Editorial Board Members' Collection Series: Advanced Oxidation Processes of Organic Contaminants

Editorial Board Members' Collection Series: Advanced Oxidation Processes of Organic Contaminants

Guest Editors

Zunyao Wang
Tiecheng Wang

MDPI

Basel • Beijing • Wuhan • Barcelona • Belgrade • Novi Sad • Cluj • Manchester

Guest Editors
Zunyao Wang
School of Environment
Nanjing University
Nanjing
China

Tiecheng Wang
College of Natural Resources
and Environment
Northwest A&F University
Yangling
China

Editorial Office
MDPI AG
Grosspeteranlage 5
4052 Basel, Switzerland

This is a reprint of the Special Issue, published open access by the journal *Toxics* (ISSN 2305-6304), freely accessible at: www.mdpi.com/journal/toxics/special_issues/DQ6MI2Z39L.

For citation purposes, cite each article independently as indicated on the article page online and using the guide below:

Lastname, A.A.; Lastname, B.B. Article Title. *Journal Name* **Year**, *Volume Number*, Page Range.

ISBN 978-3-7258-2872-2 (Hbk)
ISBN 978-3-7258-2871-5 (PDF)
https://doi.org/10.3390/books978-3-7258-2871-5

Contents

toxics

MDPI

Editorial

Advanced Oxidation Processes of Organic Contaminants

Mingzhu Liu, Shengnan Zhang and Zunyao Wang *

State Key Laboratory of Pollution Control and Resources Reuse, School of the Environment, Nanjing University, Nanjing 210023, China; 602022250029@smail.nju.edu.cn (M.L.); dg21250058@smail.nju.edu.cn (S.Z.)
* Correspondence: wangzy@nju.edu.cn

Concern is growing about the occurrence of emerging organic contaminants in various eco-environments. Most pollutants are processed in our daily life by industrial production and released into the natural environment through different channels with their consumption [1]. Furthermore, this causes environmental pollution and exposure [2,3]. Several studies have reported the occurrence of these contaminants in drinking water [4]. Their toxic effects on organisms and the ecological environment should arouse vigilance [5,6]. Many compounds have endocrine disruptors that are often considered the biggest potential problem [7]. Additionally, the presence of multiple antibiotics will promote the production of antibiotic resistance genes and antibiotic-resistant bacteria [8]. Therefore, a variety of advanced oxidation processes (AOPs) should be used to degrade organic pollutants.

Biodegradation and physical adsorption can also be used to remove organic pollutants; however, these methods are not widely used due to poor removal efficiency and high operating costs [9,10]. AOPs stand out for their efficacy in treating a wide range of refractory organic contaminants and are suitable for the removal of multiple organics. The Fenton oxidation reaction, in which materials such as iron-based catalysts activate hydrogen peroxide to produce active substances that remove contaminants, has limitations in acidic systems [11]. Electrochemical oxidation utilizes an electric field to change the functional group of organic compounds, but it requires much power and time [12]. Permanganate has been widely studied as an oxidant with good stability and relatively low cost in water purification, but its stability hinders its reactivity in some cases [13]. Therefore, some researchers choose a composite process or additives for degradation to make up for shortcomings or assist operation of single process.

The development of emerging materials has also promoted AOPs for organic pollutants. Gholami et al. [14] investigated ultrasonic degradation catalyzed by a Fe-Cu-layered double hydroxide/biochar nanocomposite for cefazolin sodium, which had a 90% higher degradation efficiency than only ultrasonic treatment. In addition, the combination of two advanced oxidation technologies could have a synergistic effect, reducing the total cost while improving the total treatment effect. Periodate and persulfate were mixed with ultraviolet radiation and titanium dioxide nanoparticles to degrade and mineralize pirimicarb insecticide in aqueous media, with mineralization rates of 35% and 46%, respectively, after 60 min [15]. As a result, many AOPs have proven efficient in removing organic contaminants and have been vigorously developed. The purpose of this Special Issue is to collect cutting-edge results from recent research, investigate the degradation of organic pollutants by advanced oxidation technologies, and explore the mechanisms and impacts on the associated environment.

Peroxonosulfate (PMS) has the advantages of high redox potential, long half-life, and wide application range, attracting much attention in the field of pollutant treatment. Shi et al. (contribution 1) confirmed the synergistic effect of PMS and ferrate(VI) on the co-degradation of butylated hydroxyanisole (BHA). When T = 25 °C, the initial pH = 8.0, $[BHA]_0$ = 100 µM, and $[PMS]_0$:$[Fe(VI)]_0$:$[BHA]_0$ = 100:1:1, the degradation efficiency could reach 92.4% within 30 min, a 29.3% improvement over the sum degradation of the PMS

Citation: Liu, M.; Zhang, S.; Wang, Z. Advanced Oxidation Processes of Organic Contaminants. *Toxics* **2024**, *12*, 579. https://doi.org/10.3390/toxics12080579

Received: 1 August 2024
Accepted: 8 August 2024
Published: 9 August 2024

system and Fe(VI) system. The main degradation pathways included hydroxylation, ring opening and coupling reactions, in which hydroxylation products were generated from •OH attacking BHA. Wang et al. (contribution 2) investigated the degradation performance of ciprofloxacin (CIP) in a magnetic graphene-oxide (MGO)-activated PMS system. The results showed that PMS could effectively remove CIP from an aquatic environment through the catalysis of stable MGO. •OH and $•SO_4^-$ generated in the MGO/PMS system were the main reaction species for oxidation, resulting in the effective destruction of CIP molecules. The Z-type heterojunction $MnO_2@g\text{-}C_3N_4$ was used by Lu et al. (contribution 3) to catalyze the oxidation of tetracycline (TC) by PMS, and the degradation efficiency increased by about 38.7% compared with the $g\text{-}C_3N_4$/PMS system. Various characterization methods showed that the addition of MnO_2 increased the absorption of $g\text{-}C_3N_4$ to visible light. Additionally, capture experiments showed that •OH and $•SO_4^-$ were still the main active species for TC degradation in the composite system.

As mentioned, sufficient studies have shown that free radicals play an important role in AOPs. Therefore, many researchers are trying to use and even produce free radicals to attack organic compounds and achieve degradation. Kang et al. (contribution 4) explored direct and indirect photolysis with free radicals (•OH, $•NO_3$, and $•SO_4^-$) as active substances of Florfenicol (FLO) in an aqueous environment. The direct photolysis of FLO involved the cleavage of C-C/C-N/C-S bonds, while indirect photolysis mainly included OH addition, NO_3 addition, and SO_4 addition on the benzene ring. Indirect photolysis of FLO was more likely to occur than direct photolysis in the natural environment. Theoretical calculations were utilized by Sun et al. (contribution 5) to study the degradation mechanism and kinetics of Metronidazole (MNZ) in the presence of •OH and $•SO_4^-$. They found that the degradation pathways were mainly O_2 addition, hydrogen abstraction, and bond breaking, and the most feasible reaction mechanism was that •OH and $•SO_4^-$ were added to the carbon atom attached to the NO_2 group in the MNZ molecule. The rate constants of MNZ with •OH and $•SO_4^-$ were calculated as $(3.54 \pm 0.42) \times 10^9$ and $(2.74 \pm 0.13) \times 10^9$ M^{-1} s^{-1}, respectively.

Plasma is also a feasible technology for effectively removing contaminants. •OH, •O, O_3, H_2O_2, and other active species are generated by high-voltage discharge between two electrodes to decompose target molecules. Yao et al. (contribution 6) confirmed that the addition of cobalt oxyhydroxide (CoOOH) significantly improved the Methylene Blue (MB) degradation performance compared with a double dielectric barrier discharge system alone. In addition, with increased CoOOH dosage and discharge voltage, the removal rate and energy efficiency improved. In the catalytic system, the active species H_2O_2 and O_3 were converted to the more oxidizing •OH, thus making the decomposition of the parent compound and intermediates more efficient. Li et al. (contribution 7) demonstrated that nanomaterial ZnO-Fe_3O_4 can effectively degrade ciprofloxacin (CIP) in chemical wastewater when combined with non-thermal plasma (NTP). The [ZnO]:[Fe_3O_4] ratio was determined as 14%:86%. The experimental results showed that •OH, $•O_2^-$, and $^1O_2$ had certain effects on degradation. The main degradation pathways revealed by liquid chromatography–mass spectrometry were hydroxyl addition, hydroxyl substitution, and piperazine ring destruction.

With so many AOPs being used for contaminant removal, different methods and contaminants should be compared for optimal degradation and reduced toxicity. Sun et al. (contribution 8) employed a variety of theoretical calculation methods to analyze the degradation mechanisms and toxicity changes of fluoroquinolones (FQs) in different AOPs. Through the application of density functional theory and comparative molecular similarity index analysis, the relationship between the bond strength and molecular structure of FQs and their degradation products was elucidated, uncovering potential differences in bond dissociation energy. The prediction of toxicity offered a unique perspective for strategically utilizing, handling, and degrading FQs to prevent the formation of highly toxic compounds.

The ultimate goal of AOP-related research is to efficiently and reasonably solve the pollution problem in the real environment. Li et al. (contribution 9) explored the migration and transformation process of polycyclic aromatic hydrocarbons (anthracene (Ant), 9-chloroanthracene (9-ClAnt), benzopyrene (BaP), and chrysene (Chr)) in soil and their interactions with native communities. By the 56th day, BaP, Chr, and Ant contents were greatly reduced, but 9-ClAnt degradation was inhibited. During the degradation process, 19 intermediate compounds were identified, including hydroxylated and carboxylated compounds. During this period, the species diversity and relative abundance of soil bacterial and fungal communities also changed, including *lysosomes*, *Bacillus*, *pseudomonas*, and *Massilia* bacteria. Amacosta et al. (contribution 10) studied the treatment of paper industry wastewater by ozone oxidation and biodegradation. In this process, ozone oxidation was used as a pretreatment to convert parent compounds into low-molecular-weight ones. Subsequently, biodegradation decomposed the initial organic mixture, achieving a higher degree of mineralization.

The ten articles in this Special Issue explain the development and exploration of AOPs from different aspects, and contribute to our understanding of AOPs and emerging organic contaminant removal. These findings promote the efficient, green, and low-cost removal of environmental pollution to ensure the healthy life of human beings and the stable operation of ecosystems.

Author Contributions: Writing—original draft preparation, M.L.; writing—review and editing, S.Z., Z.W. and M.L. All authors have read and agreed to the published version of the manuscript.

Conflicts of Interest: The authors declare no conflicts of interest.

List of Contributions:

1. Shi, P.; Yue, X.; Teng, X.; Qu, R.; Rady, A.; Maodaa, S.; Allam, A.A.; Wang, Z.; Huo, Z. Degradation of Butylated Hydroxyanisole by the Combined Use of Peroxymonosulfate and Ferrate(VI): Reaction Kinetics, Mechanism and Toxicity Evaluation. *Toxics* **2024**, *12*, 54.
2. Wang, X.; Li, Y.; Qin, J.; Pan, P.; Shao, T.; Long, X.; Jiang, D. Degradation of Ciprofloxacin in Water by Magnetic-Graphene-Oxide-Activated Peroxymonosulfate. *Toxics* **2023**, *11*, 1016.
3. Lu, G.; Li, X.; Lu, P.; Guo, H.; Wang, Z.; Zhang, Q.; Li, Y.; Sun, W.; An, J.; Zhang, Z. Z-Type Heterojunction MnO_2@g-C_3N_4 Photocatalyst-Activated Peroxymonosulfate for the Removal of Tetracycline Hydrochloride in Water. *Toxics* **2024**, *12*, 70.
4. Kang, Y.; Lu, Y.; Wang, S. Study on the Direct and Indirect Photolysis of Antibacterial Florfenicol in Water Using DFT/TDDFT Method and Comparison of Its Reactivity with Hydroxyl Radical under the Effect of Metal Ions. *Toxics* **2024**, *12*, 127.
5. Sun, J.; Chu, R.; Khan, Z.U.H. A Theoretical Study on the Degradation Mechanism, Kinetics, and Ecotoxicity of Metronidazole (MNZ) in •OH- and $SO_4^{\bullet-}$ -Assisted Advanced Oxidation Processes. *Toxics* **2023**, *11*, 796.
6. Yao, X.; Fang, Y.; Cui, X.; Cheng, X.; Cheng, Z. Dielectric Barrier Discharge Plasma Coupled with Cobalt Oxyhydroxide for Methylene Blue Degradation. *Toxics* **2023**, *11*, 763.
7. Li, A.; Wang, C.; Qian, C.; Wen, J.; Guo, H. Safe Disposal of Accident Wastewater in Chemical Industrial Parks Using Non-Thermal Plasma with ZnO-Fe_3O_4 Composites. *Toxics* **2024**, *12*, 40.
8. Sun, S.; Wang, Z.; Pu, Q.; Li, X.; Cui, Y.; Yang, H.; Li, Y. Identification and Mechanistic Analysis of Toxic Degradation Products in the Advanced Oxidation Pathways of Fluoroquinolone Antibiotics. *Toxics* **2024**, *12*, 203.
9. Li, X.; Zhang, S.; Guo, R.; Xiao, X.; Liu, B.; Mahmoud, R.K.; Abukhadra, M.R.; Qu, R.; Wang, Z. Transformation and Degradation of PAH Mixture in Contaminated Sites: Clarifying Their Interactions with Native Soil Organisms. *Toxics* **2024**, *12*, 361.
10. Amacosta, J.; Poznyak, T.; Siles, S.; Chairez, I. Sequential Treatment by Ozonation and Biodegradation of Pulp and Paper Industry Wastewater to Eliminate Organic Contaminants. *Toxics* **2024**, *12*, 138.

References

1. Wols, B.A.; Hofman-Caris, C.H.M. Review of photochemical reaction constants of organic micropollutants required for UV advanced oxidation processes in water. *Water Res.* **2012**, *46*, 2815–2827. [CrossRef] [PubMed]
2. Zhu, X.H.; Liu, S.Y.; Gao, X.F.; Gu, Y.L.; Yu, Y.; Li, M.; Chen, X.W.; Fan, M.Q.; Jia, Y.J.; Tian, L.P.; et al. Typical emerging contaminants in sewage treatment plant effluent, and related watersheds in the Pearl River Basin: Ecological risks and source identification. *J. Hazard. Mater.* **2024**, *476*, 135046. [CrossRef]

3. Harriage, S.; Harasheh, A.; Schultz, N.; Long, B.J.M. Bioconcentration of pharmaceuticals by aquatic flora in an Australian river system. *Sci. Total Environ.* **2024**, *946*, 174361. [CrossRef] [PubMed]
4. Focazio, M.J.; kolpin, D.W.; Barnes, K.K.; Furlong, E.T.; Meyer, M.T.; Zaugg, S.D.; Barber, L.B.; Thurman, M.E. A national reconnaissance for pharmaceuticals and other organic wastewater contaminants in the United States—II) Untreated drinking water sources. *Sci. Total Environ.* **2008**, *402*, 201–216. [CrossRef] [PubMed]
5. Schriks, M.; Heringa, M.B.; van der Kooi, M.M.E.; De Voogt, P.; Van Wezel, A.P. Toxicological relevance of emerging contaminants for drinking water quality. *Water Res.* **2010**, *44*, 461–476. [CrossRef] [PubMed]
6. Manamsa, K.; Crane, E.; Stuart, M.; Talbot, J.; Lapworth, D.; Hart, A. A national-scale assessment of micro-organic contaminants in groundwater of England and Wales. *Sci. Total Environ.* **2016**, *568*, 712–726. [CrossRef] [PubMed]
7. Pal, A.; Gin, K.Y.; Lin, A.Y.; Reinhard, M. Impacts of emerging organic contaminants on freshwater resources: Review of recent occurrences, sources, fate and effects. *Sci. Total Environ.* **2010**, *408*, 6062–6069. [CrossRef] [PubMed]
8. Anh, H.Q.; Le, T.P.Q.; Da Le, N.; Lu, X.X.; Duong, T.T.; Garnier, J.; Rochelle-Newall, E.; Zhang, S.; Oh, N.-H.; Oeurng, C.; et al. Antibiotics in surface water of East and Southeast Asian countries: A focused review on contamination status, pollution sources, potential risks, and future perspectives. *Sci. Total Environ.* **2021**, *764*, 142865. [CrossRef] [PubMed]
9. Min, X.; Li, W.; Wei, Z.; Spinney, R.; Dionysiou, D.D.; Seo, Y.; Tang, C.; Li, Q.; Xiao, R. Sorption and biodegradation of pharmaceuticals in aerobic activated sludge system: A combined experimental and theoretical mechanistic study. *Chem. Eng. J.* **2018**, *342*, 211–219. [CrossRef]
10. Ghattas, A.K.; Fischer, F.; Wick, A.; Ternes, T.A. Anaerobic biodegradation of (emerging) organic contaminants in the aquatic environment. *Water Res.* **2017**, *116*, 268–295. [CrossRef] [PubMed]
11. Zhang, M.H.; Dong, H.; Zhao, L.; Wang, D.X.; Meng, D. A review on Fenton process for organic wastewater treatment based on optimization perspective. *Sci. Total Environ.* **2017**, *670*, 110–121. [CrossRef] [PubMed]
12. Garcia-Segura, S.; Ocon, J.D.; Chong, M.N. Electrochemical oxidation remediation of real wastewater effluents—A review. *Process Saf. Environ. Protect.* **2018**, *113*, 48–67. [CrossRef]
13. Peng, J.; Zhou, P.; Zhou, H.; Liu, W.; Zhang, H.; Zhou, C.; Lai, L.; Ao, Z.; Su, S.; Lai, B. Insights into the Electron-Transfer Mechanism of Permanganate Activation by Graphite for Enhanced Oxidation of Sulfamethoxazole. *Environ. Sci. Technol.* **2021**, *55*, 9189–9198. [CrossRef] [PubMed]
14. Gholami, P.; Dinpazhoh, L.; Khataee, A.; Hassani, A.; Bhatnagar, A. Facile hydrothermal synthesis of novel Fe-Cu layered double hydroxide/biochar nanocomposite with enhanced sonocatalytic activity for degradation of cefazolin sodium. *J. Hazard. Mater.* **2020**, *381*, 120742. [CrossRef] [PubMed]
15. Ghavi, A.; Bagherian, G.; Rezaei-Vahidian, H. Evaluation of a hybrid advanced oxidation process for removal of pirimicarb insecticide in aqueous media: Statistical optimization and estimation of electrical energy consumption. *Environ. Prog. Sustain. Energy* **2024**, *43*, e14412. [CrossRef]

toxics

MDPI

Article

Transformation and Degradation of PAH Mixture in Contaminated Sites: Clarifying Their Interactions with Native Soil Organisms

Xiaoyu Li [1], Shengnan Zhang [1], Ruixue Guo [1], Xuejing Xiao [1], Boying Liu [1], Rehab Khaled Mahmoud [2], Mostafa R. Abukhadra [3], Ruijuan Qu [1] and Zunyao Wang [1,*]

1 State Key Laboratory of Pollution Control and Resources Reuse, School of the Environment, Nanjing University, Nanjing 210023, China; quruijuan0404@nju.edu.cn (R.Q.)
2 Faculty of Science, Beni Suef University, Beni Suef 62521, Egypt; rehabkhaled@science.bsu.edu.eg
3 Materials Technologies and Their Applications Lab, Faculty of Science, Beni Suef University, Beni Suef 62521, Egypt; abukhadra89@science.bsu.edu.eg
* Correspondence: wangzy@nju.edu.cn; Tel.: +86-025-89680358

Abstract: Soil contamination of polycyclic aromatic hydrocarbons (PAHs), especially caused by the mixture of two or more PAHs, raised great environmental concerns. However, research on the migration and transformation processes of PAHs in soils and their interactions with native communities is limited. In this work, soil samples from uncontaminated sites around the industrial parks in Handan, Hengshui, and Shanghai were artificially supplemented with three concentrations of anthracene (Ant), 9-chloroanthracene (9-ClAnt), benzopyrene (BaP), and chrysene (Chr). Ryegrass was planted to investigate the degradation of PAHs and its interaction with native soil organisms in the constructed ryegrass–microbe–soil microcosmic system. The bacterial and fungal communities in soil were affected by PAHs; their species diversity and relative abundance changed after exposure to different concentrations of PAHs, among which *Lysobacter*, *Bacillus*, *Pseudomonas*, and *Massilia* bacteria were correlated to the degradation of PAHs. On the 56th day, the contents of BaP, Chr, and Ant decreased with the degradation process, while the degradation of 9-ClAnt was limited. Nineteen intermediates, including hydroxylation and carboxylated compounds, were identified. The present research would help clarify the potential interactions between PAHs and native organisms in contaminated sites, providing fundamental information for evaluating the transformation risks of PAHs in the natural environment.

Keywords: polycyclic aromatic hydrocarbons; microbial community; ryegrass; natural degradation; transformation intermediates

Citation: Li, X.; Zhang, S.; Guo, R.; Xiao, X.; Liu, B.; Mahmoud, R.K.; Abukhadra, M.R.; Qu, R.; Wang, Z. Transformation and Degradation of PAH Mixture in Contaminated Sites: Clarifying Their Interactions with Native Soil Organisms. *Toxics* **2024**, *12*, 361. https://doi.org/10.3390/toxics12050361

Academic Editor: Lukasz Chrzanowski

Received: 18 April 2024
Revised: 8 May 2024
Accepted: 10 May 2024
Published: 13 May 2024

1. Introduction

Polycyclic aromatic hydrocarbons (PAHs) are persistent and semi-volatile compounds characterized by two or more fused aromatic rings [1]. They are emitted primarily from combustion sources, such as vehicles and coal-burning plants, through the production and use of petroleum-derived substances. PAHs can accumulate in the soil through atmospheric deposition and surface runoff [2]. Due to the low water solubility, PAHs may persist for a long time in sediments and soils, which are important sinks for PAHs in the environment [3–5]. For instance, the total concentrations of 21 PAHs in agricultural soils around coking sites were reported to vary from 294 to 1665 μg/g (dry weight) [6]. In recent years, soil contamination in industrial areas has emerged as a serious environmental problem, and the PAHs in contaminated sites might have great potential risks [7].

PAHs possess carcinogenicity, mutagenicity, and other potential toxic effects, and they have been classified as priority pollutants [8,9]. One-, two- and three-ring PAHs are demonstrated to be acutely toxic, while higher molecular weight PAHs are genotoxic [10,11]. Tong et al. showed that the possibility of carcinogenic risk was 45% if PAH exposure exceeded the acceptable threshold of 10^{-6} (TCR) [12]. When exposed to PAHs for a long time, some human organs, including lungs, skin, esophagus, colon, pancreas, bladder, and women's breasts, are prone to tumor formation [13,14]. In addition, PAHs could cause embryonic toxicity in laboratory animals by binding directly to estrogen and androgen receptors as anti-estrogens and/or anti-androgens [15]. Moreover, some research has revealed the enhanced toxicity of PAH intermediates formed under natural light irradiation and during biodegradation processes involving soil native microorganisms and plants [16,17]. Therefore, given the high environmental risks of PAHs and the great harm to the human body, it is particularly important to clarify the transformation process of PAHs in contaminated soil sites.

PAHs can exert negative effects on soil microbial diversity and may lead to the total loss of certain microbial species [18,19]. Some soil microorganisms could serve as indicators for PAH stress in soil. Moreover, microbes also play important roles in the degradation of organic pollutants, including PAHs [20]. The biodegradation of PAHs in soil is influenced by soil pH, texture, nitrate nitrogen, water content, and aeration conditions [21,22]. Due to its acceptability, sustainability, and environmental aesthetics, phytoremediation by plant cultivation is considered a promising and environmentally friendly technique to address soil pollution of recalcitrant contaminants like PAHs [23–25]. Ryegrass was generally superior to other turfgrasses in terms of bioconcentration of PAHs [26] and can be used to enrich PAHs from soil. Thus, the interaction between PAHs and soil organisms in the soil microcosmic system of plants–microbes–soil is relatively complex. However, few studies focused on the migration and transformation of PAH mixture in contaminated sites and the interactions with soil habitats.

In this work, soil samples from contaminated sites and surrounding background points in Handan, Hengshui, and Shanghai city were collected to construct a soil microcosmic system to investigate the interactions of PAHs with soil native microbial communities and plants cultivated. Four types of PAHs (including chlorinated derivatives), namely anthracene (Ant), 9-chloroanthracene (9-ClAnt), chrysene (Chr), and benzo[a]pyrene (BaP), were selected to simulate the mixed contamination of PAHs in soil. Ant, Chr, and BaP were representative PAHs containing three to five rings, and 9-ClAnt was selected as chlorinated PAHs (Cl-PAHs) to simulate the combined contamination in real soil. The growth status of ryegrass and the diversity of microbial communities were characterized. Then, the effects of plants and soil microorganisms on the degradation of mixed PAHs at different concentrations were studied. Finally, potential transformation products of PAHs in soil were identified by mass spectrometry analysis. This study is of great significance in clarifying the potential impact of PAHs on the primary microbial environment in contaminated sites and can provide a reference for assessing the potential transformation risks of PAHs in the ryegrass–microbe–soil system.

2. Materials and Methods

2.1. Chemicals and Reagents

Detailed information of chemicals and reagents can be seen in Text S1.

2.2. Soil Collection and Pretreatment

Soil samples were taken from three industrial parks in Shanghai (SH), Handan (HD), and Hengshui (HS), with the locations shown in Figure S1 and the detailed sampling sites in Figure S2. The study area is divided into different sampling units based on the actual functional areas of the factory. At the same sampling unit, five sub-samples were collected at a depth of 0–20 cm and then mixed well to provide a composite sample. According to the preliminary test results, soil in the three industrial parks was contaminated with multiple PAHs (see Text S2 for details). To facilitate inter-group comparisons and laboratory testing, surface soil samples (0–20 cm) with low background PAH values below the detection limits, which could be ignored, were collected from each of the three polluted sites for experimentation. The initial concentration of the contamination gradient was established according to the Soil Environmental Quality—Risk Control Standard for Soil Contamination of Development Land (GB36600-2018) in China [27]. Specific physicochemical property parameters of the three soils can be found in Table S1. Considering the original pollution of multiple PAHs in the three industrial parks, three concentration gradients (PAHs-1, PAHs-2, PAHs-3) with the total concentrations of the four PAHs being 50, 250, and 1000 mg/kg (dry weight), respectively, were set to simulate low, medium, and high pollution levels. The specific concentration settings of each PAH are listed in Table S2.

2.3. Pot Experiments

Details of the pot experiments, such as pot specifications, soil water content, and sampling methods, are recorded in Text S3. Unseeded pots were set as negative controls to evaluate the removal effect of PAHs by sole soil or native microbial consortium. The sterilized soil without the planting of ryegrass was used as blank control groups to evaluate natural degradation, such as leaching and adsorption. The control groups and experimental groups with PAHs ranging from low to high concentrations were named after sampling sites with the suffixes C1–C3, B1–B3, and P1–P3, respectively. For example, SH_C1, HS_B2, and HD_P3 represent unseeded and sterilized pots with 50 mg/kg PAHs in Shanghai soil, unseeded pots with 250 mg/kg PAHs in Hengshui soil, and seeded pots with 1000 mg/kg PAHs in Handan soil, respectively. The group names of different experimental groups and the experimental groups used in different analyses are organized in Table 1 for the readers' convenience and understanding. All groups were conducted in duplicate.

Table 1. Groups names, treatment of different experimental groups, and data utilization in different sections.

Group Name	Vegetation	Sterilization	Darkness	PAH Load
SH * _BK	√	×	×	×
SH_P1 (P2,P3)	√	×	×	√
SH_B1 (B2,B3)	×	×	√	√
SH_C1 (C2,C3)	×	√	×	√
	Data utilization in different sections			
Group Name	Plant Height Analysis	Microbial Diversity	Degradation Results	
SH * _BK	√	√	×	
SH_P1 (P2,P3)	√	√	√	
SH_B1 (B2,B3)	×	×	√	
SH_C1 (C2,C3)	×	×	√	

* Note: Taking the naming of soil samples from Shanghai (SH) as an example, the experimental groups set up in Handan (HD) and Hengshui (HS) were the same; numbers 1–3 represent the concentration gradients from 250 mg/kg, 500 mg/kg, and 1000 mg/kg.

2.4. Analysis Method

After being collected, the samples underwent freeze-drying at −40 °C. Larger soil particles were gently crushed with a wooden mallet to remove clumps attached to the roots of ryegrass. The remaining soil was further ground, and PAHs were extracted and analyzed from the samples within 12 h. To analyze the PAH content, 0.1 g soil was weighed into a centrifuge tube and ultrasonically extracted for 30 min by 10 mL mixture of dichloromethane and n-hexane (V:V = 1:1). At the 56th day, ryegrass shoots were rinsed three times with ultrapure water, freeze-dried, ground into powder, and weighed 0.1 g to analyze the PAH content following the same procedure as soil samples above. A total of 1 mL of supernatant was sampled for HPLC, HPLC-MS, and GC-MS analyses. The initial concentrations of PAHs in soils of different concentration gradients are listed in Table S3. Detailed information on the analysis process can be found in Text S4, Table S4 and Figure S3.

2.5. Microbial Community Analysis

Soil samples from unseeded pots were collected at the end of the pot experiments (56th day) for microbial community analysis. The control soil groups without the addition of PAHs, represented by the suffix "BK", were also analyzed for comparison. The experimental procedure for DNA extraction and PCR amplification are detailed in Text S5. The amplified products were subjected to fluorescence quantification using a BioTek (FLx800) microplate reader (Santa Clara, CA, USA) and then submitted to the Illumina platform for sequencing through Yanqu Information Technology Co., Ltd. (Hangzhou, China).

3. Results and Discussion

3.1. Effect of PAHs on Ryegrass

3.1.1. Variation in Growth Height of Ryegrass

Plant height is the most intuitive physical quantity reflecting the growth status of plants. The height of ryegrass in different soils with three contamination levels of PAH mixture was recorded as a function of growth time (Figure 1). Obvious differences in plant height were observed in different sites, among which ryegrass grew best in SH soil, followed by HD and then HS soil. HS soil took the longest time for seed germination. Ryegrass seed germinated on the third day in SH and HD soils, while the germination occurred on days 3, 5, 9, and 11 in HS soil with total PAH concentrations of 0, 50, 250, and 1000 mg/kg, respectively (the red connecting line). The cultivation of ryegrass lasted for 56 days; the plant height increased slowly, and the growth was relatively stable from the 25th day. Among the three experimental groups, plants in Shanghai soil were stout and tall, with the height ranging from 40 to 46 cm. In Handan soil, plants in the HD-P3 group grew the slowest. However, the final height of plants in the three groups showed no obvious difference, ranging from 36 to 38cm. The worst growth status of ryegrass was observed in HS soil, where the plant leaves were thinner, and the height varied between 16–30 cm. There are multiple heavy metals (e.g., Hg, 176.96 mg/kg) and other organic pollutants (e.g., CCl_4, 18.99 mg/kg) in the soil, which exceeds the risk screening values for soil contamination of development land prescribed by GB 36600-2018 (China) [27], exerting an adverse effect on plant growth.

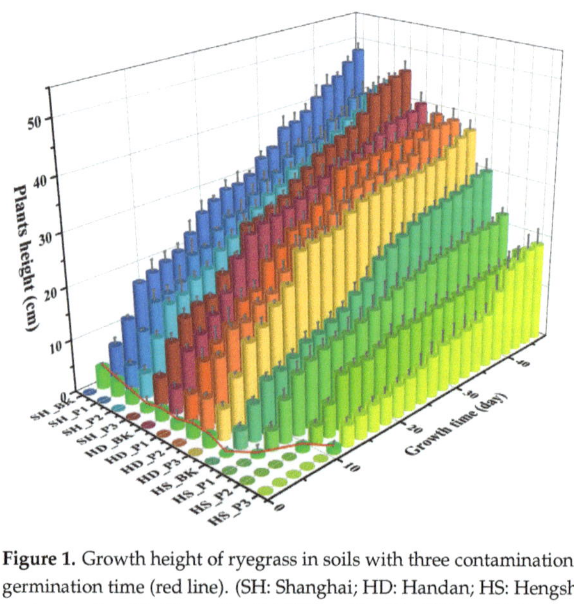

Figure 1. Growth height of ryegrass in soils with three contamination levels of PAH mixture and the germination time (red line). (SH: Shanghai; HD: Handan; HS: Hengshui; BK: Control group without contamination; P1: 50 mg/kg; P2: 250 mg/kg; P3: 1000 mg/kg).

3.1.2. Migration of PAHs into Ryegrass

Contents of PAHs in the ryegrass on the 56th day were measured. Due to the poor growth of ryegrass, it was difficult to determine the concentration of PAHs in plants in HS groups. Therefore, only the migration of PAHs into plants in HD and SH soils was studied. As the initial soil PAH content increased, the concentration of three unchlorinated PAHs in plants increased to varying degrees (Table S5). As the initial pollution concentration increased, the concentrations of Ant and BaP in HD ryegrass shoots increased from 0.36 and 0.24 mg/kg to 1.14 and 1.17 mg/kg (dry weight), respectively, representing increases of 0.78 mg/kg and 0.93 mg/kg. For SH groups, the Ant and BaP increased by 0.10 and 0.41 mg/kg from the SH_P1 group to the SH_P3 group. According to a previous study, the initial soil PAH concentration directly affects the accumulation of PAHs in ryegrass [28]. Due to the higher initial concentration in the soil, the concentration of Chr accumulated in ryegrass is higher than that of the other three PAHs. The contents of Chr in PAHs-3 groups reached 2.18 and 6.34 mg/kg in HD and SH soils, respectively. However, the content of 9-ClAnt in plant shoots increased first and then decreased in SH soil, showing a decreasing trend in HD soil. This abnormal phenomenon might be related to the properties of Cl-PAHs. The quantitative structure–activity relationship and level III fugacity model revealed that Cl-PAHs tend to be more enriched in soil and sediment instead of air and water bodies [29]. In addition, the Cl-PAHs in soil and sediment are hard to migrate to soil organisms and to be utilized by plants. The concentration of Cl-PAHs, especially those with larger molecule weight, is much higher in the sediment than in mussels of benthic organisms [30].

3.2. *Effects of PAHs on Soil Microbial Community Structure*

Microorganisms, as an important component of soil ecosystems, are affected in species composition and community structure by the stress of PAHs [31]. As previous studies have demonstrated, contaminants in soil systems could affect the composition of microbial communities [32]. In this section, the changes in soil bacterial and fungal community structure under different PAH stress conditions were studied. The effectiveness of the sequencing results is listed in Table S6. The proportion of valid sequences obtained after quality control, denoising, assembly, and chimera removal exceeds 90% for all experimental groups, except for HD_P1, which was 88.8%, indicating the validity of the sequencing results.

3.2.1. The Alpha Diversity Index Analysis

Alpha diversity, indicated by the Chao1, Shannon, Simpson, Observed Species, and Goods' Coverage indexes, was used to reveal the species richness and diversity of microbial communities in three different soils (Table S7). Chao1 and Observed Species can characterize the richness while excluding the impact of each species' proportion (Species Evenness) [33]. As the total PAH concentration increased to 50, 250, and 1000 mg/kg, the Chao1 index of bacteria in the SH soil samples changed from 418.8 (SH_BK) to 970.2, 554.0, and 530.2, respectively. The number of observed species showed a similar trend that increased from 418 (SH_BK) to 969, 553, and 528, respectively. Thus, a low level of PAH pollution greatly increased the microbial diversity of SH soil samples. The PAHs could have a dual effect on microbial communities. When the concentration of PAHs is low, it might change the dominant species in the soil, leading to an increase in the abundance of other microorganisms, especially PAH-degrading bacteria [34]. However, high concentrations of PAHs may inhibit the growth of microbial communities and reduce species abundance. Moreover, the relative abundance of species is also related to the physiochemical properties of the soil, and thus, different types of soils showed different PAH stress tolerance [35]. The Chao1 and Observed Species of bacteria in HD soil samples were higher than those in Shanghai and decreased from 6035.1 (HD_BK) to 1552.0, 1347.9, 1320.6, and from 5404 (HD_BK) to 1545, 1302 and 1316, respectively, with increasing soil PAH concentration.

In the case of fungi, the species richness had negligible variation in SH soil, of which the Chao1 increased from 95.0 (SH_BK) to 105.3, 108.9, and 101.1. However, the two indexes both showed an obvious decrease in HD soil samples, which demonstrated that HD soil was more susceptible to PAHs, resulting in a decrease in species richness. The decreased biodiversity of HD soil might exert an adverse effect on the degradation of PAHs. After exposure to 1000 mg/kg PAHs, the species richness of fungi decreased in HS soil.

The diversity of the microbial community was represented by Shannon [36] and Simpson [37] indices, which both have been traditionally used to measure diversity. The Shannon index is a measure of species richness, while the Simpson index focuses more on species evenness [38]. According to Table S7, both the Shannon and Simpson indexes indicated that the microbial diversity increased in the SH soil sample contaminated by PAHs, while opposite trends were observed in the other two kinds of soils. Good's coverage [39] was used to evaluate the coverage quality of sequencing results. This index varied from 0.98 to 1.00, suggesting that the sequencing results are representative of the real microbial communities.

3.2.2. Rank Abundance Curve Analysis

The rank abundance curve (RAC) was used to represent the distribution of OTU abundance, and the flatness of the broken line reflects the evenness of community composition. The flatter the line, the smaller the abundance difference among OTUs in the community, and the higher the evenness of community composition [40]. In SH soil samples, the curve became flatter after PAH contamination, especially for SH_P1 (Figure S4a), indicating the increased evenness. HD soil samples showed different trends, with the HD_BK group showing the flattest curve, and almost the same degree of steepening was observed in the curves of the contaminated samples (Figure S4b). Due to the differences in the community composition of the initial microorganisms (see details in Section 3.2.3), the two types of soils responded differently to the secretions of ryegrass. On the one hand, rhizosphere secretions could promote the activity of soil microorganisms [41]. On the other hand, rhizosphere secretions may cause competition by stimulating microbial activity, thereby affecting species richness and composition uniformity [42].

Thus, PAH contamination disrupted the evenness of community composition in HD soil, which had little relationship with the concentration of PAHs. Due to the low species abundance in HS soil samples, its RAC is not displayed. From the analysis of alpha diversity indices and RAC, the microbial abundance and diversity of bacteria were larger than those of the fungal community, which indicated that the bacterial community was dominant in the three soils.

3.2.3. Comparison of the Species Composition in Soils

The differences in bacterial communities among three different soils (SH, HD, HS) were analyzed via PCoA based on the OTUs of the bacterial communities [43]. In Shanghai soil samples, Axis 1 explains 63.8% of species variation (Figure S5a). The SH_P1 group was notably distant from others on Axis 1, indicating that even low PAH contamination could change Shanghai soil community composition. Combining the analysis from Section 3.2.2, the SH_P1 group exhibits higher evenness in species composition. With increasing PAH stress, there might be dominant species adapted to PAH stress. Similarly, in Handan soil, Axis 1 accounts for 49.2% of the variation, with PAH-exposed samples projecting further than blanks, indicating PAH-induced microbial community differences. In Hengshui soil, Axis 1 and Axis 2 explained 43.6% and 31.4% of the variation, respectively, with a cumulative explanatory power of 75.0%. P1, P2, and P3 are closely projected on both Axis 1 and Axis 2. Therefore, PAH mixture altered the composition of indigenous microbial communities in the soil.

The identification results of species composition analysis are listed in Table S8. HD soil samples showed the largest number of bacteria species, with 18–20 phyla, 40–56 classes, 81–111 orders, and 185–269 genera identified. Shanghai soil samples showed increased species numbers after exposure to PAHs. The lowest species numbers were found in HS soil samples, where the number of genera for bacteria and fungi ranged from 10–59 and 3–10, respectively.

The composition of bacteria in the phylum level is shown in Figure 2. The dominant bacteria in the three soils were *Actinobacteria* and *Proteobacteria*, the total proportion of which could reach even up to 92%. With the increase in PAH concentration, *Proteobacteria* abundance in Shanghai soil increased by varying degrees from 28% to 61%, 41%, and 50%, while that of *Actinobacteria* decreased from 56% to 19%, 48%, and 42%, respectively. The two microorganisms showed similar variation trends in the soil of Handan, where the appearance of *Firmicutes* was also noticed (4.2%, 3.8%, and 4.1%) after the addition of PAHs. Accompanied by a decreased abundance of *Acidobacteria*, the percentage of *Firmicutes* increased obviously in Hengshui soil. However, only the HS_P3 group showed a higher proportion of *Proteobacteria* than in uncontaminated soil. Although the specific organisms underwent different changes in different soils, the diversity of bacterial composition at the phylum level reduced with increasing PAH concentrations. Only *Firmicutes*, *Actinobacteria*, and *Proteobacteria* were found in the 1000 mg/kg PAH-contaminated Hengshui soil, which might result from the combined effect of PAH stress and soil physicochemical properties. Further analysis was conducted on the composition of bacterial communities at the genus level (Figure S6). The natural PAH-degrading bacteria in soil mainly belonged to *Pseudomonas*, *Bacillus*, *Sphingomonas*, *Lysobacter*, *Rhizobium*, and *Ohtaekwangia* [44–48]. In Shanghai soil (Figure S6a), the abundance of *Pseudomonas* increased in the SH_P1 group (32.4%) and subsequently decreased to 3.6% and 6.4% in the SH-P2 and SH-P3 groups, respectively, with that of SH_BK group being 5.1%. *Sphingomona* showed decreased abundance, which accounted for less than 0.6% of the three PAH exposure groups. For HD soil, the abundance of *Lysobacter* and *Bacillus* increased from 1.9% to 11.1%, 26.9%, and 20.0% and from 0.1% to 1.8%, 2.3%, and 2.4%, respectively with increasing PAH concentrations (Figure S6b). In both Shanghai and Handan soils, the abundance of *Massilia* increased with the increasing gradient of PAH mixture, suggesting its potentiality in PAH degradation, which was consistent with a previous study [49]. In Hengshui soil, both *Bacillus* and *Pseudomonas* increased to become the dominant population after the addition of PAHs, while the proportion of *Nocardioides* was significantly decreased (Figure S6c).

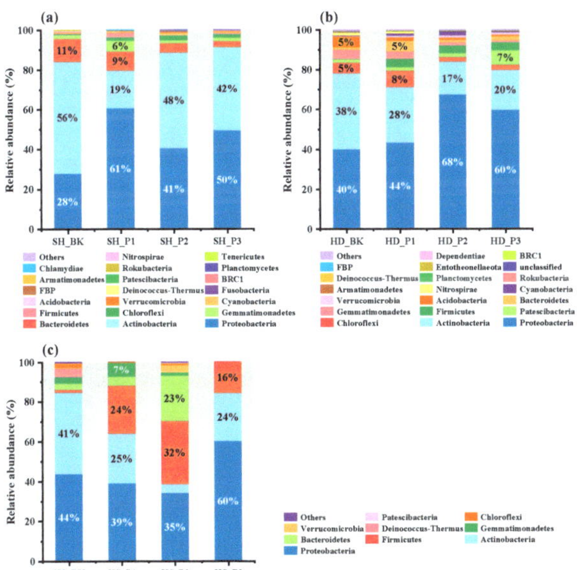

Figure 2. Relative abundance of bacteria at the phylum level in Shanghai (**a**), Handan (**b**), and Hengshui (**c**) soils with different concentrations of PAHs. (SH: Shanghai; HD: Handan; HS: Hengshui; BK: Control group without contamination; P1: 50 mg/kg; P2: 250 mg/kg; P3: 1000 mg/kg).

Fungal-mediated bioremediation of PAHs in soil has also been widely reported in recent years. Fungi do not utilize PAHs as their sole source of carbon as bacteria but co-metabolize PAHs and produce a series of oxidation products, including carbon dioxide [50]. Therefore, the abundance changes of fungal communities in different soils before and after PAH pollution were compared (Table S9). The dominant genus in SH_P1 soil was *Ascomycota*, which accounted for up to 53.0%, much higher than that in SH_BK (10.8%). However, as the loading concentration of PAHs increased, its proportion decreased to 26.1% and 16.7% in the SH_P2 and SH_P3 groups, respectively. Other fungal communities have also undergone varying degrees of changes under the influence of PAHs, as detailed in Table S8.

To sum up, PAHs could vary the composition and structure of soil microorganisms. Some bacteria and fungi that have been proven to have the ability to degrade PAHs showed increased relative abundance as PAH concentration increased. However, the abundance of these microorganisms does not always show a positive correlation with PAH concentration. Instead, their abundance increased at low concentrations of PAHs while decreasing as the PAH concentrations were further increased.

3.2.4. Comparative Cluster Analysis

In the sample clustering analysis, UPGMA clustering is performed on samples according to Euclidean Distance of species composition data, and the samples are then arranged based on the clustering results. The abundance information of the top 50 genera in the samples was selected to draw the heat map (Figures S7–S9), from which obvious differences could be found in the distribution of dominant genera. As seen in Figure S7, the dominant genera were distributed in the upper left of the heat map for HD_BK and changed by exposure to different concentrations of PAHs. *Devosia*, *Actinotalea*, *Phenylobacterium*, *Massilia*, *Sphingoaurantiacus*, *ANPR*, and *Shinella*, some of which have been proven to be quite capable of degrading PAHs in the soil or promoting plant growth, showed increased abundance with increasing PAH exposure concentrations [51,52]. By contrast, some of the bacteria, such as *Luteimonas*, *Nitrosospira*, *Longimicrobiaceae*, *Cupriavidus*, and *Anaeromyxobacter*, showed higher abundance in HD_P1 than other groups, indicating that PAH stress at a low concentration was beneficial to the growth of the bacteria to some extent.

The heat map in Shanghai soil of different PAH exposure concentrations also showed differences (Figure S8). *Georgenia*, *Actinotolea*, and *Nitrosospira* were not found in the SH_BK, but their relative abundance increased in the PAH exposure groups. Previous findings have also revealed that the stress of PAHs could lead to an increase in microbial abundance [53], which may be due to the decline of some dominant species [33]. It is noteworthy that in both Shanghai and Handan soils, *Devosia*, *Nitrosospira*, *Actinotalea*, *Luteumonas*, *Paracoccus*, and *JG30-KF-CM45* showed quite a low abundance in the control group but increased to different degrees in the PAH exposure groups. The results of heat map analysis from Hengshui soil samples (Figure S9) showed that there were significant differences in microbial abundance among the four experimental groups, and only a few genera of microbe displayed absolute predominance. The sandy soil in Hengshui was not conducive to the survival of microorganisms; therefore, the abundance of both bacteria and fungi was low.

3.3. *Effect of PAHs on Ryegrass*

The transformation of PAHs in soil is the result of many environmental factors, such as light, microorganisms, and plants. Figure 3 shows the degradation of the four PAHs in the vegetated soils, reflecting the comprehensive effects of plants, microorganisms, and light in the soil system. The residual PAHs all decreased with time in the soils studied. In general, the removal efficiency of the four PAHs was lowered with the increase in the total exposure concentration. The removal percentage of Ant, BaP, Chr, and 9-ClAnt decreased from 94%, 94%, 45%, and 45% in the HD-P1 group to 40%, 59%, 39%, and 32% in the HD-P3 group, respectively. As the total PAH concentration increased from 50 mg/kg to 1000 mg/kg, the removal of Ant decreased by 55.5%, 26.6%, and 41.25% in HD, HS, and SH soils, respectively. The initial concentration of PAHs in soil influenced microbial degradation [54]. The high concentration of PAHs may have an adverse effect on the microbial community structure and plant growth in the soil, which was unfavorable for the elimination of PAHs by soil biota, thus leading to decreased removal efficiency in PAHs-3 groups. Among the four PAHs, Ant and BaP showed better removal effects than the other two PAHs. Specifically, around 85% of Ant and BaP were removed in PAHs-1 groups of HD, HS, and SH soils on the 56th day. The removal of Chr and 9-ClAnt became much lower, which were in the range of 42–47%, revealing the higher potential environmental risk.

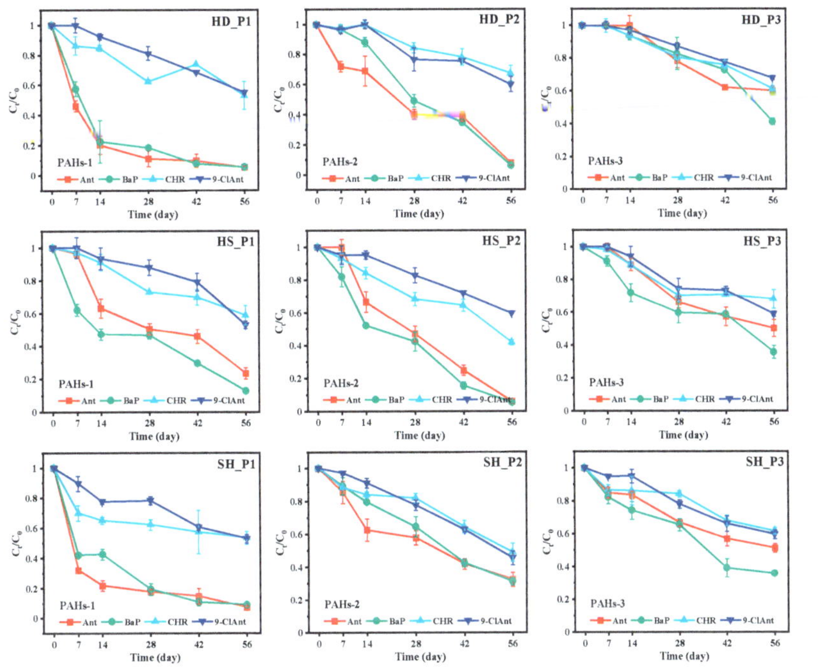

Figure 3. Effects of different PAH concentrations and different types of soils on the degradation of four PAHs in the ryegrass–microbe–soil systems under sunlight irradiation. (C_t/C_0 represents the residual concentration variation of PAHs in soil over time; P1: 50 mg/kg; P2: 250 mg/kg; P3: 1000 mg/kg).

3.3.1. Effect of Native Microorganisms on the Degradation of PAHs

Considering the degradation effects of microbial on soil PAHs [55], the soils were unplanted and kept in the dark with water content around the soil field capacity to further clarify the contribution of native microorganisms to the degradation of PAHs. The removal of PAHs in the blank control groups was mainly due to leaching and abiotic loss, such as adsorption and volatilization [27]. The four PAHs showed some degradation, among which the degradation rate decreased with increasing initial PAH concentration (Figure S10). Generally, 9-ClAnt showed the lowest degradation regardless of the type of soil and its initial concentration. It is believed that the degradation of PAHs in this experimental group of soil mainly comes from the role of microorganisms. As seen in Figure S10, the concentration of the four PAHs decreased over time in all soils. Obvious removal of Ant and BaP was observed, especially in PAHs-1 and PAHs-2 groups. The degradation percentage of Ant were 81%, 76%, and 87% in PAHs-1 groups of HD, HS, and SH, respectively. Thus, soil microorganisms could contribute greatly to soil PAH degradation.

3.3.2. Effect of Plants on the Degradation of PAHs

Further, the unplanted soils with sterilization were compared to clarify the influence of plantation (Figure S11). Compared with the unplanted group, the degradation of Chr in different soils was enhanced (Figure 4). When not planted, the removal efficiency percentages of Chr in the PAHs-1 experimental groups of Handan, Hengshui, and Shanghai soils on the 56th day were 13.5%, 20.7%, and 23.4%, respectively. In the case of ryegrass cultivation, the removal percentages were 66.0%, 42.4%, and 77.5%, respectively.

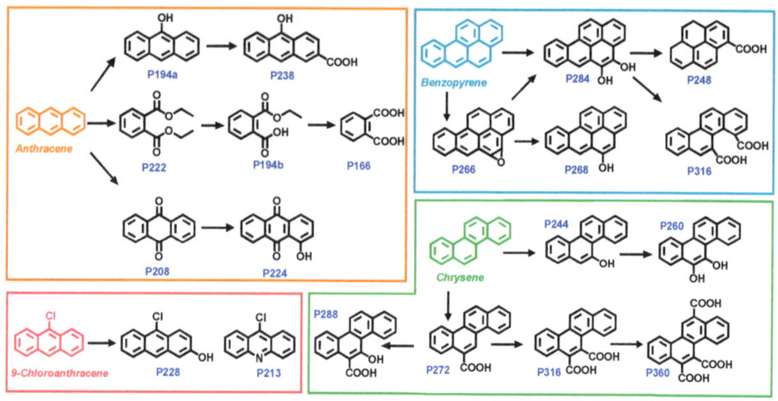

Figure 4. Proposed degradation pathways of the four PAHs. (Note that the intermediate compounds shown in the figure might have isomers.).

In addition, the degradation of 9-ClAnt was relatively low and not affected much by plant cultivation. With or without the planting of ryegrass, the removal percentages of 9-ClAnt in PAHs-1 groups of Handan, Hengshui, and Shanghai soils were 38.9% and 12.1%, 18.8% and 20.6%, 32.7%, and 36.4%, respectively. As illustrated in Section 3.1.2, the Cl-PAHs usually showed poor bioavailability, and their migration to plants was difficult. Moreover, Kamiya et al. examined the photolysis process of Cl-PAHs and found that Cl-PAHs were more stable than the corresponding parent PAHs [56]. Therefore, 9-ClAnt showed poor degradation, which was hardly affected by the vegetation of ryegrass.

To sum up, the vegetation of plants could facilitate the degradation of PAHs to some degree. In this study, the removal of Chr was improved by planting ryegrass. However, Cl-PAHs, such as 9-ClAnt, are relatively more difficult to degrade in natural environments, thus possessing higher environmental risks.

3.4. The Intermediate Products of PAHs

PAHs might be converted into other products under the combined action of plants, microorganisms, and other factors in the natural environment. The identification of intermediate products plays an important role in the environmental risk analysis of PAHs. In this study, soil and ryegrass shoot samples collected on day 56 were analyzed with HPLC/MS and GC/MS methods to identify the polar products and weak/non-polar products, respectively. A total of 19 intermediates of the four PAHs were identified. Detailed mass information on HPLC/MS analysis is shown in Table S10.

As shown in Figure 4, a hydroxylated product of ANT (anthracen-9-ol, P194a) was first generated, experiencing further carboxylation to form (10-hydroxyanthracene-2-carboxylic acid, P238). The formation of P222 (diethyl phthalate) was also observed under the effects of soil bacteria [57], which could be transformed into P194b (2-(ethoxycarbonyl)benzoic acid) and P166 (phthalic acid). Quinones are common conversion products of anthracene in the natural degradation process, which was confirmed in previous studies [58]. In addition, fungi could produce enzymes such as manganese peroxidase and laccases to convert PAHs into diphenol intermediates that are eventually oxidized to quinones [59]. In this work, anthracene-9,10-dione (P208) was identified, which could be transformed into P224 through hydroxylation.

In the degradation process of BaP, P266 was recognized through GC/MS analysis. The fungal community in the soil could produce certain intracellular enzymes, such as cytochrome P450 monooxygenase and epoxide hydrolase, which play a role in the degradation of 2–5-ring PAHs, resulting in the formation of epoxy products [60,61]. Moreover, bacteria could use BaP as a carbon source or synthesize enzymes that could transform the five-ring PAHs [10,62] Afterwards, P266 could be further transformed into the mono-hydroxylation product P268 or the di-hydroxylation product P284. Finally, P248 and P316, which were both identified by HPLC/MS, were formed from P284 after the fission process [63,64].

Continuous hydroxylation of Chr in soil could lead to hydroxylation products (P244 and P260) in sequence. Meanwhile, P272, P316, and P360, as monocarboxylic, dicarboxylic, and tricarboxylic products, respectively, were formed from sequential carboxylation processes of Chr. Then, P244 and P272 were converted to the product P288 through carboxylation and hydroxylation, respectively. As for 9-ClAnt, its hydroxylation product P228 could be identified by HPLC/MS, and the nitrogenated product P213 was found in the GC/MS analysis.

In summary, the four PAHs could undergo hydroxylation reactions to form corresponding hydroxyl addition products. For the three unchlorinated PAHs (Ant, BaP, and Chr), carboxylated products could be formed through carboxylation (e.g., P238, P272, P288, P316, and P360) or through ring opening (e.g., P166 and P316). In addition to microbial density and microbial diversity, soil properties like texture, permeability, moisture content, and organic matter content are also important factors affecting PAH degradation in soil [65]. Therefore, it is necessary to clarify the intermediate products of PAHs during natural degradation at contaminated sites as well as their potential impact on indigenous microorganisms, which are of great significance for environmental risk analysis related to PAHs.

4. Conclusions

This study examined the effects of ryegrass planting and soil native microbial communities on the degradation of PAH mixture in three types of soils, aiming to elucidate the interaction of PAHs with microbial communities and plants. The plant height was the lowest in PAH-contaminated Hengshui soil, and it took the longest time (about 5–6 days) for seed germination. Under PAH stress, the species abundance of some PAHs-degrading bacteria, such as *Pseudomonas*, *Bacillus*, *Sphingomonas*, and *Lysobacter*, changed in the soil, but the changes were not always positively correlated with the exposure concentration of PAHs. From the comparison of the degradation of PAHs in soils with and without planting of ryegrass, it was found that microorganisms in soils played an important role in the degradation of PAHs, while the plantation of ryegrass improved the degradation efficiency of chrysene. During the natural degradation process, 19 intermediates were identified by combined use of HPLC/MS and GC/MS, and hydroxylation and carboxylation were the common reaction pathways. This study is of great significance in clarifying the potential impact of PAHs on the primary microbial environment in contaminated sites and can provide valuable information for assessing the potential risks of PAHs in natural degradation processes.

Supplementary Materials: The following supporting information can be downloaded at: https://www.mdpi.com/article/10.3390/toxics12050361/s1, Text S1: Detailed information of chemicals and reagents; Text S2: Pollution of PAHs in the soil of the three industrial parks; Text S3: Detailed information of the pot experiments; Text S4: Detailed process of HPLC, HPLC-MS, and GC-MS analyses; Text S5: Detailed process of the sequencing process; Table S1: Physicochemical properties of the soil samples used in the experiment; Table S2: The concentration gradients of PAHs; Table S3: The initial concentration of PAHs after artificial contamination (mg/kg); Table S4: HPLC detection methods for the PAHs used in the study; Table S5: Contents of four PAHs in ryegrass after 56 days of growth; Table S5: The effectiveness of sequencing results; Table S6: The effectiveness of sequencing results of the vegetation groups; Table S7: Alpha diversity indices of bacteria and fungi of different soil samples; Table S8: Statistical table of species taxonomic annotation results; Table S9: The relative abundance of fungal communities in Shanghai, Handan, and Hengshui soils with different PAH concentrations; Table S10: Accurate mass measurements of the degradation products of four PAHs as determined by HPLC-TOF-MS/MS; Figure S1: Location of the three industrial parks in China; Figure S2: Sampling sites in Shanghai (a), Handan (b), and Hengshui (c); Figure S3: The standard curve of peak area and solution concentration for Ant (a), 9-ClAnt (b), BaP (c), and Chr (d).; Figure S4: The rank abundance curve of bacteria in Shanghai (a) and Handan (b) soil samples; Figure S5: PCoA profile of community dissimilarity indices (Bray—Curtis) calculated from the OUT tables of different soils; Figure S6: Relative abundance of bacteria at the phylum level in Shanghai (a), Handan (b), and Hengshui (c) soils with different concentrations of PAHs; Figure S7: Comparative heat map of bacteria in Handan soil samples; Figure S8: Comparative heat map of bacteria in Shanghai soil samples; Figure S9: Comparative heat map of bacteria in Hengshui soil samples; Figure S10: The degradation of PAHs in Handan (a–c), Hengshui (d–f), and Shanghai (g–i) soils in the dark with no planting of ryegrass; Figure S11: The loss of PAHs in Handan (a–c), Hengshui (d–f), and Shanghai (g–i) soils in the dark with sterilization and no planting of ryegrass [57,65–70].

Author Contributions: Conceptualization, X.L.; methodology, Z.W.; validation, X.L., R.Q. and S.Z.; formal analysis, X.L.; investigation, X.L.; resources, R.G.; data curation, X.L.; writing—original draft preparation, X.L.; writing—review and editing, R.G., X.X. and B.L.; visualization, R.K.M. and M.R.A.; supervision, R.Q. All authors have read and agreed to the published version of the manuscript.

Funding: This research was funded by the National Key R & D Program of China, grant number: 2020YFC1806700, and the Graduate International Exchange Key Program of Nanjing University.

Informed Consent Statement: Not applicable.

Data Availability Statement: The data presented in this study are available on request from the corresponding author.

Acknowledgments: The authors acknowledge the Researchers Supporting Project number (RSPD2024 R691), King Saud University, Riyadh, Saudi Arabia.

Conflicts of Interest: The authors declare no conflicts of interest.

References

1. Ren, Y.; Zhou, B.; Tao, J.; Cao, J.; Zhang, Z.; Wu, C.; Wang, J.; Li, J.; Zhang, L.; Han, Y.; et al. Composition and size distribution of airborne particulate PAHs and oxygenated PAHs in two Chinese megacities. *Atmos. Res.* **2017**, *183*, 322–330. [CrossRef]
2. Ma, B.; Chen, H.; He, Y.; Wang, H.; Xu, J. Evaluation of toxicity risk of polycyclic aromatic hydrocarbons (PAHs) in crops rhizosphere of contaminated field with sequential extraction. *J. Soils Sediments* **2010**, *10*, 955–963. [CrossRef]
3. El-Alawi, Y.S.; McConkey, B.J.; George Dixon, D.; Greenberg, B.M. Measurement of Short- and Long-Term Toxicity of Polycyclic Aromatic Hydrocarbons Using Luminescent Bacteria. *Ecotoxicol. Environ. Saf.* **2002**, *51*, 12–21. [CrossRef]
4. Wild, S.R.; Jones, K.C. Biological and abiotic losses of polynuclear aromatic hydrocarbons (PAHs) from soils freshly amended with sewage sludge. *Environ. Toxicol. Chem.* **1993**, *12*, 5–12. [CrossRef]
5. Zhang, P.; Chen, Y. Polycyclic aromatic hydrocarbons contamination in surface soil of China: A review. *Sci. Total Environ.* **2017**, *605*, 1011–1020. [CrossRef] [PubMed]
6. Duan, Y.; Shen, G.; Tao, S.; Hong, J.; Chen, Y.; Xue, M.; Li, T.; Su, S.; Shen, H.; Fu, X.; et al. Characteristics of polycyclic aromatic hydrocarbons in agricultural soils at a typical coke production base in Shanxi, China. *Chemosphere* **2015**, *127*, 64–69. [CrossRef]
7. Li, H.B.; Qu, C.; Cui, X.; Song, Y.; Wang, C.; Gu, C. Control, Treatment and Restoration of Contaminated Soils. *Bull. Environ. Contam. Toxicol.* **2022**, *109*, 577. [CrossRef] [PubMed]
8. Wang, J.; Liu, J.; Ling, W.; Huang, Q.; Gao, Y. Composite of PAH-degrading endophytic bacteria reduces contamination and health risks caused by PAHs in vegetables. *Sci. Total Environ.* **2017**, *598*, 471–478. [CrossRef]
9. Patel, K.; Patel, M. Improving bioremediation process of petroleum wastewater using biosurfactants producing *Stenotrophomonas sp.* S1VKR-26 and assessment of phytotoxicity. *Bioresour. Technol.* **2020**, *315*, 123861. [CrossRef]
10. Juhasz, A.L.; Naidu, R. Bioremediation of high molecular weight polycyclic aromatic hydrocarbons: A review of the microbial degradation of benzo[a]pyrene. *Int. Biodeter. Biodegr.* **2000**, *45*, 57–88. [CrossRef]
11. Sims, R.C.; Overcash, M.R. *Fate of Polynuclear Aromatic Compounds (PNAs) in Soil-Plant Systems*; Gunther, F.A., Ed.; Springer: New York, NY, USA, 1983; pp. 1–68.
12. Tong, R.; Yang, X.; Su, H.; Pan, Y.; Zhang, Q.; Wang, J.; Long, M. Levels, sources and probabilistic health risks of polycyclic aromatic hydrocarbons in the agricultural soils from sites neighboring suburban industries in Shanghai. *Sci. Total Environ.* **2018**, *616–617*, 1365–1373. [CrossRef] [PubMed]

13. Rajpara, R.K.; Dudhagara, D.R.; Bhatt, J.K.; Gosai, H.B.; Dave, B.P. Polycyclic aromatic hydrocarbons (PAHs) at the Gulf of Kutch, Gujarat, India: Occurrence, source apportionment, and toxicity of PAHs as an emerging issue. *Mar. Pollut. Bull.* **2017**, *119*, 231–238. [CrossRef]
14. Yu, H. Environmental carcinogenic polycyclic aromatic hydrocarbons: Photochemistry and phototoxicity. *J. Environ. Sci. Health C* **2002**, *20*, 149–183. [CrossRef]
15. Bolden, A.L.; Rochester, J.R.; Schultz, K.; Kwiatkowski, C.F. Polycyclic aromatic hydrocarbons and female reproductive health: A scoping review. *Reprod. Toxicol.* **2017**, *73*, 61–74. [CrossRef] [PubMed]
16. Huang, X.-D.; Dixon, D.G.; Greenberg, B.M. Impacts of UV radiation and photomodification on the toxicity of pahs to the higher plant *Lemna gibba* (duckweed. *Environ. Toxicol. Chem.* **1993**, *12*, 1067–1077.
17. Newsted, J.L.; Giesy, J.P. Predictive models for photoinduced acute toxicity of polycyclic aromatic hydrocarbons to *Daphnia magna*, strauss (cladocera, crustacea). *Environ. Toxicol. Chem.* **1987**, *6*, 445–461. [CrossRef]
18. Huang, Z.; Liu, Y.; Dai, H.; Gui, D.; Hu, B.X.; Zhang, J. Spatial distribution and source apportionment of polycyclic aromatic hydrocarbons in typical oasis soil of north-western China and the bacterial community response. *Environ. Res.* **2022**, *204*, 112401. [CrossRef] [PubMed]
19. Yan, Z.; Hao, Z.; Wu, H.; Jiang, H.; Yang, M.; Wang, C. Co-occurrence patterns of the microbial community in polycyclic aromatic hydrocarbon-contaminated riverine sediments. *J. Hazard. Mater.* **2019**, *367*, 99–100. [CrossRef] [PubMed]
20. Mahamoud Ahmed, A.; Lyautey, E.; Bonnineau, C.; Dabrin, A.; Pesce, S. Environmental Concentrations of Copper, Alone or in Mixture with Arsenic, Can Impact River Sediment Microbial Community Structure and Functions. *Front. Microbiol.* **2018**, *9*, 1852. [CrossRef]
21. Du, J.; Liu, J.; Jia, T.; Chai, B. The relationships between soil physicochemical properties, bacterial communities and polycyclic aromatic hydrocarbon concentrations in soils proximal to coking plants. *Environ. Pollut.* **2022**, *298*, 118823. [CrossRef]
22. Wu, Y.; Ding, Q.; Zhu, Q.; Zeng, J.; Ji, R.; Dumont, M.G.; Lin, X. Contributions of ryegrass, lignin and rhamnolipid to polycyclic aromatic hydrocarbon dissipation in an arable soil. *Soil Biol. Biochem.* **2018**, *118*, 27–34. [CrossRef]
23. Shahsavari, E.; Aburto-Medina, A.; Taha, M.; Ball, A.S. Phytoremediation: Management of Environmental Contaminants. In *Microbial Ecotoxicology*; Ansari, A.A., Gill, S.S., Gill, R., Lanza, G.R., Newman, L., Eds.; Springer International Publishing: Cham, Switzerland, 2016; pp. 17–38.
24. Loss, E.M.O.; Yu, J.H. Bioremediation and microbial metabolism of benzo(a)pyrene. *Mol. Microbiol.* **2018**, *109*, 433–444. [CrossRef] [PubMed]
25. Babu, A.G.; Reddy, M.S. Diversity of Arbuscular Mycorrhizal Fungi Associated with Plants Growing in Fly Ash Pond and Their Potential Role in Ecological Restoration. *Curr. Microbiol.* **2011**, *63*, 273–280. [CrossRef]
26. Ma, L.; Yao, L.; Li, Y. Bioremediation of a polycyclic aromatic hydrocarbon–contaminated urban soil: Degradation dynamics and phytotransformation pathways. *J. Soils Sediments* **2022**, *22*, 797–808. [CrossRef]
27. *GB36600-2018*; Soil Environmental Quality—Risk Control Standard for Soil Contamination of Development Land. Standardization Administration of the People's Republic of China: Beijing, China, 2018. Available online: https://www.mee.gov.cn/ywgz/fgbz/bz/bzwb/trhj/201807/W020190626596188930731.pdf (accessed on 2 May 2024).
28. Gao, Y.; Ling, W.; Zhu, L.; Shen, Q. Ryegrass-Accelerating Degradation of Polycyclic Aromatic Hydrocarbons (PAHs) in Soils. *J. Agro-Environ. Sci.* **2005**, *24*, 498–502. (In Chinese)
29. Gao, Z.; Yang, X.; Peng, Y. Analytical methods and pollution status of a new class of organic contaminants-chlorinated polycyclic aromatic hydrocarbons. *Huanjing Huaxue Environ. Chem.* **2016**, *35*, 287–296. (In Chinese)
30. Horii, Y.; Ohura, T.; Yamashita, N.; Kannan, K. Chlorinated Polycyclic Aromatic Hydrocarbons in Sediments from Industrial Areas in Japan and the United States. *Arch. Environ. Contam. Toxicol.* **2009**, *57*, 651–660. [CrossRef]
31. Pugazhendi, A.; Qari, H.; Basahi, J.M.A.-B.; Godon, J.J.; Dhavamani, J. Role of a halothermophilic bacterial consortium for the biodegradation of PAHs and the treatment of petroleum wastewater at extreme conditions. *Int. Biodeterior.* **2017**, *121*, 44–54. [CrossRef]
32. Zhou, Q.; Li, D.; Wang, T.; Hu, X. Leaching of graphene oxide nanosheets in simulated soil and their influences on microbial communities. *J. Hazard. Mater.* **2021**, *404*, 124046. [CrossRef]
33. Chao, A. Nonparametric estimation of the number of classes in a population. *Scand. Stat. Theory Appl.* **1984**, *11*, 265–270.
34. Jacquiod, S.; Cyriaque, V.; Riber, L.; Al-Soud, W.A.; Gillan, D.C.; Wattiez, R.; Sørensen, S.J. Long-term industrial metal contamination unexpectedly shaped diversity and activity response of sediment microbiome. *J. Hazard. Mater.* **2018**, *344*, 299–307. [CrossRef] [PubMed]
35. Huang, Y.M.; Michel, K.; An, S.S.; Zechmeister-Boltenstern, S. Changes in microbial-community structure with depth and time in a chronosequence of restored grassland soils on the Loess Plateau in northwest China. *J. Soil Sci. Plant Nutr.* **2011**, *174*, 765–774. [CrossRef]
36. Shannon, C.E. A mathematical theory of communication. *Bell Syst. Tech. J.* **1984**, *27*, 379–423. [CrossRef]
37. Simpson, E.H. Measurement of Diversity. *Nature* **1949**, *163*, 688. [CrossRef]
38. Bo-Ra, K.; Jiwon, S.; Robin, B.G.; Jun Hyung, L.; Doo Wan, K.; Kuk-Hwan, S.; Ju-Hoon, L.; Hyeun Bum, K. Deciphering Diversity Indices for a Better Understanding of Microbial Communities. *J. Microbiol. Biotechnol.* **2017**, *27*, 2089–2093.
39. Good, I.J. The Population frequencies of species and the estimation of population parameters. *Biometrika* **1953**, *40*, 237–264. [CrossRef]

40. Neuteboom, J.H.; Struik, P.C. Variation in rank abundance replicate samples and impact of clustering. *NJAS Wagening. J. Life Sci* **2005**, *53*, 199–222. [CrossRef]
41. Lee, S.-H.; Lee, W.-S.; Lee, C.-H.; Kim, J.-G. Degradation of phenanthrene and pyrene in rhizosphere of grasses and legumes. *J. Hazard. Mater.* **2008**, *153*, 892–898. [CrossRef] [PubMed]
42. Rentz, J.A.; Alvarez, P.J.J.; Schnoor, J.L. Benzo[a]pyrene co-metabolism in the presence of plant root extracts and exudates: Implications for phytoremediation. *Environ. Pollut.* **2005**, *136*, 477–484. [CrossRef]
43. Bray, J.R.; Curtis, J.T. An Ordination of the Upland Forest Communities of Southern Wisconsin. *Ecol. Monogr.* **1957**, *27*, 325–349. [CrossRef]
44. Li, W.; Zhang, Y.; Wu, N.; Zhao, Z.; Xu Wa Ma, Y.; Niu, Z. Colonization Characteristics of Bacterial Communities on Plastic Debris Influenced by Environmental Factors and Polymer Types in the Haihe Estuary of Bohai Bay, China. *Environ. Sci. Technol.* **2019**, *53*, 10763–10773. [CrossRef] [PubMed]
45. Wang, B.; Teng, Y.; Xu, Y.; Chen, W.; Ren, W.; Li, Y.; Christie, P.; Luo, Y. Effect of mixed soil microbiomes on pyrene removal and the response of the soil microorganisms. *Sci. Total Environ.* **2018**, *640–641*, 9–17. [CrossRef] [PubMed]
46. Wang, L.; Zhang, J.; Lu, L. Effects of different concentrations of rhamnolipid on soil polycyclic aromatic hydrocarbons removal rate and microbial community structure. *Environ. Pollut. Control* **2019**, *41*, 901–905. (In Chinese)
47. Zhao, L.; Yao, T.; Zhao, Y.; Sun, S.; Lyu, C.; Zhao, W. Reduction strategies of polycyclic aromatic hydrocarbons in farmland soils: Microbial degradation, plant transport inhibition, and their mechanistic analysis. *J. Hazard. Mater.* **2023**, *465*, 133397. [CrossRef] [PubMed]
48. Babu, A.G.; Reja, S.I.; Akhtar, N.; Sultana, M.; Deore, P.S.; Ali, F.I. Bioremediation of Polycyclic Aromatic Hydrocarbons (PAHs): Current Practices and Outlook. In *Microbial Metabolism of Xenobiotic Compounds*; Arora, P.K., Ed.; Springer: Singapore, 2019; pp. 189–216.
49. Lu, L.; Zhang, J.; Peng, C. Shift of Soil Polycyclic Aromatic Hydrocarbons (PAHs) Dissipation Pattern and Microbial Community Composition due to Rhamnolipid Supplementation. *Water Air Soil Pollut.* **2019**, *230*, 107. [CrossRef]
50. Gupta, S.; Pathak, B. *Abatement of Environmental Pollutants*; Singh, P., Kumar, A., Borthakur, A., Eds.; Elsevier: Amsterdam, The Netherlands, 2020; pp. 127–149.
51. Bao, H.; Wang, J.; Zhang, H.; Li, J.; Li, H.; Wu, F. Effects of biochar and organic substrates on biodegradation of polycyclic aromatic hydrocarbons and microbial community structure in PAHs-contaminated soils. *J. Hazard. Mater.* **2020**, *385*, 121595. [CrossRef] [PubMed]
52. Song, L.; Niu, X.; Tian, Y.; Xiao, Y. Assessment of PAH degradation potential of native species from a coking plant through identifying of the beneficial bacterial community within the rhizosphere soil. *Chemosphere* **2021**, *264*, 128513. [CrossRef] [PubMed]
53. Zhang, W.; Wang, H.; Zhang, R.; Yu, X.-Z.; Qian, P.-Y.; Wong, M.H. Bacterial communities in PAH contaminated soils at an electronic-waste processing center in China. *Ecotoxicology* **2010**, *19*, 96–104. [CrossRef]
54. Gupta, G.; Kumar, V.; Pal, A.K. Microbial Degradation of High Molecular Weight Polycyclic Aromatic Hydrocarbons with Emphasis on Pyrene. *Polycycl. Aromat. Compd.* **2019**, *39*, 124–138. [CrossRef]
55. Gupta, G.; Kumar, V.; Pal, A.K. Biodegradation of Polycyclic Aromatic Hydrocarbons by Microbial Consortium: A Distinctive Approach for Decontamination of Soil. *Soil Sediment Contam.* **2016**, *25*, 597–623. [CrossRef]
56. Kamiya, Y.; Iijima, A.; Ikemori, F.; Okuda, T.; Ohura, T. Source apportionment of chlorinated polycyclic aromatic hydrocarbons associated with ambient particles in a Japanese megacity. *Sci. Rep.* **2016**, *6*, 38358. [CrossRef] [PubMed]
57. Zhao, Z.; Chen, W.; Cheng, Y.; Li, J.; Chen, Z. *Burkholderia cepacia* immobilized onto rGO as a biomaterial for the removal of naphthalene from wastewater. *Environ. Res.* **2023**, *235*, 116663. [CrossRef]
58. Tu, Z.; Qi, Y.; Tang, X.; Wang, Z.; Qu, R. Photochemical transformation of anthracene (ANT) in surface soil: Chlorination and hydroxylation. *J. Hazard. Mater.* **2023**, *452*, 131252. [CrossRef] [PubMed]
59. Aydin, S.; Karaçay, H.A.; Shahi, A.; Gökçe, S.; Ince, B.; Ince, O. Aerobic and anaerobic fungal metabolism and Omics insights for increasing polycyclic aromatic hydrocarbons biodegradation. *Fungal Biol. Rev.* **2017**, *31*, 61–72. [CrossRef]
60. Golan-Rozen, N.; Chefetz, B.; Ben-Ari, J.; Geva, J.; Hadar, Y. Transformation of the Recalcitrant Pharmaceutical Compound Carbamazepine by *Pleurotus ostreatus*: Role of Cytochrome P450 Monooxygenase and Manganese Peroxidase. *Environ. Sci. Technol.* **2011**, *45*, 6800–6805. [CrossRef]
61. Loss, E.M.O.; Lee, M.K.; Wu, M.Y.; Martien, J.; Chen, W.P.; Amador-Noguez, D.; Jefcoate, C.; Remucal, C.; Jung, S.; Kim, S.C.; et al. Cytochrome P450 Monooxygenase-Mediated Metabolic Utilization of Benzo[a]Pyrene by Aspergillus Species. *MBio* **2019**, *10*, e00558-19.
62. Akhtar, M.N.; Boyd, D.R.; Thompson, N.J.; Koreeda, M.; Gibson, D.T.; Mahadevan, V.; Jerina, D.M. Absolute sterochemistry of the dihydroanthracene-cis- and -trans-1,2-diols produced from anthracene by mammals and bacteria. *J. Am. Chem. Soc.* **1975**, *23*, 2506–2511.
63. Mahaffey, W.; Gibson, D.; Cerniglia, C. Bacterial oxidation of chemical carcinogens: Formation of: Polycyclic aromatic acids from benz[a]anthracene. *Appl. Environ. Microbiol.* **1988**, *54*, 2415–2423. [CrossRef]
64. Schneider, J.; Grosser, R.; Jayasimhulu, K.; Xue, W.; Warshawsky, D. Degradation of pyrene, benz[a]anthracene, and benzo[a]pyrene by Mycobacterium sp. strain RJGII-135, isolated from a former coal gasification site. *Appl. Environ. Microbiol.* **1996**, *62*, 13–19. [CrossRef]

65. Chen, M.; Xu, P.; Zeng, G.; Yang, C.; Huang, D.; Zhang, J. Bioremediation of soils contaminated with polycyclic aromatic hydrocarbons, petroleum, pesticides, chlorophenols and heavy metals by composting: Applications, microbes and future research needs. *Biotechnol. Adv.* **2015**, *33*, 745–755. [CrossRef]
66. Bolyen, E.; Rideout, J.R.; Dillon, M.R.; Bokulich, N.A.; Abnet, C.C.; Al-Ghalith, G.A.; Alexander, H.; Alm, E.J.; Arumugam, M.; Asnicar, F.; et al. Reproducible, interactive, scalable and extensible microbiome data science using QIIME 2. *Nat. Biotechnol.* **2019**, *37*, 852–857. [CrossRef] [PubMed]
67. Martin, M. Cutadapt removes adapter sequences from high-throughput sequencing reads. *EMBnet. J.* **2011**, *17*, 10–12. [CrossRef]
68. Callahan, B.J.; Mcmurdie, P.J.; Rosen, M.J.; Han, A.W.; Johnson, A.J.; Holmes, S.P. Dada2: High-resolution sample inference from illumina amplicon data. *Nat. Methods* **2016**, *13*, 581–583. [CrossRef] [PubMed]
69. Katoh, K.; Misawa, K.; Kuma, K.I.; Miyata, T. Mafft: A novel method for rapid multiple sequence alignment based on fast fourier transform. *Nucleic Acids Res.* **2002**, *30*, 3059–3066. [CrossRef]
70. Price, M.N.; Dehal, P.S.; Arkin, A.P. FastTree: Computing large minimum evolution trees with profiles instead of a distance matrix. *Mol. Biol. Evol.* **2009**, *26*, 1641–1650. [CrossRef]

toxics

MDPI

Article

Identification and Mechanistic Analysis of Toxic Degradation Products in the Advanced Oxidation Pathways of Fluoroquinolone Antibiotics

Shuhai Sun [1,†], Zhonghe Wang [2,†], Qikun Pu [2], Xinao Li [2], Yuhan Cui [2], Hao Yang [2,*] and Yu Li [2,*]

1 School of Hydraulic and Environmental Engineering, Changchun Institute of Technology, Changchun 130012, China; sun18117471@163.com

2 MOE Key Laboratory of Resources and Environmental System Optimization, North China Electric Power University, Beijing 102206, China; zhonghe_wang1999@163.com (Z.W.); puqikun2000@163.com (Q.P.); lixinao921734261@163.com (X.L.); 120222232050@ncepu.edu.cn (Y.C.)

* Correspondence: yh13601614368@163.com (H.Y.); liyuxx8@hotmail.com (Y.L.)

† These authors contributed equally to this work.

Abstract: The degradation of fluoroquinolones (FQs) via advanced oxidation processes (AOPs) is a promising avenue, yet the complete mineralization of certain FQ molecules remains elusive, raising concerns about the formation of toxic by-products. This study delineates five primary AOP degradation pathways for 16 commercially available FQ molecules, inferred from existing literature. Density functional theory (DFT) was employed to calculate the bond dissociation energies within these pathways to elucidate the correlation between bond strength and molecular architecture. Subsequently, Comparative Molecular Similarity Index Analysis (CoMSIA) models were constructed for various degradation reactions, including piperazine ring cleavage, defluorination, hydroxylation, and piperazine ring hydroxylation. Three-dimensional contour maps generated from these models provide a deeper understanding of the interplay between FQ molecular structure and bond dissociation energy. Furthermore, toxicity predictions for 16 FQ molecules and their advanced oxidation intermediates, conducted using VEGA 1.2.3 software, indicate that degradation products from pathways P2 and P5 pose a heightened health risk relative to their parent compounds. Furthermore, the application of the Multwfn program to compute the Fukui function for FQ molecules discerns the disparity in degradation propensities, highlighting that N atoms with higher f0 values can augment the likelihood of piperazine ring cleavage. HOMO-LUMO distribution diagrams further confirm that methoxy substitution at the 1-position leads to a dilution of HOMOs on the piperazine ring and an increased energy gap for free radical reactions, diminishing the reactivity with hydroxyl radicals. This study elucidates the pivotal role of structural characteristics in FQ antibiotics for their degradation efficiency within AOPs and unveils the underlying mechanisms of bond dissociation energy disparities. The toxicity parameter predictions for FQ molecules and their intermediates offer unique perspectives and theoretical underpinnings for mitigating the use of high-risk FQs and for devising targeted degradation strategies to circumvent the generation of toxic intermediates in AOPs through molecular structure optimization.

Keywords: fluoroquinolones; advanced oxidation processes; degradation products; environmental and human health risk

check for updates

Citation: Sun, S.; Wang, Z.; Pu, Q.; Li, X.; Cui, Y.; Yang, H.; Li, Y. Identification and Mechanistic Analysis of Toxic Degradation Products in the Advanced Oxidation Pathways of Fluoroquinolone Antibiotics. *Toxics* **2024**, *12*, 203. https://doi.org/10.3390/toxics12030203

Academic Editor: Łukasz Chrzanowski

Received: 26 January 2024
Revised: 4 March 2024
Accepted: 4 March 2024
Published: 6 March 2024

1. Introduction

Antibiotics, as critical agents in the treatment and prophylaxis of bacterial infections, are renowned for their robust antimicrobial capabilities, rapid onset, and significant therapeutic benefits, leading to their extensive deployment in healthcare and animal husbandry [1]. The proliferation of antibiotics has been instrumental in saving countless lives.

However, this widespread usage has a significant environmental repercussion: the discharge of antibiotic-rich wastewater and waste from hospitals, aquaculture, pharmaceutical industries, and waste disposal processes, leading to grave environmental pollution [2]. Research has revealed that the concentration of ciprofloxacin (CIP) in surface waters can reach several hundred nanograms per liter [3], and in wastewater treatment plant effluents, concentrations can soar to 5 micrograms per liter [4]. Alarmingly, concentrations in wastewater from hospitals and pharmaceutical plants can reach milligram per liter magnitudes [5]. The persistence of FQs and their resistance to conventional biological degradation methods result in their frequent detection in various settings, posing insidious threats to ecological systems and human health [6]. Consequently, the development and implementation of efficacious treatment technologies for FQ-laden wastewater is of paramount importance.

The array of antibiotic removal methodologies encompasses coagulation, membrane separation, adsorption, and biodegradation [7]. However, these approaches often fall short in widespread application due to suboptimal removal efficiencies and prohibitive operational costs [7–10]. Advanced oxidation processes (AOPs) stand out for their efficacy in treating a wide array of recalcitrant organic pollutants, making them suitable for the decomposition of all antibiotic categories [11]. For example, biochar can be used as a catalyst to adsorb persulfate to oxidize NOR [11]. AOPs leverage potent oxidizing free radicals, such as hydroxyl radicals (HO·), singlet oxygen ($^1O_2$), and superoxide anion radicals (O^-), to decompose complex organic molecules into simpler substances, thereby facilitating the efficient treatment of recalcitrant organic contaminants [12]. However, the performance of AOPs in degrading different FQ antibiotics is markedly variable in complex wastewater scenarios [13]. For instance, Sciscenko et al. [14] reported variable degradation rates for different FQ antibiotics using H_2O_2/sunlight-based AOPs, with enrofloxacin, sarafloxacin, and ofloxacin showing a descending order of degradation after 120 min. Ao et al. [15] observed that ultraviolet/peracetic acid (MPUV/PAA) processes achieved removal rates of 92.1% and 96.5% for levofloxacin and norfloxacin, respectively, under 500 mJ/cm of ultraviolet irradiation. Bobu et al. [16] reported significant differences in the removal rates of enrofloxacin and ciprofloxacin using ozone/ultraviolet-based AOPs. Thus, the structural variances of FQ antibiotics significantly influence their degradability in AOPs. Hence, the structural diversity of FQ antibiotics plays a critical role in their degradation potential during AOPs. Current research trends focus on the development of novel AOPs that are both more efficient and cost-effective, with an emphasis on increasing radical concentrations to boost degradation. However, there is a scarcity of studies that investigate the relationship between the structural characteristics of antibiotics and their degradability. Investigating this relationship could be of paramount importance for theoretically underpinning the enhancement of degradation efficiency.

While AOPs have shown efficacy in breaking down antibiotics, the limited mineralization and retention of bioactive configurations could lead to degradation products with equal or greater toxicity [1]. Li et al. [17] applied QSAR methodologies to forecast the genotoxic potential of CIP and its by-products, finding increased genotoxicity in certain intermediates compared to the original compound. Wang et al. [18] developed QSAR models with high predictive accuracy for evaluating the toxicity of FQs' photodegradation products, identifying an increase in toxicity for some products. Wachter et al. [19] leveraged the ECOSAR model to estimate the acute and chronic toxicity of CIP and its intermediates on various aquatic organisms, uncovering potential chronic threats. Predictive modeling is thus essential for ecological and health risk assessments of FQs and their intermediates in AOPs, offering unique perspectives and theoretical backing for mitigating the use of FQs with high-risk profiles and for devising strategies to prevent the production of highly toxic intermediates by optimizing molecular structures.

While extensive research has been conducted on the environmental risks of fluoroquinolone (FQ) antibiotics, the environmental and health risks of their degradation intermediates remain underexplored. Therefore, summarizing the toxicity of these intermediates is crucial for reducing the use of FQ parent molecules. This study aims to provide insights and theoretical support for controlling the generation of less toxic degradation pathways in advanced oxidation processes (AOPs) through molecular structure analysis. The primary research goals are (1) to calculate bond dissociation energies for FQ molecules in various advanced oxidation pathways using density functional theory and to identify patterns between bond energy and molecular structure through cluster analysis; (2) to create a 3D-QSAR model of FQ molecules' degradation capacity in AOPs using SYBYL-X2.0, delving into how molecular structure impacts bond dissociation energy; (3) to forecast 17 toxicity parameters for 16 FQ molecules and their intermediates with VEGA 1.2.3, evaluating the health and ecological risks; (4) to construct a 2D-QSAR regression model via machine learning, linking FQ degradation potential with molecular features; and (5) to employ Fukui functions and HOMO-LUMO distribution to reveal the degradation process's bond dissociation energy mechanisms. This research will explore methods for enhancing the degradation of FQ molecules in AOPs and provide theoretical support for minimizing the use of high-risk FQ molecules and preventing the formation of highly toxic intermediates.

2. Materials and Methods

2.1. Source of FQ Molecular Structure—PubChem Retrieval Method

Existing literature reveals that 16 commercially available fluoroquinolones (FQs), including ciprofloxacin (CIP), norfloxacin (NOR), gatifloxacin (GAT), and sparfloxacin (SPA), are degradable in advanced oxidation systems (AOPs), with reaction pathways such as piperazine ring cleavage (P1), defluorination (P2), hydroxylation (P3), piperazine ring hydroxylation (P4), and decarboxylation (P5) [15,20]. Therefore, this study retrieved and obtained the 3D structures of 16 commercial FQ molecules from the "PubChem—3D Conformer" module in the National Library of Medicine database (https://pubchem.ncbi.nlm.nih.gov/ URL (accessed on 20 July 2023)) and used Gaussian 09 software to optimize the structures at the B3LYP/6-31G(d) basis set level. These optimized structures served as the initial structures for subsequent analysis of FQ degradation patterns and characteristics.

2.2. Characterization of the Degradation Capability of FQ Molecules in Advanced Oxidation Systems—DFT Coupled with Negative Index Calculation Method

To comprehensively evaluate the degradation capabilities of different FQ molecules in AOPs and to analyze the degradation patterns and characteristics of different FQs, this study first searched the Web of Science literature database for the degradation methods of FQ antibiotics in AOPs, summarized the common features of FQ molecule degradation pathways in AOPs, and inferred five main advanced oxidation degradation pathways (Figure 1). Secondly, GaussView 5.0.8 software was used to draw the structures of reactants and products in the primary degradation pathways of 16 commercial FQs. Finally, based on density functional theory at the M062X/6-31G(d) basis set level [21], the Gaussian software was used to calculate the ease of primary degradation reactions of 16 FQ molecules in AOPs (characterized by the bond dissociation energy (kJ/mol) in the advanced oxidation degradation pathway). The lower the bond dissociation energy value, the easier the internal chemical bonds of the FQ molecules break, indicating higher degradability of the FQ molecules in the AOPs [22].

Given that the lower the bond dissociation energy of the FQ molecules, the easier they are degraded in the AOPs, this study employed a negative index calculation method to standardize the bond dissociation energy values of 16 commercial FQ molecules [23]. The standardization of the bond dissociation energy of FQs in AOPs is as follows:

$$X'_{ix} = (\max\{X_{ix}\} - X_{ix})/(\max\{X_{ix}\} - \min\{X_{ix}\}) \quad (1)$$

where, i represents different commercial FQ molecules, $i = 1, 2, \ldots, 16$; x represents five distinct primary degradation pathways; X_{ix} denotes the energy required to break a chemical bond in the i FQ molecule along the x primary degradation pathway; $min\{X_{ix}\}$ and $max\{X_{ix}\}$ represent the maximum and minimum bond dissociation energies, respectively, for the i FQ molecule in the x primary degradation pathway; X'_{ix} signifies the normalized value of the energy required to break the chemical bond of the i FQ molecule in the x primary degradation pathway.

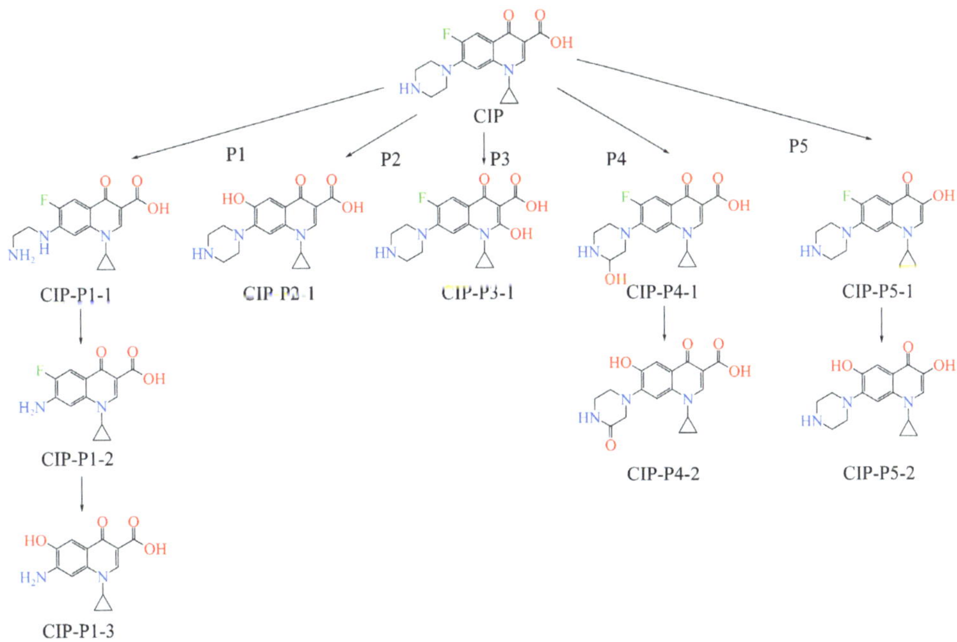

Figure 1. Inferred degradation pathways of FQ molecules in advanced oxidation systems (taking CIP as an example).

2.3. *Construction of 3D-QSAR Models for the Degradation Capability of FQ Molecules in Advanced Oxidation Systems—SYBYL Software Method*

To delve into the degradation characteristics of FQ molecules in advanced oxidation systems, this study leveraged SYBYL-X2.0 software to build a 3D-QSAR model that predicts the degradation potential of FQ molecules. The study began by importing the structural files (mol2 format) of 16 commercially available FQ molecules into the "Compute—Minimize" module of SYBYL for conformational optimization, applying Gasteiger–Hückel charges to achieve the lowest possible energy state for the FQ structures (optimization parameters: Max Iterations: 10,000 Times [24]; Min Energy Change: 0.005 kcal/mol). The optimized FQ molecules were then randomly split into a training set and a test set at a ratio of about 3:1 [22]. The study constructed four 3D-QSAR models to represent the degradation capabilities of FQs for piperazine ring cleavage, defluorination, hydroxylation, and piperazine ring hydroxylation. Each model used a specific FQ molecule as a template, with a selection of FQ molecules assigned to the training and test sets (depicted in Figure 2). Finally, the "Align Database" module was utilized to superimpose the common skeletons based on the template molecules (Figure 2).

To explore the degradation characteristics of FQ molecules in advanced oxidation systems, a 3D-QSAR model was constructed using Comparative Molecular Similarity Index Analysis (CoMSIA). The model employed partial least squares (PLS) regression to establish a quantitative relationship between the three-dimensional structural features of FQ molecules and their degradation capabilities in advanced oxidation processes, followed by a validation of the model's soundness [25]. The procedure involved the following steps: First, the leave-one-out (LOO) method was employed for cross-validation of the training set molecules to obtain the cross-validation coefficient q^2 (greater than 0.5) and the optimal number of principal components N. Next, non-cross-validation regression analysis was performed to secure the non-cross-validation coefficient r^2 (greater than 0.9), the standard error of estimation (SEE), and the test statistic F [26]. Finally, the test set underwent cross-validation to determine r^2_{pred} (greater than 0.6). Once the model parameters were verified as reasonable, the 3D-QSAR model was completed [27]. Subsequently, the "3D-QSAR—View QSAR" function in SYBYL software was used to create three-dimensional contour maps of the CoMSIA model, which illustrate the degradation capabilities of FQ molecules in advanced oxidation systems (Figure S1). These contour maps provide a visual representation of the hydrophobic, electrostatic, steric, hydrogen bond acceptor, and hydrogen bond donor fields, facilitating an analysis of the interplay between the bond dissociation energy and molecular structure of FQ molecules in advanced oxidation processes.

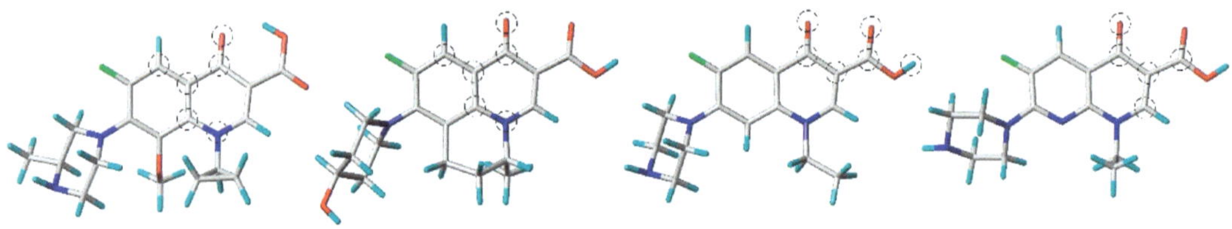

(a) (b) (c) (d)

Figure 2. Template molecules of the 3D-QSAR model for the degradation capability of FQ molecules in advanced oxidation systems: (**a**) gatifloxacin; (**b**) nalidixic acid; (**c**) norfloxacin; (**d**) enoxacin. The atoms circled in the figure represent the common skeleton overlaid in the 3D-QSAR model (O is red, N is blue, F is green).

2.4. Construction of Lasso Regression Model for the Degradation Capability of FQ Molecules in Advanced Oxidation Systems—Jupyter Notebook Tool

Machine learning (ML) is a sophisticated data mining technique that focuses on processing and distilling the interrelationships within complex datasets into mathematical representations [28]. The Lasso regression algorithm, known for constructing linear regression models with excellent stability and predictive precision [29], was selected for developing a 2D-QSAR model that predicts the bond dissociation energy of FQ molecules during advanced oxidation processes. The initial phase involved using the PaDEL 2.21-Descriptor software to forecast five principal structural property parameters across 11 commercially available FQ molecules featuring a piperazine ring structure. These parameters spanned geometric, electronic, physicochemical, spectral, and topological properties, with a comprehensive list of 1014 descriptors. In the subsequent phase, the decision tree algorithm was employed to prioritize these descriptors based on their impact on the FQ molecules' bond dissociation energy. This prioritization process culminated in the identification of nine pivotal FQ molecular feature descriptors with the most significant correlation to bond dissociation energy. The final phase entailed the application of these nine descriptors as independent variables, with the bond dissociation energy of FQ molecules as the dependent variable, to compile and execute the Lasso regression algorithm. This predictive modeling endeavor served to further dissect and corroborate the correlation between the structural features of FQs and their degradation efficacy.

2.5. Toxicity Risk Assessment of Parent FQs and Their Degradation Products in Advanced Oxidation Systems—VEGA Software Method

In an effort to comprehensively assess the potential risks posed to human health and the environment by fluoroquinolone (FQ) antibiotics and their degradation products in advanced oxidation systems, this study utilized the VEGA 1.2.3 software to forecast a suite of 17 toxicological parameters. These indicators included the modulation of androgen and estrogen receptor activities, potential for carcinogenesis, chromosomal aberrations, mutagenicity, developmental toxicity, and screening for endocrine disruptor activities. The analysis further extended to hepatotoxicity, acute toxicity, ocular and dermal irritation, skin sensitization, chronic toxicity to aquatic algae, bioaccumulation potential as indicated by the bioconcentration factor, median lethal concentrations for Daphnia magna at 48 h ($LC_{50}DM$), median lethal concentrations for fathead minnow at 96 h ($LC_{50}FM$), and the degradation half-life (kM/Half-Life). The predictive outcomes were intended to furnish a nuanced assessment of the health and ecological risks associated with FQs and their oxidative degradation intermediates.

3. Results and Discussion

3.1. Evaluation of the Degradation Capability of FQ Molecules in Advanced Oxidation Systems Based on DFT Method

In the study delineated in Section 2.2, bond dissociation energies were quantified for 16 commercially available FQ molecules undergoing primary degradation in five distinct advanced oxidation processes, with the energy values being recalibrated into kJ/mol as presented in Table 1. This investigation sought to elucidate the relative degradation difficulty of various FQ molecules when subjected to advanced oxidation systems. An analysis of the average bond dissociation energies for 11 FQ molecules with a piperazine ring structure across five advanced oxidation processes revealed that LOM and OFL molecules possess the lowest bond energies, indicating these FQs are predisposed to facile degradation in advanced oxidation contexts. In contrast, molecules such as ENO, FLE, GAT, LEV, PEF, RUF, and SPA demonstrated the highest average bond energies, indicating reduced degradability within advanced oxidation processes. The CIP and NOR molecules had average bond energies that were intermediate, suggesting a certain level of degradability within advanced oxidation frameworks. Moreover, for a holistic comparison of the degradation ease of the 16 FQ molecules, this study also calculated the average bond dissociation energies for carbon–fluorine bonds in defluorination reactions, carbon–hydrogen bonds in hydroxylation reactions, and carbon–carbon bonds in decarboxylation reactions (as shown in Table 1). The findings suggested that molecules such as CIP, CLI, ENO, LEV, LOM, MOX, NOR, and OFL, with lower average energies for these three bond types, possess a higher degradation potential in advanced oxidation systems. Conversely, molecules like FLE, GAT, NAD, PEF, RUF, SPA, TOS, and GEM, with higher average bond energies, pose a greater challenge for degradation in advanced oxidation processes, necessitating focused attention on their degradability.

Structural analysis of different FQ molecules has indicated that in hydroxylation reactions, those with oxygenated heterocycles at the 1-position and 9-position (Figure 3) of the FQ framework (such as OFL and LEV) possess lower carbon–hydrogen bond energies. In contrast, molecules with sulfur-containing or carbon-only heterocycles at these positions (e.g., RUF, NAD) have higher carbon–hydrogen bond energies. In defluorination reactions, FQ molecules with an ethyl substituent at the 9-position (e.g., LOM, NOR, ENO, PEF) demonstrate lower carbon–fluorine bond energies than those with a cyclopropyl substituent (e.g., CLI, CIP, MOX, GEM, GAT, SPA). Yet, in piperazine ring cleavage reactions, the ethyl-substituted group shows higher carbon–nitrogen bond energies within the ring. Moreover, FQ molecules with a fluorine atom at the 1-position (e.g., LOM, FLE, SPA) have lower carbon–nitrogen bond energies in the piperazine ring than those lacking a fluorine atom (e.g., NOR), shedding light on why CIP, with a cyclopropyl group, has higher carbon–nitrogen bond energies in the piperazine ring than LOM with an ethyl group. Additionally, FQs with a methoxy group at the 1-position (e.g., GAT) have higher carbon–nitrogen bond energies in the piperazine ring, which may account for the high bond energies in GAT molecules despite the presence of a cyclopropyl group. Comparing NOR and PEF molecules, it was found that FQs with a para-methyl group on the piperazine ring (e.g., PEF) have not only lower carbon–nitrogen bond energies in the ring but also lower carbon–carbon bond energies in decarboxylation reactions. However, these molecules display higher bond energies for carbon–hydrogen, carbon–fluorine, and carbon–hydrogen in the piperazine ring during defluorination, hydroxylation, and piperazine ring hydroxylation reactions, respectively. This further explains the unusual higher carbon–fluorine bond energies in PEF molecules with an ethyl group compared to CIP with a cyclopropyl group.

Table 1. Calculation of bond dissociation energies during the primary degradation process of FQ molecules in advanced oxidation systems.

FQ Molecule	Bond Dissociation Energy (kJ/mol)						
	Bond Dissociation Energy 1	Bond Dissociation Energy 2	Bond Dissociation Energy 3	Bond Dissociation Energy 4	Bond Dissociation Energy 5	Average Bond Dissociation Energy 2, 3, 5	Average Bond Dissociation Energy 1–5
CIP	315.215	475.796	428.048	345.621	420.067	441.304	396.949
CLI	-	476.809	427.754	-	417.685	440.750	-
ENO	332.520	469.497	435.749	354.007	419.707	441.651	402.296
FLE	323.916	478.663	432.779	356.372	419.095	443.513	402.165
GAT	345.581	478.644	436.300	355.487	419.518	444.821	407.106
LEV	323.651	480.971	424.094	362.482	419.069	441.378	402.053
LOM	302.263	468.812	421.597	335.379	420.174	436.861	389.645
MOX	-	474.677	430.099	-	420.721	441.832	-
NAD	-	485.717	434.137	-	422.396	447.417	-
NOR	334.738	473.383	420.770	343.804	421.537	438.563	398.846
OFL	285.316	472.396	424.136	338.739	418.284	438.272	387.774
PEF	332.362	478.256	433.633	355.115	420.883	444.257	404.050
RUF	322.860	483.039	435.182	354.429	420.621	446.281	403.226
SPA	329.582	479.695	435.563	359.746	424.528	446.595	405.822
TOS	-	481.703	439.522	-	418.397	446.541	-
GEM	-	477.578	434.822	-	417.869	443.423	-

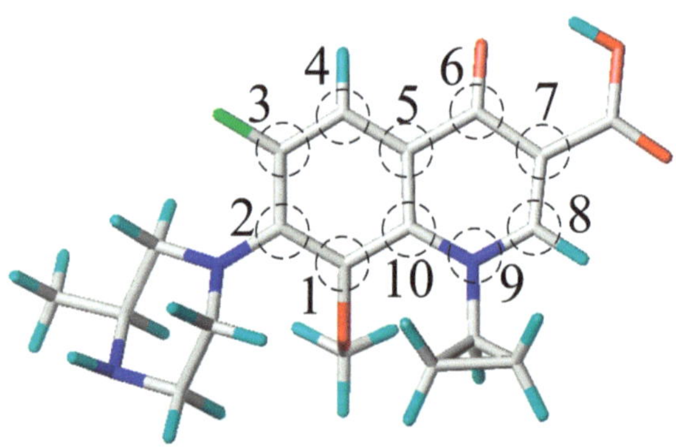

Figure 3. Basic skeleton site map of FQ molecules. (Numbers 1 to 10 represent position).

Additionally, FQ molecules with an ethyl group at the 9-position have higher bond energies than those with a fluoroethyl group at the same position (FLE). However, the dual fluorine atoms connected to the benzene ring of the FLE molecule result in lower carbon–nitrogen bond energies in the piperazine ring compared to PEF. For SPA molecules, the elevated average bond energy is attributed to the amino group at the 4-position of the FQ core. The study also found that FQ molecules with a pyridine core (position 1 as nitrogen, ENO) have lower carbon–fluorine and carbon–carbon bond energies in defluorination and decarboxylation reactions, respectively, and higher carbon–hydrogen bond energies in hydroxylation reactions compared to those with a benzene core (position 1 as carbon, NOR). The carbon–nitrogen bond energies in the piperazine ring are lower, but the carbon–hydrogen bond energies on the ring are higher.

Collectively, it can be stated that FQ molecules with an ethyl substituent at the 9-position and a methoxy substituent at the 1-position of the fundamental skeleton exhibit a lower propensity for piperazine ring cleavage reactions. FQ molecules with a para-methyl group on the piperazine ring and a fluorine atom at position 1 of the core skeleton are more likely to experience such cleavage. Molecules with a cyclopropyl substituent at the 9-position and those with a pyridine moiety in the core are more susceptible to defluorination reactions. FQ molecules with an amino substituent at the 4-position and those with sulfur-containing or carbon-only heterocycles at the 1-position and 9-position are less likely to engage in hydroxylation reactions; in contrast, those with oxygen-containing heterocycles at these positions are more conducive to hydroxylation. Furthermore, FQs with an amino substituent at the 4-position and those with a pyridine moiety are less prone to undergo ring hydroxylation reactions.

3.2. Cluster-Analysis-Based Characterization of FQ Molecules' Degradation Features in Advanced Oxidation Systems

Cluster analysis, an unsupervised machine learning technique, is adept at selecting representative features to streamline data dimensions and complexity, enabling a thorough investigation and visualization of the inherent relationships within data [30]. This approach is well suited for the current study's need to deeply explore the link between the structural features of FQs and their degradation difficulty, despite the small data volume [30]. Using Origin software's Heatmap tool, this study conducted cluster analysis on the bond dissociation energies of 11 FQ molecules across various reactions including defluorination, decarboxylation, hydroxylation, piperazine ring cleavage, and piperazine ring hydroxylation (Figure 4). The heatmap's gradient from red to blue signifies increasing bond dissociation energies. The clustering outcome divides the FQ molecules into three distinct groups based on their degradation ease in advanced oxidation systems. LOM and OFL form one group, predominantly red on the heatmap, signifying their greater ease of degradation. FLE, GAT, LEV, PEF, RUF, and SPA are clustered into a second group, marked mostly by purple and blue, denoting a lower likelihood of degradation. ENO, CIP, and NOR constitute a third group, with a mix of red and purple, suggesting their degradation difficulty is intermediate between the two groups. Additionally, the row clustering of bond dissociation energies for defluorination, hydroxylation, and decarboxylation reactions categorizes the FQ molecules into two groups. NOR, LOM, LEV, OFL, CLI, MOX, and CIP, with a heatmap dominated by red, are more prone to degradation, while SPA, NAD, RUF, PEF, TOS, GEM, GAT, FLE, and ENO, with purple and blue, are less degradable. The overall analysis indicates that the cluster analysis results align well with the conclusions of Section 3.1.

To explore the collective tendencies in the ease of bond cleavage during the primary degradation reactions of FQ molecules in advanced oxidation systems, this study conducted a column cluster analysis of the bond dissociation energies for 16 FQ molecules across these five degradation processes (illustrated in Supplementary Figure S2, with missing values filled in using the average of the column). The analysis shows that the carbon–nitrogen bonds of FQ molecules have the lowest overall energies in piperazine ring cleavage and hydroxylation reactions, indicating a higher propensity for these reactions to occur in advanced oxidation systems. The carbon–fluorine bond energies are the highest in defluorination reactions, suggesting that defluorination is the most challenging reaction to take place. The carbon–hydrogen bond energies, including those in the piperazine ring, fall between the previous two categories, implying a moderate level of difficulty for hydroxylation and piperazine ring hydroxylation reactions within advanced oxidation systems.

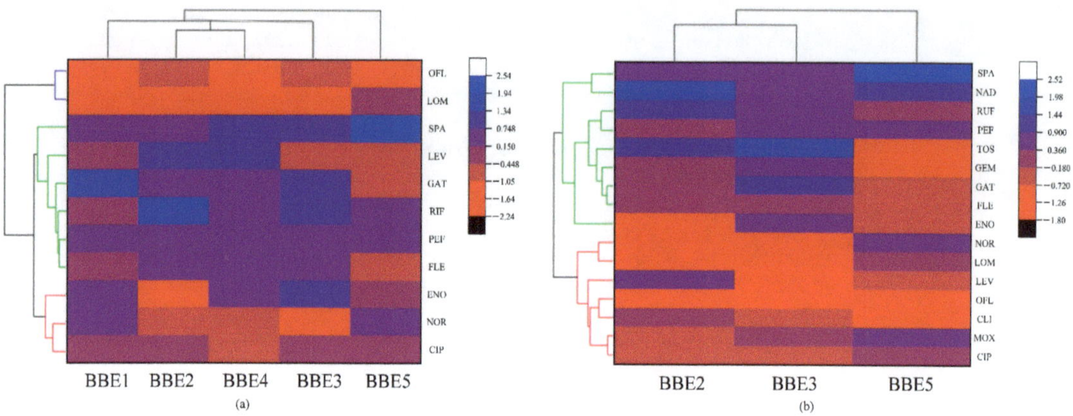

Figure 4. Cluster analysis graph of FQ molecules: (**a**) Cluster analysis of bond dissociation energies during the defluorination, decarboxylation, hydroxylation, piperazine ring breaking, and piperazine ring hydroxylation reactions of 11 FQ molecules. (**b**) Cluster analysis of bond dissociation energies during the defluorination, hydroxylation, and decarboxylation reactions of 16 FQ molecules.

3.3. *Differential Analysis of the Degradation Capability of FQ Molecules in Advanced Oxidation Systems Based on 3D-QSAR Models*

Utilizing bond dissociation energies from the primary degradation pathways of fluoroquinolone molecules, this research constructs a CoMSIA model to assess bond cleavage energies within advanced oxidation systems. It was found that the energy differences for decarboxylation among 16 commercial FQ molecules are below 2%, indicating that factors other than the core FQ structure have a minimal influence on this reaction. Attention is thus directed to CoMSIA models based on the bond dissociation energies from the other four degradation processes (evaluation parameters are shown in Table 2). The CoMSIA model for piperazine ring cleavage in FQ molecules, with a principal component of 5, exhibits a q^2 value of 0.516, indicating robust predictive capabilities [31]. The model's SEE of 0.001 and F-value of 227534.936, along with an R^2 of 1.000, underscore its stability and predictive strength [31]. The external validation coefficient r^2_{pred} of 0.972 and SEP of 0.043 further validate the model's external predictive capacity [32]. All CoMISA models meet the evaluation criteria, attesting to the reliability of these models. The three-dimensional contour maps for the piperazine ring cleavage model (Figures 5 and S2) demonstrate that the hydrogen bond acceptor, hydrophobic, and electrostatic fields collectively contribute 85.7% to the carbon–nitrogen bond, signifying their substantial influence on the cleavage reaction in FQ molecules. Detailed contour map information for the additional bond cleavage models is presented in Table 3.

Table 2. CoMSIA model evaluation parameters for the bond dissociation energies of FQ molecules in advanced oxidation systems.

CoMSIA	q^2	N	R^2	SEE	F	r^2_{pred}	SEP
Piperazine ring cleavage	0.516	5	1.000	0.001	227534.936	0.972	0.043
Defluorination	0.595	6	1.000	0.002	26018.716	0.992	0.062
Hydroxylation	0.628	7	1.000	0.008	1926.276	0.878	0.018
Piperazine ring hydroxylation	0.622	2	0.991	0.037	268.732	0.999	0.006

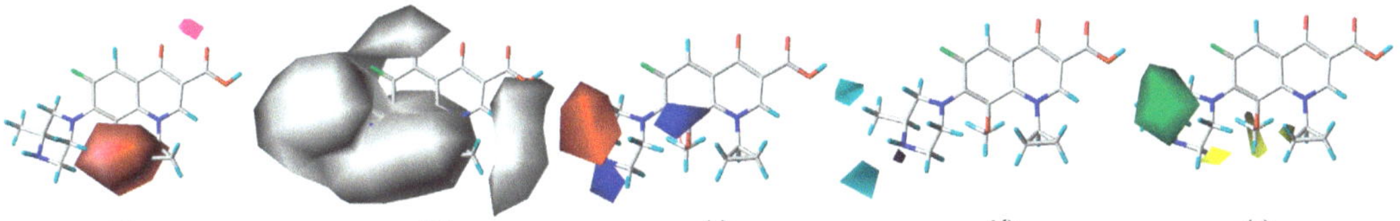

Figure 5. CoMSIA model contour maps of piperazine ring breaking in FQ molecules: (a) hydrogen bond acceptor field; (b) hydrophobic field; (c) electrostatic field; (d) hydrogen bond donor field; (e) steric field.

Table 3. Molecular force field contributions of the CoMSIA model for the bond dissociation energies of FQ molecules in advanced oxidation systems.

CoMSIA	Hydrogen Bond Acceptor Field (%)	Hydrophobic Field (%)	Electrostatic Field (%)	Hydrogen Bond Donor Field (%)	Steric Field (%)
Piperazine ring cleavage	49.30	21.00	15.40	7.30	7.00
Defluorination	43.20	22.30	18.20	8.60	7.80
Hydroxylation	25.30	24.70	24.50	6.20	19.20
Piperazine ring hydroxylation	40.60	16.90	28.50	7.70	6.30

The CoMSIA model's contour maps for the piperazine ring cleavage of FQ molecules in advanced oxidation systems (Figure 5) show that the GAT molecule, in comparison to CIP, exhibits red blocks (indicative of the hydrogen bond acceptor field) and blue blocks (indicative of the electrostatic field) around the first position of its core structure. The bond energy of GAT's piperazine ring carbon–nitrogen bond is 9.63% higher than that of CIP, suggesting increased resistance to cleavage for GAT in such oxidative environments. An analysis of the substituents at the 1-position of the core structures shows that GAT possesses a methoxy group, serving as both a hydrogen bond acceptor and an electronegative entity, unlike CIP, matching the contour map's field patterns. Furthermore, the hydrophobic field contour map comparison between CIP and NOR shows white blocks around the 9-position of CIP's core structure, and its piperazine ring's carbon–nitrogen bond has a 5.83% lower bond energy than NOR's, suggesting a greater ease of cleavage for CIP in advanced oxidation systems. The substituent analysis at the 9-position shows CIP's cyclopropyl group to be more hydrophilic than NOR's ethyl group, consistent with the hydrophobic field contour map.

In the advanced oxidation systems, the defluorination CoMSIA model's hydrogen bond acceptor field contour map (Figure S2) for FQ molecules reveals magenta blocks surrounding the 1-position of the ENO molecule's basic skeleton. The carbon–fluorine bond energy in ENO is 0.82% lower than that in NOR, suggesting an increased likelihood of defluorination for ENO. An analysis of the substituents at the first position of the core structures of these FQ molecules shows that ENO has a nitrogen atom that serves as a hydrogen bond acceptor, unlike NOR. This is in agreement with the hydrogen bond acceptor field observed in the contour maps. Furthermore, the hydrophobic field contour map comparison between GAT and LOM reveals yellow blocks around the 1-position of LOM's core structure, with LOM's carbon–fluorine bond energy being 2.05% lower than GAT's, suggesting LOM is more susceptible to defluorination. The substituent analysis at the 1-position shows that LOM's ethyl group is more hydrophobic than GAT's cyclopropyl group, consistent with the hydrophobic field contour map.

The hydrophobic field contour map for the hydroxylation CoMSIA model of FQ molecules in advanced oxidation systems (Figure S2) shows that SPA molecules exhibit yellow blocks at the 4-position of their basic skeleton, in contrast to LOM molecules. The carbon–hydrogen bond energy in SPA is 3.31% higher than that in LOM, suggesting a decreased likelihood of hydroxylation for SPA. Further comparison of the characteristics of the groups at the 4-position of the basic skeleton of the two aforementioned FQ molecules reveals that the SPA molecule possesses a more hydrophilic amino group, while the one at the 4-position of the LOM molecule possesses a more hydrophobic H atom, which coincides with the color-blocking pattern of the hydrophobic field in the contour maps. In the piperazine ring hydroxylation CoMSIA model for FQ molecules (Figure S2), SPA exhibits magenta blocks at the fourth position of its basic skeleton when compared to LOM, and the carbon–hydrogen bond energy in SPA is 7.27% higher than that in LOM, indicating a greater resistance to piperazine ring hydroxylation. The presence of an amino group in SPA, acting as a hydrogen bond donor, compared to the lack of such a group in LOM, matches the hydrophobic field patterns observed in the contour maps.

The analysis leads to the conclusion that FQ molecules with either a hydrogen bond donor or a positive group at the 1-position and a hydrophilic group at the 9-position of their basic skeleton have an increased likelihood of carbon–nitrogen bond cleavage within the piperazine ring. Molecules with a hydrogen bond donor or hydrophobic group at the 1-position are more inclined to participate in defluorination reactions. Those with hydrophobic groups at the 4-position are more amenable to hydroxylation reactions. FQ molecules with a hydrogen bond acceptor group at the 4-position of their basic skeleton are more readily subject to piperazine ring hydroxylation.

3.4. Toxicity Risk Assessment of Degradation Products of FQ Molecules in Advanced Oxidation Systems Based on Toxicokinetic Models

Although advanced oxidation processes demonstrate significant efficacy in the removal of FQ molecules, their constrained mineralization ability results in the formation of numerous intermediates with potential toxicity, leading to ecological concerns [1]. Studies have identified that out of 30 potential degradation products of ciprofloxacin, 28 pose risks of mutagenicity and are challenging to biodegrade, with all products carrying certain levels of toxicity and carcinogenicity [33]. Likewise, the 25 degradation products of norfloxacin are characterized by mutagenicity, toxicity, carcinogenicity, and biodegradation resistance [33]. The objective of this study is to investigate the toxicity variation of intermediate products during the degradation of 16 commercial FQ molecules, aiming to circumvent pathways with toxic intermediates and reduce the potential toxicity risks associated with FQs and their degradation products. The research employs VEGA 1.2.3 software for the risk assessment of human health and the ecological impact of these FQ molecules and their intermediates (Table S1).

3.4.1. Human Health Risk Assessment of Degradation Products of FQ Molecules in Advanced Oxidation Systems

Table S1 indicates that the 16 commercial FQ molecules are assessed as inactive for both estrogenic and androgenic receptor-mediated effects, as well as for endocrine-disrupting activity, suggesting no associated risk from these FQ molecules in these areas. Further risk analysis for carcinogenicity, mutagenicity, and chromosomal aberrations among the 16 FQ molecules reveals that, apart from LOM and MOX, the rest exhibit potential risks in one or more of these categories. PEF and GEM, in particular, are identified as having all three potential genotoxic "triad" effects, corroborating findings from research by Segalin et al. [33]. Evaluations of developmental toxicity and hepatotoxicity risks show that all 16 FQ molecules are implicated, with 12 of them—CIP, CLI, ENO, FLE, GAT, MOX, NAD, NOR, PEF, RUF, SPA, and TOS—facing risks in both categories. Risk assessments for eye irritation, skin sensitization, and skin irritation among the 16 commercial FQ molecules reveal risks of eye irritation or skin sensitization for all molecules, with no skin irritation risks identified. The acute toxicity assessment, based on LD50 predictions, reveals significant variability, identifying RUF as the molecule with the highest acute toxicity risk and ENO as the least risky. Furthermore, the study employs a negative index standardization approach to process acute toxicity data, calculating a comprehensive human health risk index for each FQ molecule, with PEF presenting the highest risk and LOM the lowest.

An analysis of Table S1 for the 16 commercial FQ molecules and their degradation products indicates that the risk assessments for estrogenic and androgenic receptor-mediated effects, as well as endocrine-disrupting activities, are consistently inactive. This demonstrates that the degradation intermediates of the 16 FQ molecules do not exhibit a significant change in risk levels for these effects and activities compared to the parent FQ molecules. Further scrutiny reveals that, except for MOX, the degradation intermediates of the other 15 FQ molecules all exhibit potential carcinogenic, mutagenic, or chromosomal aberration risks. To explore the trends of these risks during degradation, the study assigns a value of 0 for no risk and 1 for risk presence, calculating a composite index for the "triad" risks of the FQ molecules and their intermediates (Table S1). The results indicate that the degradation intermediates of CIP, CLI, ENO, FLE, LEV, LOM, NAD, RUF, and SPA have an increased average "triad" risk index compared to their parent molecules, suggesting an overall elevation in these risks during the degradation process. The developmental toxicity and hepatotoxicity risk assessments indicate that all intermediates of the 16 FQ molecules carry these risks, with the exception of GEM and some intermediates of FLE, all other intermediates exhibit both risks. The composite index for these risks (Table S1) shows that the degradation intermediates of LEV, LOM, and OFL have increased developmental toxicity and hepatotoxicity risks compared to the parent molecules, while the risks for the intermediates of other FQ molecules remain the same or decrease. Moreover, all FQ molecule degradation intermediates, except for some intermediates of FLE, GAT, and LEV, present risks of eye irritation, skin sensitization, or skin irritation. Notably, only two intermediates of SPA do not pose a risk of skin irritation. The composite index for these risks (Table S1) reveals that the degradation intermediates of SPA, RUF, and GEM show an increasing trend in these risks compared to the parent molecules, while the risk levels for the intermediates of NAD and TOS remain unchanged, and the risks for the intermediates of the other 11 FQ molecules decrease. The acute toxicity risk assessment results indicate that, apart from LEV, OFL, RUF, and TOS, the acute toxicity of the degradation intermediates of the other FQ molecules generally shows an upward trend (e.g., the primary degradation products of ENO show an average increase in toxicity of 25.42% over the parent molecule). In addition, the overall human health risk index for the degradation intermediates of the 16 FQ molecules shows that, except for LEV, LOM, NAD, RUF, SPA, and GEM, the human health risks for the remaining 10 FQ molecules decrease in varying degrees during degradation.

To assess the human health risks of intermediates from various FQ degradation pathways, this study adds up the overall human health risk indices of intermediates from the same pathway and computes their average (Table S1, excluding FQ molecules with missing data). The findings indicate that the overall human health risk index for degradation products from pathway P1 is reduced by 11.13%, which is the largest reduction among all primary degradation pathways. As advanced oxidation systems continue to break down primary degradation products, there is a trend of increasing human health risks for later-stage degradation products, with the overall risk index for tertiary degradation products surpassing the parent molecules by 0.48%. Moreover, the overall human health risk indices for primary degradation products from pathways P3 and P4 are lower than the parent molecules, but the secondary degradation products from pathway P4 show a higher risk than the parent molecules (an increase of 2.19% in the overall index). Compared to the parent molecules, degradation products from pathways P2 and P5 show increased human health risks (with overall indices rising by 3.69% and 4.08%, respectively). A detailed analysis of the degradation products with increased human health risks compared to the parent molecules during FQ degradation shows that these products consistently contain hydroxyl groups, suggesting that hydroxyl groups may enhance the potential human health risks of the degradation products. In summary, the use of parent molecules with substantial human health risks (like PEF) and molecules whose risks increase during the advanced oxidation degradation process (such as LEV, LOM, NAD, RUF, SPA, and GEM) should be minimized to mitigate potential health hazards.

3.4.2. Ecological Environmental Risk Assessment of Degradation Products of FQ Molecules in Advanced Oxidation Systems

To conduct a holistic evaluation of the ecological and environmental risks posed by FQ molecules and their degradation intermediates, this study integrates five key assessment indicators: chronic toxicity to algae, 48 h LC_{50} for Daphnia magna, 96 h LC50 for Pimephales promelas, bioconcentration factor, and half-life. Since lower values for the first three indicators correlate with higher ecological risks, these are treated as negative indicators. Conversely, larger values for bioconcentration factor and half-life indicate higher ecological risk, so these are treated as positive indicators. The bio-toxicity risk index is derived by summing the processed values for the first three indicators, and the total ecological and environmental risk index is calculated by summing the processed values for all five indicators, as detailed in Table S1. Upon analysis, it is found that of the 16 parent FQ molecules, TOS exhibits the highest bio-toxicity risk, while ENO presents the lowest. The bioconcentration factor assessment indicates that TOS has the highest potential for bioaccumulation, whereas LEV, NOR, and OFL have the lowest. The half-life assessment shows that MOX has the longest half-life, and FLE has the shortest. The comprehensive ecological and environmental risk index suggests that TOS carries the highest risk, while ENO carries the lowest.

A comparative analysis of the ecological and environmental risks of the 16 parent FQ molecules and their degradation products indicates that, with the exception of CLI, MOX, and SPA, the bio-toxicity risk index for the intermediates of the remaining FQ molecules decreased during degradation when compared to the parent molecules (for example, the primary degradation products of ENO showed a 22.67% reduction in average bio-toxicity compared to the parent molecule). In terms of bioaccumulation, except for the intermediates of CIP, which exhibited no significant change, the overall bioaccumulation potential of the degradation products of the other 15 FQ molecules decreased (for instance, the primary degradation products of CLI showed a 22.73% reduction in average bioaccumulation potential compared to the parent molecule). The half-life assessment indicated that, except for NAD, which had an increased half-life, the half-lives of the other 15 FQ molecules were reduced during degradation (for example, the primary degradation products of ENO showed a 39.44% reduction in average half-life compared to the parent molecule). The comprehensive ecological and environmental risk index showed a decrease for all 16 FQ molecules during degradation (for instance, the overall ecological and environmental risk index for the primary degradation products of PEF decreased by 16.76% compared to the parent molecule). Overall, the production or widespread use of molecules with the highest ecological and environmental risks (such as TOS), those with the longest half-lives (such as MOX), those with increased combined toxicity during degradation (such as CLI, MOX, and SPA), and those with extended half-lives (such as NAD) should be avoided. Additionally, to investigate the ecological and environmental risks of intermediates from different FQ degradation pathways, this study aggregates the overall ecological and environmental risk indices of intermediates from the same pathway and calculates their average. The results indicate that the primary degradation products from all five pathways pose a lower ecological and environmental risk than before degradation, with pathway P2 yielding the lowest risk (a 16.58% decrease in the overall index). In pathways P1, P4, and P5, the ecological and environmental risks of intermediates further decrease as degradation progresses.

Upon comprehensive evaluation, it is evident that the parent molecules of PEF, TOS, and MOX present significant risks to human health or the ecological environment. The degradation intermediates of CLI, MOX, LEV, LOM, NAD, RUF, SPA, and GEM exhibit increased risks to human health or the ecological environment compared to their respective parent molecules. Consequently, the production or extensive use of these compounds should be avoided. Furthermore, degradation products resulting from the piperazine ring cleavage reaction have the lowest aggregate index for human health risk, while those from defluorination and decarboxylation reactions have the highest. Therefore, it is recommended to minimize the occurrence of these two reactions during advanced oxidation degradation processes.

3.5. Mechanistic Analysis of the Differential Degradation Capability of FQ Molecules in Advanced Oxidation Systems

3.5.1. Differential Analysis of the Degradation Capability of FQ Molecules in Advanced Oxidation Systems Based on 2D-QSAR Model

To delve deeper into the causes of the variability in the degradation capabilities of FQ molecules, this study selected 11 FQ molecules with piperazine rings to construct Lasso regression models for defluorination, hydroxylation, piperazine ring cleavage, and piperazine ring hydroxylation reactions. Initially, a comprehensive set of molecular descriptors was calculated for the 11 molecules, including geometric, electronic, physicochemical, spectral, and topological parameters (a total of 1014 feature parameters), to establish a molecular feature database for the Lasso regression models. Previous research has indicated that a larger number of relevant feature parameters can enhance the reliability of model predictions to some extent, but may also lead to low training efficiency or excessive prediction bias [34]. Therefore, this study employed a decision tree method to identify the importance of the 1014 feature parameters for the 11 commercial FQ surrogate molecules. The top nine feature parameters, prioritized by importance, served as independent variables, and the bond-breaking energy (BBE) during the advanced oxidation degradation processes of the FQs served as the dependent variable. The datasets were split into training and testing sets in a 7:3 ratio to construct Lasso regression models for piperazine ring cleavage, defluorination, hydroxylation, and piperazine ring hydroxylation reactions (Table 4). The results showed that the Lasso regression model for piperazine ring cleavage had an R^2 of 0.731, the defluorination model had an R^2 of 0.770, the hydroxylation model had an R^2 of 0.885, and the piperazine ring hydroxylation model had an R^2 of 0.825. These findings demonstrate that the Lasso regression models for the advanced oxidation degradation of the 11 commercial FQ surrogate molecules meet the requirements for significance testing and possess good predictive capabilities [35]. The model equations are as follows:

$$\begin{gathered}\mathrm{BBE1} = 20.785(\mathrm{ALogP})^2 + 39.707\mathrm{MATS2m} - 98.611\mathrm{maxsssCH} + 6.559\mathrm{GATS5c} \\ +0.530\mathrm{ATS6s} + 1878.111\mathrm{MATS1m} - 50.654\mathrm{SP} - 2 - 48.891\mathrm{SM1\_Dze} \\ +275.920\mathrm{AVP} - 1 + 568.480\end{gathered} \tag{2}$$

$$\begin{gathered}\mathrm{BBE2} = 0.137\mathrm{ECCEN} - 66.598\mathrm{MATS8i} - 62.187\mathrm{MATS3c} + 1.367\mathrm{nHaaCH} \\ -19.339\mathrm{SpMin3}_{\mathrm{Bhm}} - 5.192\mathrm{SpMin8}_{\mathrm{Bhi}} + 0.136\mathrm{EE}_{\mathrm{Dzp}} + 3.848\mathrm{MAXDP} \\ + 397.123\end{gathered} \tag{3}$$

$$\begin{gathered}\mathrm{BBE3} = 7.067(\mathrm{ALogP})^2 + 48.210\mathrm{GATS6e} + 1.091\mathrm{SP} - 2 + 21.735\mathrm{GATS3i} \\ -1.089\mathrm{SpMAD\_Dt} - 24.991\mathrm{GATS5i} + 4.360\mathrm{C1SP2} + 368.843\end{gathered} \tag{4}$$

$$\begin{gathered}\mathrm{BBE4} = 7.224\mathrm{MLFER\_BH} + 5.962\mathrm{SP} - 2 - 183.529\mathrm{GATS6i} + 0.525\mathrm{Sare} \\ -5.857\mathrm{SpMax8\_Bhe} + 10.888\mathrm{SP} - 5 - 24.674\mathrm{GATS6e} + 0.004\mathrm{ATS0m} \\ -0.003\mathrm{ATS3i} + 620.923\end{gathered} \tag{5}$$

As illustrated in Equation (2), the nine most significant feature parameters in the Lasso regression model for piperazine ring cleavage include ALOGP descriptors $(\mathrm{ALogP})^2$, autocorrelation descriptors (MATS2m, GATS5C, ATS6s, MATS1m), atomic-type electrotopological state descriptors (maxsssCH), Chi path descriptors (SP-2, AVP-1), and matrix-based descriptors (SM1_Dze). The positive coefficients associated with $(\mathrm{ALogP})^2$, GATS5C, ATS6s, MATS2m, and AVP-1 demonstrate a positive correlation with the bond energy of the carbon–nitrogen bonds in the piperazine rings of FQ molecules, indicating that an increase in these feature parameters correlates with an increase in bond energy. Conversely, the coefficients for MATS1m, maxsssCH, SP-2, and SM1_Dze are negative, suggesting an inverse relationship with the bond energy of the carbon–nitrogen bonds in the piperazine rings, meaning that as the values of these feature parameters increase, the bond energy tends to decrease. Among these, the feature importance of $(ALogP)^2$ is the highest (0.713) compared to the other eight parameters. ALogP represents a parameter for molecular lipophilicity, with higher values indicating more pronounced lipophilic properties of the molecule [36]. The analysis suggests a significant positive correlation between the bond energy of the carbon–nitrogen bonds in the FQ molecules' piperazine rings and the lipophilicity of the FQ molecules. That is, the higher the ALogP value of an FQ molecule, the greater the bond energy of the carbon–nitrogen bonds in its piperazine ring, making the piperazine ring cleavage reaction less likely to occur in AOPs. This finding is consistent with research by Liu et al., who observed that antibiotics with higher hydrophobicity were more susceptible to degradation through activated carbon ultrasonic degradation [37].

Table 4. Feature importance of FQ molecular feature parameters selected by decision tree screening algorithm.

Implicit Variable	Feature Parameter	Feature Importance
Bond dissociation energy 1	$(AlogP)^2$	0.713
	MATS2m	0.187
	maxsssCH	0.069
	GATS5c	0.022
	ATS6s	0.005
	MATS1m	0.002
	SP-2	0.002
	SM1_Dze	<0.001
	AVP-1	<0.001
Bond dissociation energy 2	ECCEN	0.737
	MATS8i	0.132
	MATS3c	0.065
	nHaaCH	0.006
	SpMin3_Bhm	0.003
	SpMin8_Bhi	0.001
	EE_Dzp	<0.001
	MAXDP	<0.001
Bond dissociation energy 3	ALogp2	0.847
	GATS6e	0.105
	SP-2	0.023
	GATS3i	0.001
	SpMAD_Dt	<0.001
	GATS5i	<0.001
	C1SP2	<0.001
Bond dissociation energy 4	MLFER_BH	0.749
	SP-2	0.119
	GATS6i	0.085
	Sare	0.037
	SpMax8_Bhe	0.005
	SP-5	0.003
	GATS6e	0.002
	ATS0m	<0.001
	ATS3i	<0.001

Figure S2 illustrates that the presence of white coloration around the ninth position of the FQ molecule signifies that FQ molecules with hydrophilic groups at this position are more prone to piperazine ring carbon–nitrogen bond cleavage. This observation allows for the Lasso regression model for piperazine ring cleavage in advanced oxidation systems to be corroborated by the CoMSIA model for the same reaction. As shown in Equation (3), in the defluorination Lasso regression model, the eight most significant feature parameters include the eccentric connectivity index descriptor (ECCEN), autocorrelation descriptors (MATS8i, MATS3c), electrotopological state atomic descriptors (nHaaCH, MAXDP), charge-modified eigenvalue descriptors (SpMin3_Bhm, SpMin8_Bhi), and the Barysz matrix descriptor (EE_Dzp). The positive coefficients for ECCEN, nHaaCH, EE_Dzp, and MAXDP indicate a direct correlation with the bond energy of the carbon–fluorine bonds in FQ molecules, meaning that as these feature parameters increase, the bond energy of the carbon–fluorine bonds tends to strengthen. Conversely, the negative coefficients for MATS8i, MATS3c, SpMin3_Bhm, and SpMin8_Bhi suggest an inverse relationship, indicating that as these feature parameters increase, the bond energy of the carbon–fluorine bonds tends to decrease. Among these, ECCEN has the highest feature importance (0.737) compared to the other seven parameters. ECCEN is a topological descriptor that characterizes the topological properties of a chemical molecule by combining information about the distances and adjacency of atoms within the molecule [38]. Specifically, the ECCEN descriptor calculates the eccentricity (the maximum distance from an atom to all other atoms in the molecule) for each atom, multiplies it by the atom's connectivity (number of adjacent atoms), and then sums these values to obtain the ECCEN value for the entire molecule. This indicates that the bond energy of the carbon–fluorine bonds in FQ molecules is significantly positively correlated with the overall eccentricity of the molecule, meaning that the greater the overall eccentricity of an FQ molecule, the higher the bond energy of its carbon–fluorine bonds. FQ molecules with lower ECCEN values (CIP, ENO, LOM, NOR) tend to have a simpler basic skeleton and fewer substituents on the piperazine ring, correlating with lower carbon–fluorine bond energy. Therefore, FQ molecules with fewer substituents and simpler structures are more likely to undergo defluorination reactions in advanced oxidation systems. Tian et al. studied the degradation of recalcitrant organic compounds such as humic acids using a combined process (UASB-A/O-USSB) and found that organic compounds with simple molecular structures are preferentially degraded through anaerobic digestion and short-range denitrification [39]. Zhu et al. investigated the degradation of organic dyes using magnetic catalysts and found that smaller and simpler molecular structures are more easily degraded [40].

As can be seen from Equation (4), within the hydroxylation Lasso regression model, the seven most significant feature parameters include ALOGP descriptors $(ALogP)^2$, autocorrelation descriptors (GATS6e, GATS3i, GATS5i), Chi path descriptors (SP-2), Detour matrix descriptors (SpMAD_Dt), and carbon-type descriptors (C1SP2). The coefficients for $(ALogP)^2$, GATS6e, SP-2, GATS3i, and C1SP2 are positive, indicating a positive correlation with the bond energy of the carbon–hydrogen bonds in FQ molecules, suggesting that as these feature parameters increase, the bond energy of the carbon–hydrogen bonds tends to strengthen. Conversely, the negative coefficients for SpMAD_Dt and GATS5i imply an inverse relationship, indicating that as these feature parameters increase, the bond energy of the carbon–hydrogen bonds tends to decrease. Among these, $(ALogP)^2$ has the highest feature importance (0.847), signifying a significant positive correlation between the bond energy of the carbon–hydrogen bonds in FQ molecules and the lipophilicity of the FQ molecules; that is, the higher the ALogP value of an FQ molecule, the greater the bond energy of its carbon–hydrogen bonds. As shown in Figure S2, the presence of white patches around the 1-position of the FQ molecule indicates that FQ molecules with hydrophilic groups at this site have carbon–hydrogen bonds that are more susceptible to cleavage. This finding allows for the hydroxylation Lasso regression model for FQ molecules in advanced oxidation systems to be corroborated by the CoMSIA model for the hydroxylation of FQ molecules in the same systems.

As demonstrated in Equation (5), within the Lasso regression model for hydroxylation of the piperazine ring, the nine most significant feature parameters include MLFER descriptors (MLFER_BH), Chi path descriptors (SP-2, SP-5), autocorrelation descriptors (GATS6i, GATS6e, ATS0m, ATS3i), constitutional descriptors (Sare), and charge-modified eigenvalue descriptors (SpMan8_Bhe). The coefficients for MLFER_BH, SP-2, Sare, SP-5, and ATS0m are positive, indicating a positive correlation with the bond energy of the carbon–hydrogen bonds in the piperazine rings of FQ molecules, indicating that as these feature parameters increase, the bond energy of the carbon–hydrogen bonds tends to strengthen. In contrast, the negative coefficients for GATS6i, SpMan8_Bhe, GATS6e, and ATS3i suggest a negative correlation, indicating that as these feature parameters increase, the bond energy of the carbon–hydrogen bonds tends to decrease. Among these, MLFER_BH, which is a descriptor that characterizes the hydrogen bond basicity of a molecule, has the highest feature importance (0.749). A higher MLFER_BH value indicates a stronger ability of the molecule to act as a hydrogen bond acceptor and form intermolecular hydrogen bonds with other hydrogen bond donors [41]. This finding is consistent with a significant positive correlation between the bond energy of the carbon–hydrogen bonds in the piperazine rings of FQ molecules and the molecules' ability to form hydrogen bonds as acceptors; the stronger the ability of an FQ molecule to form hydrogen bonds as an acceptor, the higher the bond energy of the carbon–hydrogen bonds in its piperazine ring. Figure S2 shows purple patches around the 4-position and 8-position of the FQ molecule, indicating that FQ molecules with hydrogen bond donor groups at these positions have piperazine ring carbon–hydrogen bonds that are more susceptible to cleavage. This indicates that the Lasso regression model for piperazine ring hydroxylation in advanced oxidation systems can be validated against the CoMSIA model for the hydroxylation of the piperazine ring in advanced oxidation systems.

In the realm of advanced oxidation systems, this research has elucidated several key insights regarding FQ molecules. Firstly, FQ molecules that possess hydrophilic groups at the 9-position exhibit a heightened susceptibility to the cleavage of carbon–nitrogen bonds within the piperazine ring. Secondly, FQ molecules characterized by a paucity of substituents and a simplistic structure are more likely to undergo defluorination reactions. Thirdly, the presence of hydrophilic groups at the 1-position of FQ molecules correlates with an increased propensity for carbon–hydrogen bond cleavage. Fourthly, FQ molecules that feature hydrogen bond donor groups at the 4-position and 8-position are more prone to the disruption of carbon–hydrogen bonds in the piperazine ring. The analytical conclusions derived from the Lasso regression models for chemical bond cleavage in FQ molecules are in concordance with the findings from 3D-QSAR models, indicating a robust consistency between these modeling approaches.

3.5.2. Mechanistic Analysis of the Differential Degradation Capability of FQ Molecules in Advanced Oxidation Systems Based on Fukui Function

The Fukui function is a commonly used real-space function for predicting potential reactive sites within molecules, extensively applied across the fields of biology, medicine, chemistry, and environmental science. Predictive outcomes are typically categorized into three types: nucleophilic attack (f^+), electrophilic attack (f^-), and radical attack (f^0), with higher predictive values indicating stronger reactivity of atoms within the molecule [42]. In this study, the Fukui function is employed to assess the relative ease of degradation of different FQ molecules within advanced oxidation systems, thereby investigating the mechanistic differences in the advanced oxidation degradation capabilities of FQ molecules (Table S2). To investigate the mechanistic impact of specific functional groups on molecular degradation capabilities based on the Fukui function, this study selected two molecules with highly similar core structures for comparative analysis. These molecules are NOR and CIP, which differ in their substituents at the 9-position; ENO and NOR, which have different atoms at the first position; and LOM and SPA, which vary in the substituents at the 4-position (Figure 6).

An analysis of the Fukui function predictive values for the NOR molecule reveals that atoms such as O20, C15, C3, and N8 are most susceptible to radical attack. Bao et al. have also confirmed through the Fukui function and electrostatic potential maps that the piperazine ring is more prone to radical attack relative to other parts of the NOR molecule [15,43]. For the CIP molecule, atoms such as O20, N1, C3, and C10 are most affected by radicals, indicating these atoms are more likely to be attacked by radicals in advanced oxidation processes. Notably, the N8 atom in the NOR molecule and the N1 atom in the CIP molecule correspond to the nitrogen atoms on their respective piperazine rings. A comparison of their f^0 values shows that the f^0 value for the N8 atom in NOR is 18.37% higher than that for the N1 atom in CIP, suggesting that the nitrogen atom on the piperazine ring of CIP is more susceptible to radical attack compared to that of NOR. Additionally, the bond energy of the carbon–nitrogen bond in the piperazine ring of CIP is 5.83% lower than that of NOR, indicating that a higher f^0 value for the nitrogen atom may increase the likelihood of piperazine ring cleavage in advanced oxidation systems. Furthermore, a comparison of the molecular structural features of CIP and NOR reveals that the substituent at the 9-position in CIP is a cyclopropyl group, while in NOR, it is an ethyl group. This suggests that compared to the ethyl group, the cyclopropyl group at this position of FQ molecules may enhance the reactivity of the nitrogen atom on the piperazine ring, thereby making the ring more susceptible to cleavage.

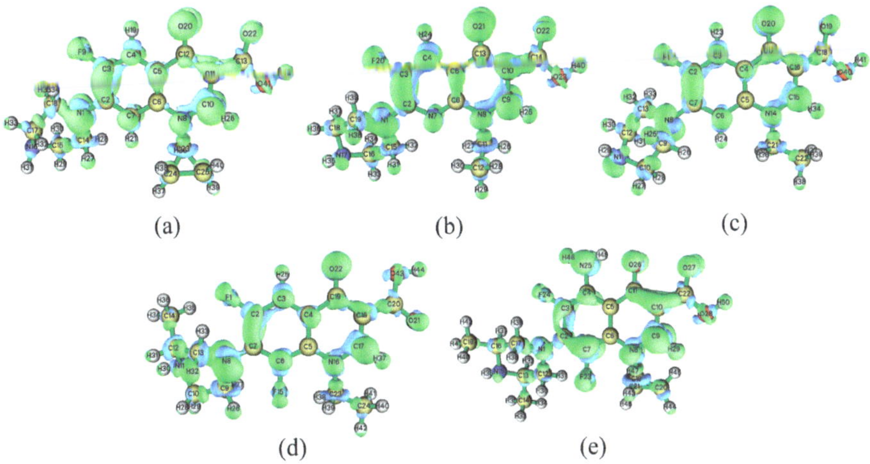

Figure 6. Fukui function isosurface maps of FQ molecules: (**a**) CIP; (**b**) ENO; (**c**) NOR; (**d**) LOM; (**e**) SPA.

An analysis of the Fukui function predictive values for the NOR molecule reveals that atoms such as O20, C15, C3, and N8 are most susceptible to radical attack. Bao et al. have also confirmed through the Fukui function and electrostatic potential maps that the piperazine ring is more prone to radical attack relative to other parts of the NOR molecule [15,43]. For the CIP molecule, atoms such as O20, N1, C3, and C10 are most affected by radicals, indicating these atoms are more likely to be attacked by radicals in advanced oxidation processes. Notably, the N8 atom in the NOR molecule and the N1 atom in the CIP molecule correspond to the nitrogen atoms on their respective piperazine rings. A comparison of their f^0 values shows that the f^0 value for the N8 atom in NOR is 18.37% higher than that for the N1 atom in CIP, suggesting that the nitrogen atom on the piperazine ring of CIP is more susceptible to radical attack compared to that of NOR. Additionally, the bond energy of the carbon–nitrogen bond in the piperazine ring of CIP is 5.83% lower than that of NOR, indicating that a higher f^0 value for the nitrogen atom may increase the likelihood of piperazine ring cleavage in advanced oxidation systems. Furthermore, a comparison of the molecular structural features of CIP and NOR reveals that the substituent at the 9-position in CIP is a cyclopropyl group, while in NOR, it is an ethyl group. This suggests that compared to the ethyl group, the cyclopropyl group at this position of FQ molecules may enhance the reactivity of the nitrogen atom on the piperazine ring, thereby making the ring more susceptible to cleavage.

An analysis of the Fukui function predictive values for the ENO molecule indicates that atoms such as O21, C3, C4, C9, and N1 are most susceptible to radical attack. Ao et al. also utilized the Fukui function to determine that the carbon atoms connected to the fluorine atoms in the LEV molecule are relatively prone to radical attacks [15]. The C3 atom in ENO and the C2 atom in NOR are the carbon atoms in the basic FQ skeleton connected to fluorine atoms. A comparison of their f0 values reveals that the f^0 value for the C3 atom in ENO is 26.28% higher than that for the C2 atom in NOR, suggesting that the fluorine atom on the basic skeleton of ENO is more susceptible to radical attack compared to that of NOR. Moreover, the bond energy of the carbon–fluorine bond in ENO is 0.82% lower than that in NOR, indicating that the higher the f^0 value of the carbon atom connected to the fluorine atom in FQ molecules, the greater the likelihood of defluorination reactions occurring in advanced oxidation systems. Further structural analysis of ENO and NOR reveals that the basic skeleton of ENO contains a pyridine ring (with the nitrogen atom at the 1-position), while the basic skeleton of NOR contains a benzene ring (with the carbon atom at the 1-position). This suggests that, compared to a benzene ring structure, a pyridine structure on the basic skeleton of FQ molecules enhances the radical reactivity of the carbon atom connected to the fluorine atom, thereby increasing the propensity for defluorination in FQ molecules.

The Fukui function predictive values for the LOM molecule indicate that atoms such as N8, C17, and O22 are most susceptible to radical attack, while in the SPA molecule, atoms like C7, C9, N25, O26, and O27 are most prone to radical attack. Notably, the C13 atom in LOM and the C17 atom in SPA are carbon atoms on their respective piperazine rings. A comparison of the f^0 values for C13 and C17 reveals that the f^0 value for the C13 atom in LOM is 130.61% higher than that for the C17 atom in SPA, suggesting that the carbon atom on the piperazine ring of LOM is more readily attacked by radicals compared to that of SPA. Additionally, the carbon–hydrogen bond energy in LOM's piperazine ring is 7.27% lower than that in SPA, suggesting that a higher f^0 value for carbon atoms increases the likelihood of piperazine ring hydroxylation in FQ molecules. Further analysis of the substituent characteristics of LOM and SPA reveals that the 4-position in SPA is occupied by an amino group, while in LOM, it is a hydrogen atom. This suggests that, compared to an amino group, a hydrogen atom at the 4-position of FQ molecules may enhance the reactivity of carbon atoms in the piperazine ring, leading to a higher likelihood of hydroxylation reactions on the piperazine ring.

In summary, this study finds that in the advanced oxidation process of FQ molecules, atoms with higher f^0 values facilitate the cleavage reactions of chemical bonds associated with those atoms. Moreover, adding a cyclopropyl group at the 9-position of FQ molecules, replacing the benzene ring structure on the basic skeleton with a pyridine structure, and avoiding the addition of an amino group at the 4-position are conducive to enhancing the reactivity of atoms involved in advanced oxidation reactions. Consequently, these modifications increase the propensity of FQ molecules to undergo piperazine ring cleavage, defluorination, and hydroxylation.

3.5.3. Mechanistic Analysis of the Differential Degradation Capability of FQ Molecules in Advanced Oxidation Systems Based on Fukui Function

In an effort to decode the reactivity trends of FQ molecules with radicals within advanced oxidation systems, this research focused on CIP and GAT molecules, which, while structurally similar, differ in the substituents at the 1-position. The Multiwfn 3.8 software [44] was employed to calculate the highest occupied molecular orbital (HOMO) and the lowest unoccupied molecular orbital (LUMO) energies of 16 optimized FQ molecules (Figure 7), analyzing their reactivity with radicals. It was found that the strongest reactivity between two reactants occurs when they exhibit the most favorable interaction energy, a phenomenon closely associated with the frontier molecular orbitals of the interacting substances [45]. Therefore, considering the HOMO-LUMO energy gap (ΔE) of FQ molecules and radicals is crucial, as a smaller energy gap indicates stronger interactions between the orbitals [45]. The hydroxyl radical, due to its prevalent and highly reactive nature in advanced oxidation systems, was selected to probe the reactivity of FQ molecules. The

LUMO energy value of the optimized hydroxyl radical structure was calculated using the Multiwfn 3.8 software.

An analysis of the HOMO and LUMO distributions of CIP and GAT molecules reveals that the HOMOs of CIP are primarily located on the two ring structures of its core skeleton (atoms C2, C3, C5, C6, C11, C10, N8, etc.), the piperazine ring (atoms N1, C18, C14, C17, C15, etc.), and the atoms F9 and O20, indicating that these structural parts of the CIP molecule have a strong electron-donating capability. The LUMOs are mainly distributed over the two ring structures of the core skeleton (atoms C2, C3, C4, C6, C7, N8, C10, C11, C12, etc.), the piperazine ring (atom N1), the carboxyl group (atoms C13, O22, O41), and the atoms F9 and O20, suggesting that these structural parts of the CIP molecule have a strong electron-accepting capability. For GAT, the HOMOs are mainly located on the two ring structures of its core skeleton (atoms C2, C3, C4, C6, etc.), the piperazine ring (atoms N8, C13, C9, N11, etc.), the methoxy group (atom O14), and the atoms F1 and O22, indicating that these structural parts of the GAT molecule have a strong electron-donating capability. The LUMOs are mainly distributed over the two ring structures of the core skeleton (atoms C2, C3, C4, C6, C7, N16, C17, C19, etc.), the piperazine ring (atom N8), the methoxy group (atom O14), and the atom O22, suggesting that these structural parts of the GAT molecule have a strong electron-accepting capability. This is in line with the analysis by Ao et al. on the LEV molecule, which found a rich distribution of HOMOs near the piperazine ring and core skeleton, and LUMOs concentrated around the core skeleton and carboxyl group. Further comparison of the HOMO distribution maps of CIP and GAT molecules shows that the HOMOs of CIP are more densely distributed over the piperazine ring, the carbon atoms involved in defluorination, and the carbon atoms involved in hydroxylation reactions than those of GAT, indicating that in advanced oxidation systems, CIP is more likely than GAT to undergo piperazine ring cleavage, defluorination, hydroxylation, and hydroxylation of the piperazine ring.

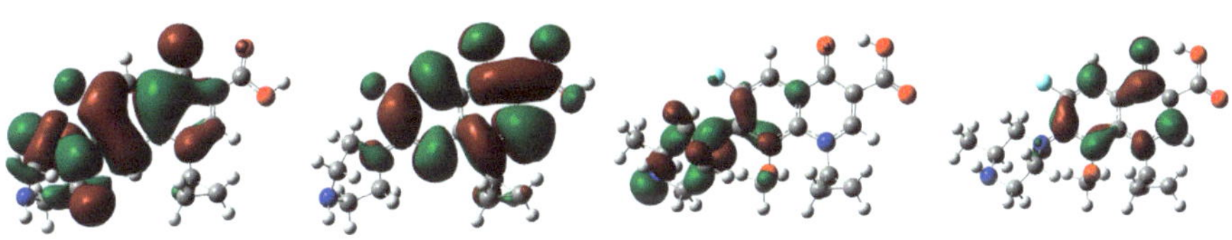

(a) (b) (c) (d)

Figure 7. HOMO and LUMO distribution maps of FQ molecules: (**a**) HOMO distribution map of CIP molecule; (**b**) LUMO distribution map of CIP molecule; (**c**) HOMO distribution map of GAT molecule; (**d**) LUMO distribution map of GAT molecule.

Furthermore, a comparison of the ΔE between CIP and GAT molecules reveals that the ΔE of CIP is 5.19% lower than that of GAT, indicating stronger interactions between the orbitals of CIP and the hydroxyl radical and thus a higher reactivity of CIP with the hydroxyl radical compared to GAT. Asghar et al. found that the interaction between Acid Blue dye and ozone was the strongest by calculating the HOMO of the dye and the LUMO of various oxidizing agents [45]. A structural comparison between CIP and GAT molecules shows that the 1-position of GAT is occupied by a methoxy group, while in CIP, it is a hydrogen atom. This suggests that the addition of a methoxy group at the 1-position of FQ molecules can cause the highest occupied molecular orbital to be further removed from the carbon atoms involved in the piperazine ring, defluorination, and hydroxylation reactions, leading to an increase in the ΔE between FQ molecules and the hydroxyl radical and consequently a decrease in their reactivity. This finding is consistent with the conclusions drawn in Section 3.5.1, which determined through bond dissociation energy calculations that "FQ molecules with a methoxy group connected at the 1-position of the basic skeleton are less likely to undergo piperazine ring cleavage reactions."

Overall, the addition of a methoxy group at the 1-position of FQ molecules can cause a sparser distribution of the HOMO over the carbon atoms involved in the piperazine ring, defluorination, and hydroxylation reactions, thereby weakening the overall reactivity of FQ molecules with radicals. Moreover, the addition of a methoxy group at the 1-position leads to an increase in the HOMO-LUMO energy gap between FQ molecules and radicals, resulting in decreased reactivity of FQ molecules with the hydroxyl radical.

4. Conclusions

In this research, density functional theory (DFT) was employed to quantify the oxidative degradation potential of 16 commercially available fluoroquinolone (FQ) molecules within advanced oxidation systems. The study elucidates a significant correlation between the molecular structural characteristics of FQ molecules and their degradation capabilities across five identified advanced oxidation pathways. Additionally, three-dimensional quantitative structure–activity relationship (3D-QSAR) contour maps were utilized to discern the influence of specific functional groups on the advanced oxidation degradation of FQ molecules. Moreover, the study utilized VEGA software to evaluate the toxicity risks posed by FQs and their degradation by-products, proposing restrictions on FQ molecules that present a high risk and strategies to circumvent degradation reactions that could heighten the toxicity of degradation products. The intent of this research was to provide a data-driven foundation for thorough and rational health risk assessments of FQ molecules, to contribute novel perspectives for directing the efficient and sustainable degradation of FQs in advanced oxidation systems, and to offer theoretical support for the design of eco-friendly FQ alternatives.

Supplementary Materials: The following supporting information can be downloaded at https://www.mdpi.com/article/10.3390/toxics12030203/s1: Figure S1: Equipotential diagram of FQs molecule CoMSIA model (a) hydrogen bond receptor field of defluorinated CoMSIA model; (b) Hydrophobic field of defluorinated CoMSIA model; (c) Defluorinated CoMSIA model electrostatic field; (d) Defluorinated CoMSIA model hydrogen bond donor field; (e) Defluorinated CoMSIA model three-dimensional field; (f) Hydroxylated CoMSIA model hydrogen bond receptor field; (g) Hydrophobic field of hydroxylated CoMSIA model; (h) Hydroxylated CoMSIA model electrostatic field; (i) Hydroxylated CoMSIA model hydrogen bond donor field; (j) Hydroxylated CoMSIA model stereo field; (k) CoMSIA model hydrogen bond receptor field for hydroxylation of piperazine ring; (l) Hydrophobic field of CoMSIA model for hydroxylation of piperazine ring; (m) CoMSIA model electrostatic field for hydroxylation of piperazine ring; (n) CoMSIA model hydrogen bond donor field for hydroxylation of piperazine ring; (o) CoMSIA model for hydroxylation of piperazine ring in three-dimensional field; Figure S2: FQs Molecular column clustering analysis chart; Table S1: The risk assessment of human health and the ecological impact of these FQ molecules and their intermediates; Table S2: Nucleophilic attack, electrophilic attack and free radical attack index of different FQs molecules; Table S3: LUMO-HOMO between different FQs molecules and hydroxyl radicals.

Author Contributions: Conceptualization, S.S. and Z.W.; methodology, Q.P.; software, X.L.; validation, H.Y., Y.C. and Y.L.; formal analysis, Z.W.; investigation, Z.W.; resources, S.S.; data curation, S.S.; writing—original draft preparation, S.S.; writing—review and editing, H.Y.; visualization, Q.P.; project administration, S.S. All authors have read and agreed to the published version of the manuscript.

Funding: This research was funded by the Research and Development of Key Technologies and Equipment for Catalytic Oxidation of Iron-based Nanomaterials and in Situ Joint Remediation of Petroleum Hydrocarbon Contaminated Soil by Dry Rotary Jet Pile Process project (20200403027SF), and by the Study on Advanced Oxidation Process of Persulfate Catalyzed by Iron-based Nanomaterials to Repair Petroleum Polluted Soil project (JJKH20191270KJ).

Institutional Review Board Statement: Not applicable.

Informed Consent Statement: Not applicable.

Data Availability Statement: Data are available upon request.

Conflicts of Interest: The authors declare no conflicts of interest.

References

1. Li, S.; Wu, Y.; Zheng, H.; Li, H.; Zheng, Y.; Nan, J.; Chang, J.S. Antibiotics degradation by advanced oxidation process (AOPs): Recent advances in ecotoxicity and antibiotic-resistance genes induction of degradation products. *Chemosphere* **2023**, *311*, 136977. [CrossRef] [PubMed]
2. Focazio, M.J.; Kolpin, D.W.; Barnes, K.K.; Furlong, E.T.; Meyer, M.T.; Zaugg, S.D.; Thurman, M.E. A national reconnaissance for pharmaceuticals and other organic wastewater contaminants in the United States—II) Untreated drinking water sources. *Sci. Total Environ.* **2008**, *402*, 201–216. [CrossRef] [PubMed]
3. Tello, A.; Austin, B.; Telfer, T.C. Selective pressure of antibiotic pollution on bacteria of importance to public health. *Environ. Health Perspect.* **2012**, *120*, 1100–1106. [CrossRef]
4. Rosal, R.; Rodríguez, A.; Perdigón-Melón, J.A.; Petre, A.; García-Calvo, E.; Gómez, M.J.; Fernández-Alba, A.R. Occurrence of emerging pollutants in urban wastewater and their removal through biological treatment followed by ozonation. *Water Res.* **2010**, *44*, 578–588. [CrossRef] [PubMed]
5. Larsson, D.J.; de Pedro, C.; Paxeus, N. Effluent from drug manufactures contains extremely high levels of pharmaceuticals. *J. Hazard. Mater.* **2007**, *148*, 751–755. [CrossRef]
6. Akbari, M.Z.; Xu, Y.; Lu, Z.; Peng, L. Review of antibiotics treatment by advance oxidation processes. *Environ. Adv.* **2021**, *5*, 100111. [CrossRef]
7. Wang, J.; Zhuan, R. Degradation of antibiotics by advanced oxidation processes: An overview. *Sci. Total Environ.* **2020**, *701*, 135023. [CrossRef]
8. Wang, J.; Wang, S. Microbial degradation of sulfamethoxazole in the environment. *Appl. Microbiol. Biotechnol.* **2018**, *102*, 3573–3582. [CrossRef]
9. Wang, J.; Wang, S. Preparation, modification and environmental application of biochar: A review. *J. Clean. Prod.* **2019**, *227*, 1002–1022. [CrossRef]
10. Zhuang, S.; Liu, Y.; Wang, J. Covalent organic frameworks as efficient adsorbent for sulfamerazine removal from aqueous solution. *J. Hazard. Mater.* **2020**, *383*, 121126. [CrossRef]
11. Avramiotis, E.; Frontistis, Z.; Manariotis, I.D.; Vakros, J.; Mantzavinos, D. On the performance of a sustainable rice husk biochar for the activation of persulfate and the degradation of antibiotics. *Catalysts* **2021**, *11*, 1303. [CrossRef]
12. Wang, X.; Brigante, M.; Dong, W.; Wu, Z.; Mailhot, G. Degradation of Acetaminophen via UVA-induced advanced oxidation processes (AOPs). Involvement of different radical species: HO, SO^{4-} and HO^2/O^{2-}. *Chemosphere* **2020**, *258*, 127268. [CrossRef] [PubMed]
13. Bermúdez, L.A.; Pascual, J.M.; Martínez, M.D.M.M.; Poyatos Capilla, J.M. Effectiveness of advanced oxidation processes in wastewater treatment: State of the art. *Water* **2021**, *13*, 2094. [CrossRef]
14. Sciscenko, I.; Hắng, H.T.M.; Escudero-Oñate, C.; Oller, I.; Arques, A. Fluorescence spectroscopy and chemometrics: A simple and easy way for the monitoring of fluoroquinolone mixture degradation. *ACS Omega* **2021**, *6*, 4663–4671. [CrossRef] [PubMed]
15. Ao, X.; Zhang, X.; Li, S.; Yang, Y.; Sun, W.; Li, Z. Comprehensive understanding of fluoroquinolone degradation via MPUV/PAA process: Radical chemistry, matrix effects, degradation pathways, and toxicity. *J. Hazard. Mater.* **2023**, *445*, 130480. [CrossRef]
16. Bobu, M.; Yediler, A.; Siminiceanu, I.; Zhang, F.; Schulte-Hostede, S. Comparison of different advanced oxidation processes for the degradation of two fluoroquinolone antibiotics in aqueous solutions. *J. Environ. Sci. Health A* **2013**, *48*, 251–262. [CrossRef] [PubMed]
17. Li, M.; Wei, D.; Zhao, H.; Du, Y. Genotoxicity of quinolones: Substituents contribution and transformation products QSAR evaluation using 2D and 3D models. *Chemosphere* **2014**, *95*, 220–226. [CrossRef]
18. Wang, D.; Ning, Q.; Dong, J.; Brooks, B.W.; You, J. Predicting mixture toxicity and antibiotic resistance of fluoroquinolones and their photodegradation products in *Escherichia coli*. *Environ. Pollut.* **2020**, *262*, 114275. [CrossRef]
19. Wachter, N.; Aquino, J.M.; Denadai, M.; Barreiro, J.C.; Silva, A.J.; Cass, Q.B.; Rocha-Filho, R.C. Electrochemical degradation of the antibiotic ciprofloxacin in a flow reactor using distinct BDD anodes: Reaction kinetics, identification and toxicity of the degradation products. *Chemosphere* **2019**, *234*, 461–470. [CrossRef]
20. Anjali, R.; Shanthakumar, S. Insights on the current status of occurrence and removal of antibiotics in wastewater by advanced oxidation processes. *J. Environ. Manage.* **2019**, *246*, 51–62. [CrossRef]
21. Aydogdu, S.; Hatipoglu, A. Aqueous degradation of 6-APA by hydroxyl radical: A theoretical study. *J. Mol. Model.* **2023**, *29*, 222. [CrossRef] [PubMed]
22. Wang, Z.; Pu, Q.; Li, Y. Bidirectional selection of the functional properties and environmental friendliness of organophosphorus (OP) pesticide derivatives: Design, screening, and mechanism analysis. *Sci. Total Environ.* **2023**, *879*, 163043. [CrossRef] [PubMed]
23. Munkhdalai, L.; Munkhdalai, T.; Park, K.H.; Lee, H.G.; Li, M.; Ryu, K.H. Mixture of activation functions with extended min-max normalization for forex market prediction. *IEEE Access* **2019**, *7*, 183680–183691. [CrossRef]
24. Clark, M.; Cramer, R.D., III; Van Opdenbosch, N. Validation of the general purpose tripos 5.2 force field. *J. Comput. Chem.* **1989**, *10*, 982–1012. [CrossRef]
25. Pu, Q.; Han, Z.; Li, X.; Li, Q.; Li, Y. Designing and screening of fluoroquinolone substitutes using combined in silico approaches: Biological metabolism–bioconcentration bilateral selection and their mechanism analyses. *Green Chem.* **2022**, *24*, 3778–3793. [CrossRef]

26. Li, Q.; Qiu, Y.; Li, Y. Molecular design of environment-friendly PAE derivatives based on 3D-QSAR assisted with a comprehensive evaluation method combining toxicity and estrogen activities. *Water Air Soil Pollut.* **2020**, *231*, 1–13. [CrossRef]
27. Liu, Y.; Li, X.; Pu, Q.; Fu, R.; Wang, Z.; Li, Y.; Li, X. Innovative screening for functional improved aromatic amine derivatives: Toxicokinetics, free radical oxidation pathway and carcinogenic adverse outcome pathway. *J. Hazard. Mater.* **2023**, *454*, 131541. [CrossRef]
28. Haghighatlari, M.; Hachmann, J. Advances of machine learning in molecular modeling and simulation. *Curr. Opin. Chem. Eng.* **2019**, *23*, 51–57. [CrossRef]
29. Dong, Y.; Georgakis, C.; Santos-Marques, J.; Du, J. Dynamic response surface methodology using Lasso regression for organic pharmaceutical synthesis. *Front. Chem. Sci. Eng.* **2021**, *16*, 221–223. [CrossRef]
30. Patel, P.S.; Pandya, D.M.; Shah, M. A holistic review on the assessment of groundwater quality using multivariate statistical techniques. *Environ. Sci. Pollut. Res.* **2023**, *30*, 85046–85070. [CrossRef] [PubMed]
31. Zhao, Y.Y.; Li, Y. Design of environmentally friendly neonicotinoid insecticides with bioconcentration tuning and Bi-directional selective toxic effects. *J. Clean. Prod.* **2019**, *221*, 113–121. [CrossRef]
32. Salahinejad, M.; Ghasemi, J.B. 3D-QSAR studies on the toxicity of substituted benzenes to Tetrahymena pyriformis: CoMFA, CoMSIA and VolSurf approaches. *Ecotoxicol. Environ. Saf.* **2014**, *105*, 128–134. [CrossRef] [PubMed]
33. Segalin, J.; Arsand, J.B.; Jank, L.; Schwalm, C.S.; Streit, L.; Pizzolato, T.M. In silico toxicity evaluation for transformation products of antimicrobials, from aqueous photolysis degradation. *Sci. Total Environ.* **2022**, *828*, 154109. [CrossRef]
34. Drgoňa, J.; Tuor, A.R.; Chandan, V.; Vrabie, D.L. Physics-constrained deep learning of multi-zone building thermal dynamics. *Energy Build.* **2021**, *243*, 110992. [CrossRef]
35. Khalifa, N.M.; Srour, A.M.; Abd El-Karim, S.S.; Saleh, D.O.; Al-Omar, M.A. Synthesis and 2D-QSAR study of active benzofuran-based vasodilators. *Molecules* **2017**, *22*, 1820. [CrossRef]
36. Ghose, A.K.; Pritchett, A.; Crippen, G.M. Atomic physicochemical parameters for three dimensional structure directed quantitative structure-activity relationships III: Modeling hydrophobic interactions. *J. Comput. Chem.* **1988**, *9*, 80–90. [CrossRef]
37. Liu, P.; Wu, Z.; Lee, J.; Cravotto, G. Sonocatalytic degrading antibiotics over activated carbon in cow milk. *Food Chem.* **2024**, *432*, 137168. [CrossRef]
38. Sharma, V.; Goswami, R.; Madan, A.K. Eccentric connectivity index: A novel highly discriminating topological descriptor for structure− property and structure− activity studies. *J. Chem. Inf. Comput. Sci.* **1997**, *37*, 273–282. [CrossRef]
39. Tian, Z.; Li, G.; Bai, M.; Hou, X.; Li, X.; Zhao, C.; Zhu, Q.; Du, C.; Li, M.; Liu, W.; et al. Microbial mechanisms of refractory organics degradation in old landfill leachate by a combined process of UASB-A/O-USSB. *Sci. Total Environ.* **2022**, *848*, 157737. [CrossRef] [PubMed]
40. Zhu, Y.; Zhou, X.; Chen, D.; Li, F.; Xue, T.; Farag, A.S. Ternary Fe 3 O 4@ PANI@ Au nanocomposites as a magnetic catalyst for degradation of organic dyes. *Sci. China Technol. Sci.* **2017**, *60*, 749–757. [CrossRef]
41. Platts, J.A.; Butina, D.; Abraham, M.H.; Hersey, A. Estimation of molecular linear free energy relation descriptors using a group contribution approach. *J. Chem. Inf. Comput. Sci.* **1999**, *39*, 835–845. [CrossRef]
42. Pucci, R.; Angilella, G.G.N. Density functional theory, chemical reactivity, and the Fukui functions. *Found. Chem.* **2022**, *24*, 59–71. [CrossRef]
43. Bao, C.; Zhao, J.; Sun, Y.; Zhao, X.; Zhang, X.; Zhu, Y.; Xing, B. Enhanced degradation of norfloxacin by Ce-mediated Fe-MIL-101: Catalytic mechanism, degradation pathways, and potential applications in wastewater treatment. *Environ. Sci. Nano* **2021**, *8*, 2347–2359. [CrossRef]
44. Lu, T.; Chen, F.W. Comparison of computational methods for atomic charges. *Acta Phys.-Chim. Sin.* **2012**, *28*, 1–18. [CrossRef]
45. Asghar, A.; Bello, M.M.; Raman, A.A.A.; Daud, W.M.A.W.; Ramalingam, A.; Zain, S.B.M. Predicting the degradation potential of Acid blue 113 by different oxidants using quantum chemical analysis. *Heliyon* **2019**, *5*, e02396. [CrossRef]

toxics

Article

Sequential Treatment by Ozonation and Biodegradation of Pulp and Paper Industry Wastewater to Eliminate Organic Contaminants

Jessica Amacosta [1], Tatyana Poznyak [1,*], Sergio Siles [1] and Isaac Chairez [2,*]

[1] Escuela Superior de Ingeniería e Industrias Extractivas, Instituto Politécnico Nacional of México, Ciudad de México 07738, Mexico; amacosta@ipn.mx (J.A.); ssilesa@ipn.mx (S.S.)
[2] Institute of Advanced Materials for Sustainable Manufacturing, Tecnologico de Monterrey, Zapopan 45201, Mexico
* Correspondence: tpoznyak@ipn.mx (T.P.); isaac.chairez@tec.mx (I.C.)

Abstract: In this research, the decomposition of toxic organics from pulp and paper mill effluent by the sequential application of ozonation and biodegradation was studied. Ozonation, as a pre-treatment, was executed to transform the initial pollutants into less toxic compounds (such as organic acids of low molecular weights). Biodegradation was executed during three days with acclimated microorganisms that were able to complete the decomposition of the initial organic mixture (raw wastewater) and to achieve a higher degree of mineralization (85–90%). Experiments were performed under three different conditions: (a) only ozonation of the initial contaminants, (b) only biodegradation of residual water without previous treatment by ozone and (c) ozonation followed by biodegradation performed by acclimated microorganisms. In the case of 72 h of biodegradation, the mineralization efficiency reached 85% and 89% after 30 and 60 min of ozonation, respectively. The no significant difference in this parameter coincided with the calculated generalized microorganisms' consortia specific growing rate μ_{max} that was reduced from 2.08×10^{-3} h^{-1} to 6.05×10^{-4} h^{-1} when the ozonation time was longer. The identification of the organics composition by gas chromatography with mass detector (GC-MS) before and after treatments confirmed that the proposed combined process served as a more efficient alternative to secondary and tertiary treatments (mineralization degree between 60 and 80% in average) of the paper industry wastewater.

Keywords: pulp and paper; toxic compounds; residual water; combined treatment; ozonation-biodegradation

Citation: Amacosta, J.; Poznyak, T.; Siles, S.; Chairez, I. Sequential Treatment by Ozonation and Biodegradation of Pulp and Paper Industry Wastewater to Eliminate Organic Contaminants. *Toxics* **2024**, *12*, 138. https://doi.org/10.3390/toxics12020138

Academic Editors: Zunyao Wang and Tiecheng Wang

Received: 27 December 2023
Revised: 19 January 2024
Accepted: 24 January 2024
Published: 8 February 2024

1. Introduction

The principal product of the pulp and paper industry (PPI) is cellulose pulp. This product is derived from the Kraft process applied to wood (cellulose 45–50 wt%, hemicelluloses 20–25 wt%, lignin 20–25 wt% and additional extractable organic compounds 5–25 wt%). The rest of the final elements of the Kraft process are various fibers and recycled material [1]. Cellulose is chemically bleached to remove some organic materials from the pulp. The resulting liquid stream comprises organic chemicals such as lignin and its derivatives: phenols, steroids, tannins terpenes and resin acids and dioxins and their derivative compounds, which are usual components of wood-based materials. These chemicals are removed or converted to chlorinated organic molecules during the bleaching phase of the Kraft process [2]. The major features of pulp and paper mills following the bleaching stage include high color, a very large content of suspended particulates, unregulated levels of chemical oxygen demand (COD) and biological oxygen demand (BOD), plentiful quantities of chlorinated chemicals and an unstable pH range [3–5].

These residuals dissolved in PPI wastewater have been considered a serious environmental problem. This is especially concerning because this business is a major buyer of natural raw materials and a significant generator of harmful pollutants (solids, liquids and gaseous effluents). Overall, 72 tons of highly contaminated liquid effluents are generated for every ton of paper produced. This proportion can potentially create irreversible pollution of rivers, lakes, seas and soil [6]. Some plants and animals are poisonous, mutagenic and genotoxic due to all of these substances [7,8]. They do contain lignin and its derivatives. Furthermore, this type of waste effluent frequently contains a complicated mixture of inorganic chemicals that are very recalcitrant and resistant to biological treatment [9,10]. Various sewage treatments for these effluents have been devised over the previous two decades, including physical, biological and chemical approaches or mixtures [11–19]. BOD, COD, and total organic carbon (TOC) measurements are widely used to regulate all treatments.

Chemical treatments promote faster decomposition of many different highly concentrated toxic compounds. For example, chemical oxidation by ozone is based on the molecular reaction mechanism or on the indirect reaction aided with the generation of hydroxyl radicals ($^{\bullet}OH$) by the decomposition of gas or with UV radiation [20,21]. N. Kaushalya et al. investigated the ozone-induced discoloration of wastewater samples from the Kraft process [22,23], where the reaction parameters were set to a 240 min reaction time and two different pH values (3 and 10). Under these conditions, there was a substantial decline in phenolic compounds before and after treatment (70% and 52%, respectively). These approaches were effective but can be expensive if mineralization is the main aim of treatment. The sequential application of flocculation precipitation and ozonation (using the flocculation precipitation-like pre-treatment to remove the undissolved organic and inorganic matter by decantation) produces water clarification. With continuous oxidation of dissolved byproducts by ozone after chemical precipitation, the biodegradability was also incremented [20,23].

On the other hand, biological treatment is cheaper and more environmentally friendly, but it requires extended treatment time to eliminate pollutants and exhibits microorganisms' inhibition when high concentrations of contaminants and/or accumulation of toxic metabolites occur. In the textile and paper industries, different mixtures of microorganisms (bacteria, fungi and algae) have been used to fulfil the biological treatments: *Pseudomonas mendocina*, *Pseudomonas alcaligenes*, *Curvulariainaequalis*, *Phanerochaetechrysosporium*, *Pleurotussajorcaju*, *Pleurotusostreatus* and many others to eliminate organic matter [24–26]. Investigators have researched how an *Aspergillus awamori* strain was successful in biodegrading 1 g/dm^3 of toxic contaminants such as phenol, catechol, 2,4-dichlorophenol and 2,6-dimethoxyphenol and their derivatives. This study confirmed their decomposition after 3, 5, 7 and 8 days of treatment, correspondingly [23,24,27].

Because the wastewater that is created throughout the bleaching cycle contains suspended solids, the primary treatment uses several filters. Secondary and tertiary treatments are typically carried out using conventional methods. Still, in the study, the elimination of soluble organic matter was inefficient because the values of BOD and COD in the treated wastewater sample remained above the reported levels in various local and national legislations [28]. This ineffective result encourages the concept of ozonation followed by biodegradation as a feasible option for treating the complex variety of organic substances prevalent in the pulp and paper industry's effluent wastewater. This type of treatment technique, comprised of ozonation and biodegradation, was previously evaluated in previous investigations [12,29–31].

Decomposition of toxic contaminants was carried out firstly by ozonation to reduce the toxicity of initial organic matter. This fact was justified because ozone promoted the formation of less toxic simple organic mixtures [14,32–34]. However, in the mentioned publications, only different model compounds (analytical degree) or actual wastewater were looked at, with no detection of contaminants before and following treatment. Furthermore, the final toxicity of the organic mixture was not determined.

The aim of the present research was to evaluate the effectiveness of organic compounds' degradation from the real wastewater samples of the bleaching step in the Kraft process by the sequential application of ozonation and then biodegradation. As a secondary objective, adequate operational conditions for both processes were determined in order to reduce the treatment time and to achieve a higher mineralization degree. Additionally, identification of the organic compounds before and after treatment by mass chromatography was performed. This information was used to evaluate the efficiency of the proposed sequential treatment for the elimination of toxic organic contaminants.

2. Materials and Methods

All the experiments reported in this section were performed under three different conditions: (a) only ozonation of the initial contaminants, (b) only biodegradation of residual water without previous treatment by ozone and (c) ozonation followed by biodegradation performed by acclimated microorganisms.

2.1. Wastewater Samples Preparation

The raw wastewater sample was taken from the bleaching wastewater of a standard Kraft operation. Samples were collected from an unnamed private enterprise in Morelia, Michoacan, Mexico. White liquor generated by sodium hydroxide (NaOH) and sodium sulfide (Na_2S) was used in the initial phase of this unique Kraft process for cooking wood in digesters at high pressures and temperatures.

The digested wood was removed and cleaned to remove the cooking chemicals and dissolved organics from the fibers. These fibers were then treated in the order listed below. The wastewater treatment consisted of a single cycle of chlorination, followed by two rounds of extraction and dioxin application.

All samples were collected and kept at 4 °C. The aforementioned samples were sterilized in a conventional sterilizer for 20 min at a temperature of 120 °C and a pressure of 0.1 MPa. To prevent the typical foaming of real wastewater and to facilitate sample analysis, the untreated residual water was diluted 1:10 with distilled water and utilized in all tests. Consider that this dilution was not a major issue because the ozone stoichiometric in its reaction with the raw wastewater sample was not affected (therefore, increasing the ozone concentration 10 times solved the effect of dilution).

2.2. Ozonation Procedure

Ozone was produced from extra dried oxygen with a corona discharge ozone generator HTU500G "Azco" Industries Limited(Langley, Canada) (marked with letter i in Figure 1). The ozone/oxygen mixture flow of 0.5 dm^3/min with an initial ozone concentration of 10 mg/L was injected through a ceramic porous filter in the semi-batch glass reactor (0.25 dm^3) with samples of 0.10 dm^3 (marked with ii in Figure 1) for 30 and 60 min. The Ozone Analyzer BMT 963 "S" (BMT Messtechnik, Berlin, Germany) (marked with iii in Figure 1) provided the ozone concentration determination in the gas phase at the reactor output for the ozone monitoring (via a data acquisition card). A self-designed software based on the Graphical User Interface Development Environment was developed to perform the on-line monitoring task (marked with iv in Figure 1). This software configured and connected the acquisition board and performed the plotting and recording of the ozone concentration.

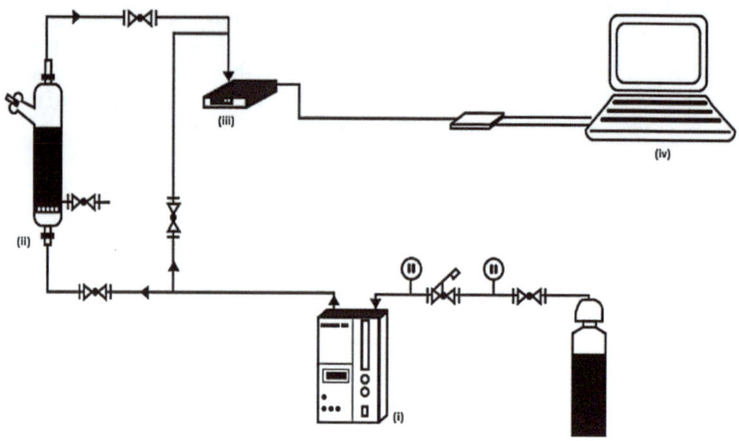

Figure 1. General diagram of ozonation process: **(i)** ozone generator, **(ii)** ozonation reactor, **(iii)** ozone sensor in the gaseous phase, **(iv)** personal computer for capturing ozone concentration.

2.3. Biodegradation Procedure

Operative conditions for the biodegradation stage were fixed using information from a preliminary investigation [35]. The biodegradation stage was performed in a batch reactor with a constant agitation fixed to 300 rpm using a magnetic steering system, with a fixed pH of 7.0 at the ambient temperature (Figure 2). The reaction used supplemented oxygen at a concentration of 21% with a flowrate of 0.5 L min^{-1}. Erlenmeyer flasks of 0.25 dm^3 encompassing 0.1 dm^3 of mineral culture broth (1.0 dm^3 of solution contained: 3 g $(NH_4)_2SO_4$, 0.6 g KH_2PO_4, 2.4 g K_2HPO_4, 1.5 g $MgSO_4{\bullet}7H_2O$, 0.15 g $CaSO_4$ and 0.03 g $FeSO_4$) were added with:

- Inoculum of the pre-acclimated consortium (with Ph and A, 1:1—Ph:A) to the sample without previous ozonation.
- Inoculum of the pre-acclimated consortium (with oxalic and formic acids—OFA) to the ozonated samples.

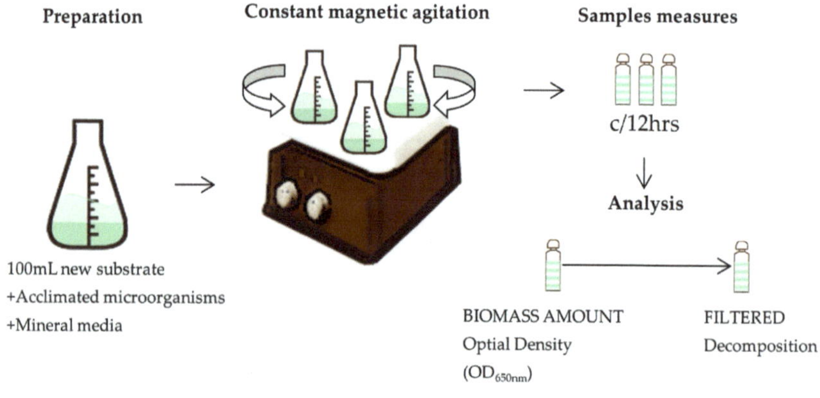

Figure 2. General diagram of biodegradation process.

The biodegradation lasted for a period of 168 h. To attain the best possible consortium acclimation conditions, the method of fill-and-draw (F&D) [30] was used. In this study, a complex microbial consortium performed biodegradation that was previously identified [35,36] by extracting DNA from samples using the Easy-DNATM Kit (Invitrogen, Waltham, MA, USA).

The prevalent microorganisms for this consortium are the following bacteria: *Xanthomomna* ssp., *Rhodopseudomona* ssp. and *Ancylobacter* spp. that can degrade chlorohexanes [37], 2-chloroethanol and 2-cloroetylvynyl ether [38], 3-chlorobenzoate [39], etc. During all biological procedures, organic deprivation was determined with respect to time.

2.4. Analytical Methods

The color units were determined using the Pt-Co index [28]. The index was created to analyze the pollution levels in wastewater. It is a popular method of comparing the strength of yellow-tinted samples. It is specific to the color yellow and is derived from dilutions of a 500 ppm platinum cobalt solution. The color produced by one milligram of platinum cobalt dissolved in one liter of water is defined as one unit of color on the platinum-cobalt scale. ASTM Designation D1209, "Standard Test Method for Color of Clear Liquids (Platinum-Cobalt Scale)" provides a full description and methodology.

The COD analysis was performed only in the samples with 5 and 60 min of ozonation time by chemical digestion according to method 410.4 (USEPA). The standard COD analysis kit was used with the ready-to-use vials. Low-range Hanna (HI94754A) reagent vials were used to digest the samples. The samples were filtered with a 0.2 µm pore filter after previous digestion for 2 h at 150 °C. After digestion, the samples were measured in the UV-Vis spectrophotometer PerkinElmer Lambda 25 at 600 nm (Waltham, MA, USA).

On the other hand, the TOC analysis was determined by direct injection in the Torch IQOQ equipment (Teledyne-Tekmar, Mason, OH, USA). The solutions were previously filtered by 0.2 µm membrane. These experiments used tert-butanol (TBA) as •OH scavenger with 200 mg L^{-1}.

UV-Vis spectroscopy was used to obtain the information of the treatment efficiency and monitoring the organic compound decomposition in ozonation (at 210 nm and 260 nm). These wavelengths were selected because they corresponded to the characteristic maximum absorbance for lignin and its derivatives (260 nm), as well as simple organic acids (210 nm).

The same technique was used to measure the discoloration (at 465 nm) and biodegradation efficiency of the organic source (at 210 nm) of raw wastewater, as well as microbial growth (at 600 nm). The starting compositions of the organic compound mixture (obtained as ozonation process effluent), the byproducts and final products formed in the ozonation procedures, and the biodegradation phases were partly recognized using high-performance liquid chromatography (HPLC, PerkinElmer, USA, Series 200) with UV detector at 210 and 260 nm under different analysis conditions. The Pt-Co scale was determined using this strategy.

In Table 1, the HPLC analysis conditions for the identification of intermediates and final products of the ozonation and biodegradation processes are detailed. The five selected methods of analysis were used to obtain a better separation of the complex mixture contained in the samples of residual water. So, methods I and III were used to obtain the time variation for the concentration of phenolic compounds, and method II was specifically used to obtain the characterization of phenolic compounds and acids. Finally, methods IV and V were specifically used to separate and characterize organic acids' distribution.

To determine the efficiency of the contaminant's degradation, the identification of the raw material, as well as the products of the ozonation and biodegradation, was carried out by a gas chromatograph with a mass detector, the GC-MS Claurus600 PerkinElmer, with helium as a carrier gas, an 0.8 cm^3/min feed flow, split ratio of 1:20 and an analysis temperature of 230 °C. In Table 2, the GC-MS analysis conditions are presented.

A more comprehensive characterization of the original wastewater sample was obtained through successive chloroform (CH_3Cl) extraction, acid hydrolysis and classification using the Carrez II method. The extraction was performed using the Soxhlet system with a volume of 0.25 L with a liquid volume of 0.1 L. The extraction liquid was CH_3Cl. The details of this procedure are shown in Table 3.

Table 1. HPLC analysis conditions for the identification of intermediates and products of ozonation and biodegradation.

	Ozonation					
	I	II		III		IV
Column		Nova Pack C-18, 250 × 4.6 mm: 10 μm				Prevail Organic Acid, 150 × 4.6 mm
Mobile phase	CH_3OH-H_2O (50:50)	C_2H_3N-H_2O (30:70)		C_2H_3N-H_2O (70:30)		H_2O-KH_2PO_4 25 mmole, pH = 2.5
Flow rate, cm^3/min	1.0	1.0		1.0		1.0
Injection volume, μL	30	30		30		30
λ, nm	210	260	210	260	210	210

	Biodegradation
	V
Column	Prevail Organic Acid, 150 × 4.6 mm
Mobile phase	H_2O-KH_2PO_4, 25 mmole, pH = 2.5
Flow rate, cm^3/min	1.0
Injection volume, μL	30
λ, nm	210

Numbers I to V refer the selected methods of analysis that we used to obtain a better separation of the complex mixture contained in the samples of residual water. The correspondign details of these methods are described above this table.

Table 2. GC-MS analysis conditions.

Column	**Elite 5 MS; 30 m × 0.32 mm × 0.25 mm**
Mobile phase	Helium
Sample volume, μL	0.5
Split	20%
Flow rate, cm^3/min	0.8
Injector temperature	230 °C
Temperature program	60 °C after 8 min; 60–240 °C, 30 °C/min; 240 °C after 10 min
Source temperature	230 °C
Transfer temperature	230 °C
Ionization energy, eV	70
Mass, m/z	25 to 400

Three treatment strategies were investigated: ozonation, biodegradation (without prior ozonation treatment or WoPT) and a combination of treatments (a sequential system with ozonation followed by biodegradation). The raw wastewater sample was chosen as a control experiment for comparison purposes.

A set of studies was carried out to evaluate conventional biodegradation in the absence of preceding ozonation. Two distinct ozonation tests were carried out: the first was discontinued after 30 min (M30O) of reaction, and the second after 60 min (M60O). These two precise reaction times were selected because, under the given conditions of the experiment, the byproducts generated in ozonation reached their maximum concentration (measured by HPLC) after 30 min of ozonation but were nearly undetectable (almost degraded) after 60 min. The outcomes of each treatment are detailed below.

Table 3. Composition of raw wastewater obtained by GC-MS.

Wastewater			Extract by CH3Cl			Acid Hydrolysis H+			Carrez Classification II		
Retention Time	Formula/MW		Retention Time	Formula/MW		Retention Time	Formula/MW		Retention Time	Formula/MW	
1.620	CH_3OH	32	1.620	CH_3OH	32	1.62	CH_3OH	32	1.61	CH_3OH	32
3.33	C_2Cl_4	164	1.72	C_5H_{12}	72	3.29	C_2Cl_4	164	4.817	$C_5H_8O_4$	132
4.633	C_5H_8O	84	2.25, 2.37, 2.48	$C_5H_{12}O$	88	1.83	C_6H_{12}	84	1.782	$C_5H_{10}O_2$	102
4.450	$C_5H_{10}O_3$	118	5.12	C6H8O	96	2.28	C7H16	100	2.08	C_6H_{12}	84
1.700	C_5H_{12}	72	2.8, 2.92, 3.19, 3.68	$C_6H_{10}O$	98	6.82	$C_{12}H_{16}$	160	5.08	$C_6H_{10}O_3$	130
5.620	$C_6H_4Cl_2$	146	4.95	C_6H_{12}	84	9.49	$C_{12}H_{24}$	168	3.066	$C_6H_{12}O$	100
2.100	C_6H_6	78	1.77, 1.8, 1.88, 1.99	C_6H_{14}	86	13.95	$C_{14}H_{12}O_4$	244	6.418	$C_6H_{12}O2$	116
2.075	C_6H_{10}	82	2.48	C_7H_{16}	100	10.64	$C_{16}H_{22}O_4$	278	4.917	$C_6H_{14}O_2$	118
1.98, 2.12	C_6H_{12}	84	3.13, 5.27	$C_7H_{14}O$	114	11.77	$C_{18}H_{38}$	254	1.82, 1.85, 1.87, 1.95	C_6H_{14}	86
2.42, 3.15	$C_6H_{10}O$	98	2.53	$C_7H_{16}O$	116				2.28	C_6H_8	80
4.650	$C_6H_{12}O3$	132	8.75, 8.84	$C_8H_{12}O_3$	156				4.82, 4.92 5.08	C_7H_{16}	100
1.75, 1.78, 1.8, 1.87, 1.97	C_6H_{14}	86	3.62	$C_8H_{18}O$	130				4.583	$C_7H_{14}O$	114
2.849	C_7H_8	92	6.32	$C_{11}H_{24}$	156				7.085	$C_7H_{14}O_2$	130
3.62, 4.25	C_7H_{14}	98	7.001	$C_{12}H_{26}$	170				1.948	$C_7H_{16}O$	116
5.2	$C_7H_{14}O_3$	146	9.85	$C_{13}H_{28}$	184				7.651	$C_8H_{16}O_2$	144
2.15, 2.22, 2.43, 3.45, 4.24, 4.97	C_7H_{16}	100	6.77	$C_{15}H_{32}$	212				2.933	C_9H_{20}	128
11.27	$C_8H_{14}O_4$	174	9.39	$C_{15}H_{28}O_2$	240				12.57	$C_{11}H_{24}$	156
3.660	C_8H_{16}	112	13.28, 14.02	$C_{20}H_{42}$	282				15.27	$C_{18}H_{38}$	254

Table 3. *Cont.*

Wastewater			Extract by CH3Cl			Acid Hydrolysis H+			Carrez Classification II		
Retention Time	Formula/MW		Retention Time	Formula/MW		Retention Time	Formula/MW		Retention Time	Formula/MW	
2.50, 2.58, 2.75, 2.8, 2.88, 3.2	C_8H_{18}	114	18.67	$C_{26}H_{54}$	366				13.64	$C_{20}H_{42}$	282
4.68, 5, 5.08, 5.17, 5.25, 5.42, 5.67	C_9H_{12}	120									
4.967	$C_9H_{16}O$	140									
4.57, 4.77,	$C_9H_{18}O$	142									
4.38, 6.751, 6.98	C_9H_{20}	128									
6.98, 8.22	$C_{10}H_8$	128									
5.317	$C_{10}H_{20}$	140									
4.833, 6.30, 8.068	$C_{10}H_{22}$	142									
5.37, 6.15, 6.20, 6.45, 6.70,	$C_{10}H_{24}$	144									
7.65, 7.35, 7.75	$C_{11}H_{10}$	142									
7.070	$C_{11}H_{16}$	148									
8.068	$C_{12}H_{10}$	154									
5.7, 5.8, 6.02, 6.07, 7.07	$C_{11}H_{24}$	156									
8.22, 8.29, 8.3, 8.4, 8.47	$C_{12}H_{12}$	156									
8.59, 8.64	$C_{13}H_{12}$	168									
7.55, 11.29	$C_{12}H_{26}$	170									

Table 3. *Cont.*

Wastewater			Extract by CH3Cl		Acid Hydrolysis H+		Carrez Classification II	
Retention Time	Formula/MW		Retention Time	Formula/MW	Retention Time	Formula/MW	Retention Time	Formula/MW
8.69, 8.80–8.97	$C_{13}H_{14}$	170						
9.60	$C_{14}H_{12}$	180						
9.09, 9.23	$C_{14}H_{14}$	182						
9.42	$C_{14}H_{16}$	184						
10.62	$C_{14}H_{18}O_4$	250						
16.94	$C_{16}H_{22}O_4$	278						
10.42	$C_{17}H_{34}O_2$	270						
9.84	$C_{18}H_{38}$	254						
11.47	$C_{19}H_{36}O_2$	296						
14.52, 16.24	$C_{26}H_{54}$	366						
16.51	$C_{28}H_{34}O$	286						

A total of three rounds of studies were carried out with the acclimatization technique in mind. The first utilized a sample that had not been pre-treated with ozone, while this acclimated microbe consortium combination was used for comparison (using the WoPT experiments). The second set of studies employed an ozone-treated sample for 30 min (M30O). The third experiment employed an ozone-treated sample for 60 min (M60O). In both cases, the consortium of acclimated microorganisms was dubbed OFA.

The short first-lag phase in microorganism development prevents us from using a standard modeling technique to quantify the effect of acclimation on biomass growth. Furthermore, no specific biomass quantification method was used. As a result, as a suitable technique for characterizing the biomass growth procedure, the biomass accumulation in the 10th cycle of acclimation was modeled as pseudo-monomolecular kinetics [38]. The selected mathematical model was $A = A_{max}\ (1 - e^{-bt})$, where A is the absorbance measured at the wavelength of 600 nm, A_{max} is its relating maximum absorbance and t is the reaction time, while the inverse of b (h^{-1}) denotes an approximation of the reaction rate.

In regard to the kinetics of the biological process, the specific growing rate (SGR or μ_{max}) was determined using a Monod reaction rate [38].

3. Results and Discussion

The primary physicochemical properties of the residual water after dilution were pH = 8, SDT = 3.5 g/dm^3, color unit of 1400 mg Pt/m^3, Pt/determined by the Pt-Co technique [28], COD of 670 mg O_2/dm^3 and conductivity of 1.0 mS/cm. All of the results were presented for a raw wastewater sample (using the inverse factor of dilution). The diluted sample has been created to be studied using normal spectroscopic and chromatographic procedures. When considering real wastewater, one may notice that this is a common strategy because the levels of color units, as well as concentrations of diluted organics, can be over the maximum level of detectability in the specialist equipment.

An initial GC-MS analysis was made to characterize the organics composition of untreated wastewater. The untreated sample of wastewater contained a wide variety of organic compound groups such as aliphatic, aromatic, chlorinated and polar (acids, esters, ketones, etc.). In general, about 80 organic compounds (C_1–C_{28}) were identified (see

Table 3). These organics were chosen from the chromatographs considering the peaks with maximal area.

3.1. Ozonation

To determine the ozone consumed in the chemical reaction, the previously proposed parameter named as the saturation constant (k_{sat} $[s^{-1}]$) was used to characterize the ozone mass transfer in the ozone–water heterogeneous system. This parameter was calculated according to the method presented in [40,41].

The sample discoloration (obtained at the wavelength of $\lambda_{max} = 465$ nm) was determined along the first 60 min of the ozonation stage. A reduction in color from 140.0 to 30.4 mg Pt/dm^3 (experimental data are not presented) was obtained (78%).

In Figure 3, the time variation in lignin as well as its derivatives and simple organic acids in ozonation (obtained by the UV absorbance measured at 260 nm and 210 nm, respectively) are shown. The absorbance of the lignin derivatives was reduced by 67% after the first 3 min of ozonation. On the other hand, the organic acids' absorbance was reduced by 83% after the first 2 min of ozonation.

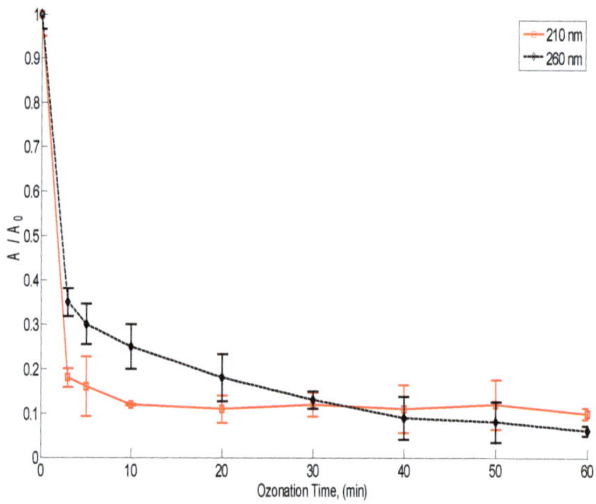

Figure 3. Normalized behavior of absorbance in ozonation during 60 min process. Simple organic acids at 210 nm wavelength and lignin and their derivatives measured at 260 nm wavelength.

On the other hand, the HPLC analysis of intermediates and final products proved the existence of oxalic, fumaric, maleic and formic acids, as well as some phenolic compounds (pseudocumene; 3-methoxy-methyl-ester-propanoic acid; catechol, hydroquinone, dichlorobenzene, decanoic-acid-methyl-ester).

The behavior of some recognized hazardous chemicals generated during ozonation is depicted in Figure 4. Catechol and hydroquinone were virtually completely eliminated (99.0%) after 20 and 30 min of ozonation, respectively, because their ozonation reaction constants were larger than 3.0×10^5 dm^3/(mole s). All other discovered byproducts (except oxalic acid) followed the same dynamics and were degraded during the reaction time.

After 30 min of ozonation, the lignin and its derivatives were almost decomposed with a transformation into simple organic acids (94%). The maleic acid accumulated during the first 5 min and decomposed practically completely after 8 min of ozonation. The oxalic acid, as a recalcitrant compound with a low ozonation constant ($k > 40 \times 10^{-2}$ dm^3/(moles)), accumulated during the 30 min of ozonation from 2.4 mg/dm^3 in the initial residual water sample to 60.0 mg/dm^3 (Figure 4b). This result is in same range of values found by some other authors [42–44]. In this particular case, its concentration decreased by 10% during the next 30 min. This behavior can be explained by the effect of OH radicals formed in the case when the reaction is performed at pH > 7 (Figure 4b).

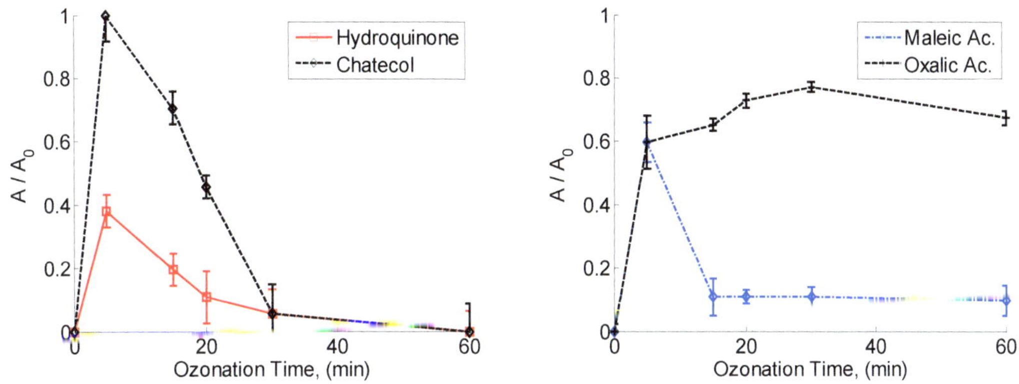

(**a**) (**b**)

Figure 4. Decomposition behavior of the main organic compounds obtained by HPLC: (**a**) hydroquinone and chatecol, and (**b**) maleic acid and oxalic acid.

3.2. Inoculum Acclimation Prior to Biodegradation

The biodegradation was applied after a preliminary acclimation of microorganisms that was important to improve the mineralization degree of organic matter. This strategy was proposed to make a fair comparison between the results obtained in the ozonation treatment, as well as the sequential process.

This stage used a specific carbon source and carried out a series of ten biodegradation phases (5 days each) for consortium maintenance [29]. Because of the amount of culture cycles, the microorganisms were compelled to consume 10 times more substrate during this phase. As a result, their quantity increased, as well as the possibility of increasing the consumption of organic sources. Figure 5a–c describe the growth of microorganisms during the acclimation phase in the three different experiments described above. When the first cycle of 5 days (Figure 5) was evaluated, the biomass growth was 12 times smaller than the one observed when the tenth cycle of acclimation was considered. This condition was obtained with and without the application of a preliminary ozonation process and appeared no matter what the preliminary ozonation time was.

The effect of pre-ozonation was evident on the biomass accumulation. In the case with no previous treatment with ozone, biomass growth was slower than the growth observed in the experiments with pre-ozonation. However, after 10 cycles of acclimation (Figure 5), despite the prior treatment with ozone, the same trend of the biomass accumulation was observed for the sets of experiments when ozonation was used as a pretreatment. The acclimation procedure was efficient to improve the capability of microorganisms to use the specific ozonation byproducts mixture (mainly organic acids) as a carbon source. Moreover, this strategy provided a more adequate form to compare the decomposition efficiency results for the pulp and paper industry. For comparison purposes, the treatment without the acclimation process and without prior ozonation was also investigated. In this case, the constant b was 0.48 day^{-1}. If the set of microorganisms was acclimated during 10 cycles, this kinetic parameter was modified to 1.13 day^{-1}. This variation obeyed the selective growing of microorganisms that were able to degrade the compounds found in the water samples. A similar increment of the same parameter was determined for the cases when the pretreatment with ozone was carried out. Similar variations in parameter b were observed when preozonated samples were analyzed.

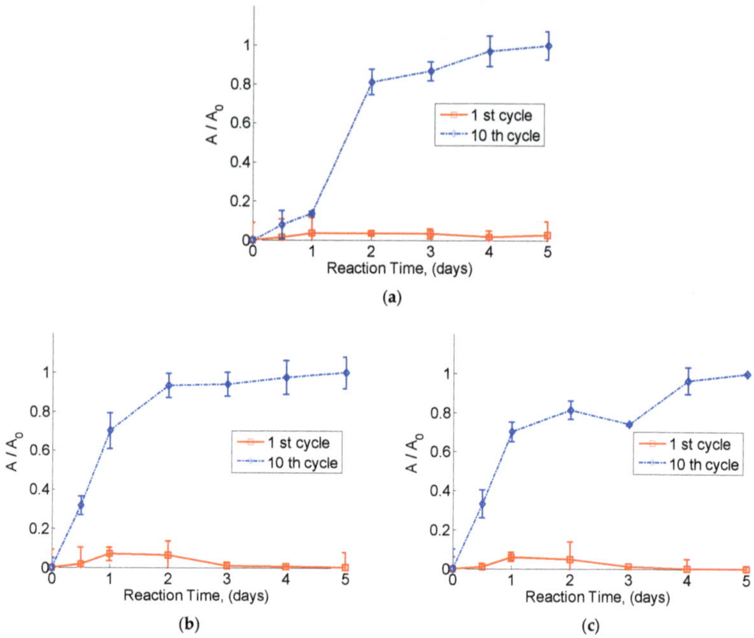

Figure 5. Microorganism's growth dynamics in the 1st and the 10th cycles of acclimation with the specific organic source: (**a**) without previous ozonation process (WoPT), (**b**) sample previously ozonated by 30 min (M30O) and (**c**) sample previously ozonated by 60 min (M60O).

3.3. Biological Treatment without Ozonation

Biodegradation with the acclimated microorganisms showed an incomplete decomposition of the organic compounds considering the complexity and heterogeneity of the wastewater samples' composition. In general, after 120 h of treatment, the increment of optical density (OD at 600 nm) indicated only partial decomposition of the organic compounds, at 20%:10% in the first 20 min and 10% additional after 50 min of bioprocess, without variation until the end of the experiment. The lines named WoPT in Figure 6a,b demonstrated these facts. These experiments also proved that despite the acclimation procedure, the microorganism's effectiveness to decompose organic matter was reduced. This fact is confirmed by the total TOC obtained after biodegradation without pre-ozonation.

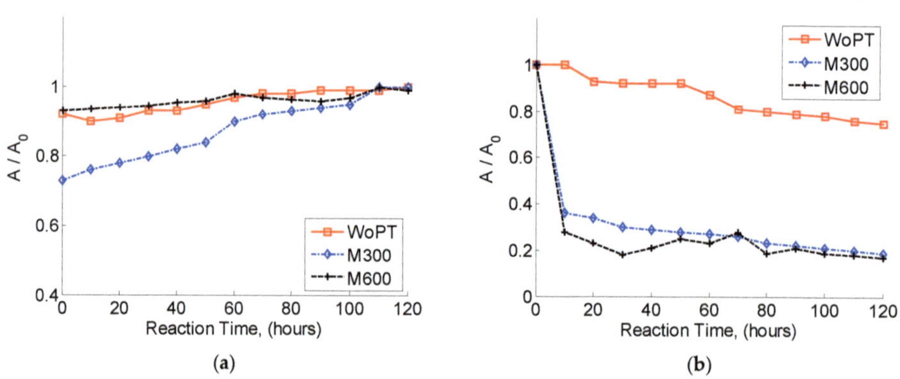

Figure 6. Microbial growth (**a**) and biodegradation of organic matter (**b**) for the WoPT and pre-treated samples of M30O and M60O at the initial pH = 7.

3.4. Combined Treatment: Ozonation before Biodegradation

The total amount of organics' decomposition obtained as a result of the combined sequential treatment is also shown in Figure 6. In both tests with 30 and 60 min of ozonation, the UV-vis spectrums at $\lambda_{max} = 210$ nm, corresponding to the concentration of simple acids, and the OD at 600 nm were utilized as general markers of the effectiveness of the combined treatment. Following ozonation, most of the organic matter degraded into oxalic acid and other low-molecular-weight acids, which were identified using the HPLC analytical technique.

Following biodegradation, all these organic acids were converted to CO_2, H_2O and a few other non-toxic metabolites. In the two treated samples, because of the presence of these chemicals, a sequential treatment based on ozonation, and biodegradation was used. After the first 12 h of the bioprocess, the pollutant level in the two treated samples was reduced by 72% for M30O and 78% for M60O. As can be seen, all organic components began to degrade during the first few minutes of the process. If the organic matter was pre-treated with ozone for 60 min, it was reduced by 78% during the first 30 h, but its concentration did not alter from 30 to 120 h without demonstrating a significant effect of incrementing ozonation time twice. This characteristic can be ascribed to the development of particular metabolites that restrict microbial growth.

At the final stage of the bioprocess (120 h), the total organic matter decrease was 82% for M30O and 83% for M60O, practically without a difference. However, the different ozonation times promoted a significant difference in the biomass dynamics accumulation, which confirms the effect of the initial substrate composition. Notice that the smaller ozonation time (30 min) had the same decomposition effectiveness after 50 h of the combined treatment and maintained this for 120 h. This effect may be used as a partial optimization of this treatment system by reducing the ozonation time. This is regarded as a partial optimization of pulp and paper wastewater treatment. The SGR was estimated without pre-ozonation as $\mu_{max} = 8.33 \times 10^{-4}$ h^{-1}. After performing preliminary ozonation for 30 min, the SGR augmented to 2.08×10^{-3} h^{-1}. This significant rise in SGR was attributed to the influence of ozone, which reduces the complexity of pollutants while increasing their biodegradability.

The SGR was lowered to 6.05×10^{-4} h^{-1} after 60 min of preparatory ozonation, which can be explained by a reduction in the microorganism's ability to remove the polar molecules (C_5–C_{17}) generated after this ozonation time. Because of the presence of these polar molecules (organic acids), the SGR was reduced to a level lower than that attained in normal biodegradation. It was established in this example that oxidized products were not appealing to microbes as carbon and energy sources.

The primary refractory product generated during ozonation was oxalic acid, as previously determined by HPLC examination of the treated samples (Table 1). As a result, the dynamics of its decomposition were investigated using the same three sets of experimental circumstances described above (biodegradation, ozonation and the sequential process). The mineralization of oxalic acid was the overall result of biodegradation.

The concentration of oxalic acid was lowered in all three systems, with and without ozonation. The final oxalic acid concentration was reduced by 18% in the non-ozonated sample, while it decreased by 90% in both pretreatment samples. However, when the ozonation period was reduced by half, the best result was obtained when the wastewater sample was ozonated for 30 min (M30O). The mineralization of oxalic acid with other intermediates was validated by measuring TOC, which dropped by 80% after 30 min of ozonation (from 680 to 150 mg C/dm^3 in diluted water samples).

As can be seen from the results of the GC-MS analysis of the initial untreated water, the water sample had a very complicated composition and contained 70 different organics, including organic acids, fatty acids, esters and ketones, among others, as well as aliphatics and aromatics (Table 3) with a maximal molecular weight of 386. After ozonation, 32 and 40 organics were eliminated from the water samples, depending on the ozonation time (30 and 60 min, correspondently), resulting in degradation efficiencies of 46 and 57% for this step. Moreover, the GC-MS data demonstrated that following ozonation, the water samples comprised different chemical components to the initial sample (Table 4). On the other hand, the treated water samples contained ozonation products which were decomposed in the biodegradation stage (Table 4). As observed, after biodegradation, only 15 compounds remained in the treated samples, which was 21% of the initial composition of organic compounds.

Table 4. Composition of wastewater after ozonation and after biodegradation. RT stands for retention time.

No	Compounds	MW	M30O		M60O	
			RT before Biodegradation	**RT after Biodegradation**	**RT before Biodegradation**	**RT after Biodegradation**
1	CH_4O	32	1.30	1.32	1.30	1.27
2	$C_5H_{10}O_3$	118	4.47	5.68	4.47	
3	C_6H_6	78	2.09			
4	$C_6H_{10}O$	98		3.19		
5	$C_6H_{12}O$	100		2.58		2.58
6	C_6H_{12}	84		1.98 2.13		
7	C_6H_{14}	86		1.87		
8	$C_6H_{12}O_2$	116		3.37		
9	$C_6H_{12}O_3$	132		3.98	4.67	3.98
10	C_7H_8	91	2.82			
11	C_7H_{14}	98	2.03 2.25			
12	$C_7H_{12}O_4$	160			4.78	5.23
13	$C_7H_{14}O_2$	130			2.13	
14	$C_7H_{14}O_3$	146		5.28		
15	C_7H_{16}	100	2.12 2.28 2.50			2.87
16	C_8H_{10}	106	3.97 4.07 4.33		3.97	
17	C_8H_{18}	114	2.22 2.43 2.52 2.90 3.32 6.42		4.33 2.43	3.37
19	$C_8H_{14}O$	126			2.52	
20	C_9H_{12}	120	5.01 5.08 5.27 5.45	4.68	4.40 5.00 5.08 5.42 5.45	4.68
21	$C_9H_{18}O$	142			4.78	

Table 4. *Cont.*

No	Compounds	MW	M30O		M60O	
			RT before Biodegradation	RT after Biodegradation	RT before Biodegradation	RT after Biodegradation
22	C_9H_{20}		4.40		2.89	
23		134	5.68 6.48	5.40	5.70 5.98 6.48	5.40
24	$C_{10}H_{18}O$	154			4.88	
25	$C_{10}H_{22}$	142	5.46			
26	$C_{11}H_{10}$	142	7.27		7.65	
27	$C_{11}H_{16}$	148	6.68		6.63	
28	$C_{11}H_{24}$	156	6.30		6.30	
29	$C_{12}H_{10}$	154	8.07		8.07	
30	$C_{12}H_{12}$	168	8.38		8.40	
31	$C_{13}H_{14}$	170	8.97		8.95	
32	$C_{13}H_{28}$	184	7.55		11.27	
33	$C_{14}H_{10}$	178				10.09
34	$C_{14}H_{30}$	198	6.98		6.98	
35	$C_{14}H_{18}O_4$	250			10.64	
36	$C_{17}H_{34}O_2$	270	5.80		10.42	
37	$C_{27}H_{56}$	380	9.11 9.22 10.42 11.27			9.13 10.40 14.44
38	$C_{28}H_{58}$	394		15.01 16.29 18.58		15.72 16.41 18.59
Total number of compounds			38	15	30	15

Figure 5 demonstrates the evolution of microorganisms' biomass considering the presence of the pre-ozonation with different periods. This comparison demonstrates in an integrated manner the effect of ozonating the mixture before biodegradation with different periods. Moreover, the data shown in Table 4 establish the effect on the accumulated byproducts after ozonation. The significant difference in the accumulated profiles of these byproducts justifies the variation in the biomass profiles shown in Figure 5.

The results of the GC-MS analysis results on the original and treated wastewater samples are summarized in Table 5, showing that the compounds were, in general, eliminated. Indeed, after 30 min of ozonation, 23 compounds out of 38 (60%) were removed during the bioprocess. Organic acids, fatty acids, esters and ketones were among the components that were removed. A buildup of heavy alkanes (C_{27}, C_{28}) and polar molecules was discovered at the end of the tests, which can be described as probable results of bacterial metabolism. Even if there is no precise identification of byproducts generated during successive treatment, a significant variation in organics composition following treatment may occur. Furthermore, there may be a significant decrease in the average molecular weight of the byproducts. This is an indirect confirmation that during biomass growth, a significant amount of organic material with part high-molecular-weight organic compounds are consumed [45,46].

Table 5. Summary of the results obtained by the GC-MS technique of the raw and treated wastewater samples (Compound content, wt%).

Compound	Original Wastewater	After Ozonation M30O	After Biodegradation	After Ozonation M60O	After Biodegradation
Chlorinated Hydrocarbons	4.4	-	-	-	-
Alkanes	17,7 (C_5–C_{12}, C_{18}, C_{26})	47.4 (C_7–C_{14}, C_{28})	53.3 (C_{28})	13.3 (C_{27}, C_{28})	53.3 (C_{27}, C_{28})
Unsaturated Hydrocarbons	33.3	50.0 (C_{10}–C_{14})	-	53.4 (C_{11}–C_{14})	20.0
Polar Hydrocarbons	31.1 (C_5–C_{28})	-	46.7 (C_5–C_6)	33.3 (C_5–C_{17})	26.7
Aromatic Hydrocarbons	13.5	2.6	-	-	-
Total	100	100	100	100	100

The metabolism of the previously identified microorganisms that form the consortia showed low toxic metabolites because they were low-weight acids that coincide to some preliminary studies [35]. Therefore, the sequential treatment proposed in this study can be considered successful according to these results.

4. Conclusions

At the base of the experimental results and discussion, the following conclusions can be suggested:

Following the bleaching phase of the Kraft process, the suggested sequential combination of ozonation and biodegradation of hazardous compounds in pulp and paper mill effluent produced remarkable results. In general, after 30 min of ozonation, a significant quantity of initial harmful pollutants (83%) was reduced, which were subsequently changed into non-toxic organic acids that were easily destroyed at rates of up to 85% after 72 h of biodegradation.

Acclimatization of microorganisms was an important step in the bioprocess treatment since it stimulated the development of microorganisms. After acclimation, the related pseudo-monomolecular constant b associated with biomass growth increased from 0.48 day^{-1} up to 1.13 day^{-1}.

Ozonation time played a key role in the enhancement of biomass growth and the decrease in the concentration of toxic organic compounds. In general, without the previous ozonation, the specific growth rate was 8.33×10^{-4} h^{-1} and it increased up to 2.08×10^{-3} h^{-1} when the sample was ozonated for 30 min. If the preliminary ozonation was carried out for 60 min, this reaction constant was reduced to 6.05×10^{-4} h^{-1}.

During experimental settings, a 30 min ozonation period was regarded the best among the evaluated cases since the mineralization degree after 72 h of biodegradation was nearly identical for both preozonated systems (89% vs. 85%) but the ozonation duration was cut in half.

Author Contributions: Conceptualization, T.P. and I.C.; methodology, J.A.; software, I.C.; validation, J.A., S.S. and I.C.; formal analysis, T.P. and I.C.; investigation, J.A. and S.S.; resources, T.P.; data curation, I.C.; writing—original draft preparation, J.A. and S.S.; writing—review and editing, T.P.; visualization, I.C.; supervision, T.P. and S.S.; project administration, I.C.; funding acquisition, T.P. All authors have read and agreed to the published version of the manuscript.

Funding: The paper was prepared with the general financial support of Consejo Nacional de Ciencia y Tecnología (Project No. A1-S-21608), Instituto Politécnico Nacional (Projects No. 20220758, No. 20230273) the Tecnologico de Monterrey, Institute of Advanced Materials for Sustainable Manufacturing under the grant Challenge-Based Research Funding Program 2022 number I006-IAMSM004-

C4-T2-T; and PROGRAMA PARA LA VINCULACIÓN DE EMPRESAS CON INSTITUCIONES DE EDUCACIÓN SUPERIOR Y CENTROS DE INVESTIGACIÓN COMECYT-EDOMÉX project No 2023/009.

Data Availability Statement: All data used or obtained during the development of this study may be available under appropriate request to the Corresponding Authors.

Conflicts of Interest: The authors declare no conflicts of interest.

References

1. Rowell, R. *Handbook of Wood Chemistry and Wood Composites*, 2nd ed.; CRC Press Taylor & Francis Group, EEUU: Boca Raton, FL, USA, 2012.
2. Pokhrel, D.; Viraraghavan, T. Treatment of pulp and paper mill wastewater—A review. *Sci. Total Environ.* **2004**, *333*, 37–58. [CrossRef]
3. Freire, R.S.; Kunz, A.; Durán, N. Remediation and toxicity removal from Kraft E1 paper mill effluent treatment with ozone. *Environ. Technol.* **2000**, *21*, 717–721. [CrossRef]
4. Peng, G.; Roberts, J.C. Solubility and toxicity of resin acids. *Water Res.* **2000**, *34*, 2779–2785. [CrossRef]
5. Catalkaya, E.C.; Kargi, F. Color, TOC and AOX removals from pulp mill effluent by advanced oxidation processes: A comparative study. *J. Hazard. Mater.* **2007**, *139*, 244–253. [CrossRef]
6. Kansal, S.; Singh, M.; Sud, D. Effluent quality at kraft/soda agro-based paper mills and its treatment using a heterogeneous photocatalytic system. *Desalination* **2008**, *228*, 183–190. [CrossRef]
7. Johnsen, K.; Tana, J.; Lehtinen, K.-J.; Stuthridge, T.; Mattsson, K.; Hemming, J.; Carlberg, G.E. Experimental Field Exposure of Brown Trout to River Water Receiving Effluent from an Integrated Newsprint Mill. *Ecotoxicol. Environ. Saf.* **1998**, *40*, 184–193. [CrossRef] [PubMed]
8. Tsang, Y.; Hua, F.; Chua, H.; Sin, S.; Wang, Y. Optimization of biological treatment of paper mill effluent in a sequencing batch reactor. *Biochem. Eng. J.* **2007**, *34*, 193–199. [CrossRef]
9. Karrasch, B.; Parra, O.; Cid, H.; Mehrens, M.; Pacheco, P.; Urrutia, R.; Valdovinos, C.; Zaror, C. Effects of pulp and paper mill effluents on the microplanktonans microbial self-purification capabilities of the Bibio River. *Chile Sci. Total Environ.* **2006**, *359*, 194–208. [CrossRef] [PubMed]
10. US EPA. Available online: http://www.epa.gov/waterscience/methods/pollutants.htm (accessed on 17 September 2023).
11. Trapido, M.; Hirvonen, A.; Veressinina, Y.; Hentunen, J. Ozonation, Ozone/UV and UV/H2O2 Degradation of Chlorophenols. *Ozone Sci. Eng.* **1996**, *18*, 75–95.
12. Poznyak, T.; Vivero, J. Degradation of Aqueous Phenol and Chlorinated Phenols by Ozone. *Ozone Sci. Eng.* **2005**, *27*, 447–458. [CrossRef]
13. Beltrán, F.J.; Rivas, J.; Alvarez, P.M.; Alonso, M.A.; Acedo, B.A. Kinetic Model for Advanced Oxidation Processes of Aromatic Hydrocarbons in Water: Application to Phenanthrene and Nitrobenzene. *Ind. Eng. Chem. Res.* **1999**, *38*, 4189–4199. [CrossRef]
14. Yordanov, R.; Melvin, M.; Law, S.; John, L.J.; Lamb, A. Effect of ozone pre-treatment of colored Upland water on some biological parameters of sand filters. *Ozone Sci. Eng.* **1999**, *21*, 615–626. [CrossRef]
15. Griffini, O.; Bao, M.L.; Barbieri, K.; Burrini, D.; Santianni, D.; Pantani, F. Formation and Removal of Biodegradable Ozonation By-Products during Ozonation-Biofiltration Treatment: Pilot-Scale Evaluation. *Ozone Sci. Eng.* **1999**, *21*, 79–98. [CrossRef]
16. Kitis, M.; Adams, C.D.; Kuzhikannil, J.; Daigger, G.T. Effects of Ozone/Hydrogen Peroxide Pretreatment on Aerobic Biodegradability of Nonionic Surfactants and Polypropylene Glycol. *Environ. Sci. Technol.* **2000**, *34*, 2305–2310. [CrossRef]
17. Adams, C.D.; Cozzens, R.A.; Kim, B.J. Effects of ozonation on the biodegradability of substituted phenols. *Water Res.* **1997**, *31*, 2655–2663. [CrossRef]
18. Beltrán, F.J.; García-Araya, J.F.; Álvarez, P.M. Wine Distillery Wastewater Degradation. 1. Oxidative Treatment Using Ozone and Its Effect on the Wastewater Biodegradability. *J. Agric. Food Chem.* **1999**, *47*, 3911–3918. [CrossRef] [PubMed]
19. Rodríguez Fernández, A.; Letón García, P.; Rosal García, R.; Dorado Valiño, M.; Villar Fernández, S.; Sanz García, J. Tratamientos Avanzados de Aguas Residuales Industriales. In *Informe de Vigilancia Tecnológica*; Madrid, Spain, 2006; Available online: https://www.madrimasd.org/ (accessed on 17 September 2023).
20. De los Santos Ramos, W.; Poznyak, T.; Chairez, I. Remediation of lignin and its derivatives from pulp and paper industry wastewater by the combination of chemical precipitation and ozonation. *J. Hazard. Mater.* **2009**, *169*, 428–434. [CrossRef]
21. Amat, A.; Arques, A.; Miranda, M.; López, F. Use of ozone and/or UV in the treatment of effluents from board paper industry. *Chemosphere* **2005**, *60*, 1111–1117. [CrossRef]
22. Herath, N.K.; Ohtani, Y.; Ichiura, H. Color and phenolic compound reduction of Kraft Pulp Mill effluent by ozonation with some pretreatments. *Am. J. Sci. Ind. Res.* **2011**, *2*, 798–806. [CrossRef]
23. Tünay, O.; Erdeml, E.; Kabdaşli, I.; Ölmez, T. Advanced treatment by chemical oxidation of pulp and paper effluent from a plant manufacturing hardboard from waste paper. *Environ. Technol.* **2008**, *29*, 1045–1051. [CrossRef]
24. Belém, A.; Panteleitchouk, A.V.; Duarte, A.C.; Rocha-Santos, T.A.P.; Freitas, A.C. Treatment of the effluent from a kraft bleach plant with white rot fungi: Pleurotussajorcaju and Pleurotusostreatus. *Glob. NEST J.* **2008**, *10*, 426–431.

25. Ruiz-Dueñas, F.J.; Martínez, Á.T. Microbial degradation of lignin: How a bulky recalcitrant polymer is efficiently recycled in nature and how we can take advantage of this. *Microb. Biotechnol.* **2009**, *2*, 164–177. [CrossRef]
26. Qayyum, H. Peroxidase mediated decolorization and remediation of wastewater containing industrial dyes: A review. *Rev. Environ. Sci. Biotechnol.* **2010**, *9*, 117–140.
27. Stoilova, I.; Krastanov, A.; Stanchev, V.; Daniel, D.; Gerginova, M.; Alexieva, Z. Biodegradation of high amounts of phenol, catechol, 2,4-dichlorophenol and 2,6-dimethoxyphenol by Aspergillusawamori cells. *Enzym. Microb. Technol.* **2006**, *39*, 1036–1041. [CrossRef]
28. Sridhar, R.; Sivakumar, V.; Immanuel, V.P.; Maran, J.P. Treatment of pulp and paper industry bleaching effluent by electrocoagulant process. *J. Hazard. Mater.* **2011**, *186*, 1495–1502. [CrossRef]
29. Ratledge, C. *Biochemistry of Microbial Degradation*; Kluwer Academic Publishers: Kingston upon Hull, UK, 1994; pp. 233–277.
30. Ali, M.; Sreekrishnan, T.R. Aquatic toxicity from pulp and paper mill effluents: A review. *Adv. Environ. Res.* **2001**, *5*, 175–196. [CrossRef]
31. Assalin, M.R.; Almeida, E.d.S.; Durán, N. Combined System of Activated Sludge and Ozonation for the Treatment of Kraft E_1 Effluent. *Int. J. Environ. Res. Public Health* **2009**, *6*, 1145–1154. [CrossRef]
32. Laari, A.; Korhonen, S.; Kallas, J.; Tuhkanen, T. Selective Removal of Lipophilic Wood Extractives from Paper Mill Water Circulations by Ozonation. *Ozone Sci. Eng.* **2000**, *22*, 585–605. [CrossRef]
33. Guerra, P.; Amacosta, J.; Poznyak, T.; Siles, S.; García, A.; Chairez, I. Aerobic biodegradation coupled to preliminary ozonation for the treatment of model and real residual wáter. In *Biodegradation, Chapter 14*; Intech, Inc.: Seoul, Republic of Korea, 2013; pp. 365–387. ISBN 978-953-51-1153-5.
34. *NMX-AA-045-SCFI-2001*; Análisis de Agua—Determinación de Color Platino Cobalto en Aguas Naturales, Residuales y Residuales Tratadas. Norma Nacional Mexicana: México City, México, 2001.
35. García-Peña, E.I.; Zarate-Segura, P.; Guerra-Blanco, P.; Poznyak, T.; Chairez, I. Enhanced Phenol and Chlorinated Phenols Removal by Combining Ozonation and Biodegradation. *Water Air Soil Pollut.* **2012**, *223*, 4047–4064. [CrossRef]
36. Vazquezrodriguez, G.; Youssef, C.; Waissmanvilanova, J. Two-step modeling of the biodegradation of phenol by an acclimated activated sludge. *Chem. Eng. J.* **2006**, *117*, 245–252. [CrossRef]
37. Manickam, N.; Misra, R.; Mayilraj, S. A novel pathway for the biodegradation of γ-hexachlorocyclohexane by a *Xanthomonas* sp. strain ICH12. *J. Appl. Microbiol.* **2007**, *102*, 1468–1478. [CrossRef] [PubMed]
38. van den Wijngaard, A.J.; Prins, J.; Smal, A.J.; Janssen, D.B. Degradation of 2-chloroethylvinylether by *Ancylobacter aquaticus* AD25 and AD27. *Appl. Environ. Microbiol.* **1993**, *59*, 2777–2783. [CrossRef] [PubMed]
39. Kamal, V.S.; Wyndham, R.C. Anaerobic phototrophic metabolism of 3-chlorobenzoate by Rhodopseudomonaspalus-trisWS17. *Appl. Environ. Microbiol.* **1990**, *56*, 3871–3873. [CrossRef] [PubMed]
40. Poznyak, T.I.; Manzo, A.; Mayorga, J.L. Elimination of Chlorinated Unsaturated Hydrocarbons from Water by Ozonation. Simulation and Experimental Data Comparison. *J. Mex. Chem. Soc.* **2003**, *47*, 58–65.
41. Poznyak, T.; Chairez, I.; Poznyak, A. Application of a neural observer to phenols ozonation in water: Simulation and kinetic parameters identification. *Water Res.* **2005**, *39*, 2611–2620. [CrossRef]
42. Hoigné, J.; Bader, H. Rate constants of reactions of ozone with organic and inorganic compounds in water-II. *Water Res.* **1983**, *17*, 185–194. [CrossRef]
43. Munter, R.; Preis, S.; Kamenev, S.; Siirde, E. Methology of Ozone Introduction into water and wastewater treatment. *Ozone Sci. Eng.* **1993**, *15*, 149–165. [CrossRef]
44. Xiangbin, L.; Ortega, R.; Hongye, S.; Jian, C. Identification of Nonlinearly Parameterized Nonlinear Models: Application to Mass Balance Systems. In Proceedings of the Joint 48th IEEE Conference on Decision and Control and 28th Chinese Control Conference, Shanghai, China, 16–18 December 2009.
45. Wu, S.; Yang, Z.; Zhou, Z.; Li, X.; Lin, Y.; Cheng, J.J.; Yang, C. Catalytic activity and reaction mechanisms of single-atom metals anchored on nitrogen-doped carbons for peroxymonosulfate activation. *J. Hazard. Mater.* **2023**, *459*, 132133. [CrossRef] [PubMed]
46. Wu, S.; Shen, L.; Lin, Y.; Yin, K.; Yang, C. Sulfite-based advanced oxidation and reduction processes for water treatment. *Chem. Eng. J.* **2021**, *414*, 128872. [CrossRef]

toxics

MDPI

Article

Study on the Direct and Indirect Photolysis of Antibacterial Florfenicol in Water Using DFT/TDDFT Method and Comparison of Its Reactivity with Hydroxyl Radical under the Effect of Metal Ions

Yue Kang, Ying Lu and Se Wang *

Collaborative Innovation Center of Atmospheric Environment and Equipment Technology, Jiangsu Key Laboratory of Atmospheric Environment Monitoring and Pollution Control, School of Environmental Science and Engineering, Nanjing University of Information Science and Technology, Nanjing 210044, China; 202083280094@nuist.edu.cn (Y.K.); 20211248120@nuist.edu.cn (Y.L.)

* Correspondence: wangse@nuist.edu.cn; Tel./Fax: +86-25-58731090

Abstract: Florfenicol (FLO) is a widely used antibacterial drug, which is often detected in the environment. In this paper, the photolysis mechanism of FLO in water was investigated using density functional theory (DFT) and time-dependent density functional theory (TDDFT). The focus of the study is to elucidate the direct photolysis mechanism of FLO in the water environment and the indirect photolysis of free radicals (·OH, $·NO_3$, and $·SO_4^-$) as active species. The effect of metal ions $Ca^{2+}/Mg^{2+}/Zn^{2+}$ on the indirect photolysis was also investigated. The results show that the direct photolysis of FLO involves C–C/C–N/C–S bond cleavage, the C5–S7 bond cleavage is most likely to occur, and the C17–C18 cleavage reaction is not easy to occur during the direct photodegradation of FLO. The indirect photolysis of FLO is more likely to occur in the environment than direct photolysis. The main indirect photolysis involves OH-addition, NO_3-addition, and SO_4-addition on benzene ring. The order of difficulty in the indirect photolysis with ·OH is C2 > C3 > C4 > C5 > C6 > C1, Ca^{2+} can promote the indirect photolysis with ·OH, and Mg^{2+}/Zn^{2+} has a dual effect on the indirect photolysis with ·OH. In other words, Mg^{2+} and Zn^{2+} can inhibit or promote the indirect photolysis with ·OH. These studies provide important information for theoretical research on the environmental behavior and degradation mechanism of drug molecules.

Keywords: antibacterials; florfenicol; photolysis; metal ion influence; DFT

check for updates

Citation: Kang, Y.; Lu, Y.; Wang, S. Study on the Direct and Indirect Photolysis of Antibacterial Florfenicol in Water Using DFT/TDDFT Method and Comparison of Its Reactivity with Hydroxyl Radical under the Effect of Metal Ions. *Toxics* **2024**, *12*, 127. https://doi.org/10.3390/toxics12020127

Academic Editors: Zunyao Wang and Tiecheng Wang

Received: 7 January 2024
Revised: 25 January 2024
Accepted: 1 February 2024
Published: 3 February 2024

1. Introduction

Florfenicol (FLO) is a broad-spectrum chloramphenicol antibiotic widely used in the prevention and treatment of bacterial diseases, which can inhibit or eliminate the growth of microorganisms [1–5]. During aquaculture, FLO, as a pharmaceutical feed, is usually leached directly or released into the aquatic environment with feces [6,7]. The average and maximum concentration of FLO in the water around the farm is 0.42 and 2.84 μg/L, and the maximum concentration in the Huangpu River is 46.6 ng L^{-1} [8,9]. Its residues are likely to lead to the development and spread of drug-resistant bacteria/resistance genes, which in turn limits the treatment options for human infections [10,11].

Since FLO exists in water for a long time and may pose a threat to the aquatic environment and food safety, there is an urgent need to investigate its chemical behavior in water to evaluate its risks [12,13]. The degradation of FLO in water mainly includes hydrolysis, biodegradation, and photolysis [14–16]. Photolysis can be divided into direct photolysis and indirect photolysis. Direct photolysis means that substances directly absorb light energy for decomposition reactions. During the indirect photolysis process, FLO can react with ·OH, O_3, $^1O_2$, and other reactive oxygen species in the water environment [17].

Direct and indirect photolysis play an indispensable role in the environmental behavior of drugs [14,18].

In recent years, some studies have focused on the degradation behavior of FLO under UV/H_2O_2, UV/Fe (II), and UV/$Na_2S_2O_3$ conditions [18,19]. Xu et al. [20] studied the degradation behavior of FLO and its metabolites under different soil and natural conditions based on a simple method. In the past, photocatalysis has been shown to be an efficient treatment for the removal of aquaculture antibiotics, and it is necessary to consider changes in toxicity due to FLO decomposition [21,22]. Liu et al. [18] concluded that the toxicity of final products of FLO by UV irradiation was lower than FLO. UV irradiation-activated persulfate process can effectively control the toxicity of FLO degradation [19]. Meanwhile, photolysis is dramatically affected by environmental factors such as the type of light, pH, or antibiotic initial concentration [21,23,24]. Jiang et al. [25] considered that the superior stability and low concentration of C–F bonds in FLO hindered its degradation, and the degradation products may be as active/toxic as FLO. GE et al. [24] studied the effects of Cl^-, humic acid (HA), and other water components on the photodegradation of antibiotics; under UV–visible irradiation, Cl^- can promote the formation of $^1O_2$ and accelerate the photodegradation of phenols, while under simulated solar irradiation, its photodegradation is not related to Cl^-; HA inhibits photolysis under UV–visible irradiation, but it can undergo photosensitive degradation under simulated solar irradiation. Laboratory experiments are time-consuming, laborious, and costly, and it is difficult to decipher all the pathways of organic pollutant degradation in a short time. More importantly, theoretical calculations can find the intermediate products of the reaction process, which is difficult to achieve with laboratory experiments. As a result, there is a lack of computational research on the water environment behavior of FLO, especially the photodegradation behavior in water environment, which has guiding significance and verification function for the laboratory study.

By the means of quantum chemical calculation, this study intends to use density functional theory (DFT) and time density functional theory (TDDFT) to explore the photochemical transformation process of FLO in water environment [26–30]. The focus of the study is to elucidate the direct photolysis mechanism of FLO in water and the indirect photolysis of free radicals ($\cdot OH$, $\cdot NO_3$, $\cdot SO_4^-$) as active species. In addition, the effects of different metal ions (Ca^{2+}/Mg^{2+}/Zn^{2+}) on the indirect photodegradation of FLO and $\cdot OH$ were also studied. The computational methods employed in this research provide valuable insights into the photodegradation pathways and reactive intermediates involved in the photolysis of FLO in aqueous environments. Additionally, these findings contribute to a better understanding of the environmental fate and the potential remediation strategies of FLO contaminants.

2. Computational Methods

The calculations in this study were performed using the Gaussian 16 quantum chemistry software package [31,32]. The DFT/TDDFT has been extensively tested and validated against experimental data and has been shown to provide reliable predictions for the photolysis behavior of a variety of organic compounds [33–36]. Based on the B3LYP/6-311+G (d,p) level, DFT and TDDFT were used to optimize the stable structures of FLO in the ground state (S_0) and the lowest excited triplet state (T_1), respectively [37–39]. The intrinsic reaction coordinate (IRC) was used to analyze and verify the specified reactants (R) and products (P) for each transition state (TS) linkage [40]. In all calculations, the solvent effect in water is taken into account, and the integral equation in the form of the polarized continuum model (IEFPCM) is used [39]. At the B3LYP/6-311+G(d,p) level, TDDFT theory is used to calculate the electronic absorption spectra of monomer FLO and metal ion complexes FLO–Ca^{2+}/FLO–Mg^{2+}/FLO–Zn^{2+} in water [41,42]. The photodegradation mechanism of FLO in aqueous solution was analyzed by calculating the atomic charge and electron spin density based on natural bond orbital (NBO) at the B3LYP/6-311+G (d,p) level of theory. It mainly includes direct photodegradation (C–C bond cleavage, C–N bond cleavage, C–S bond cleavage), indirect photodegradation with free radicals ($\cdot OH$, $\cdot NO_3$, $\cdot SO_4^-$) in water, and indirect photodegradation with $\cdot OH$ under the effect of metal $Ca^{2+}/Mg^{2+}/Zn^{2+}$.

The structures of Rs (reactants), TSs (transition states), and Ps (products) during the direct and indirect photolysis were optimized at the B3LYP/6-311+G(d,p) level. As T_1 state has been found to be a precursor of photochemical reactions that exists for a long time, the calculation of photodegradation processes in this paper is carried out in T_1 state [16]. Through the calculation and analysis of TS vibration, it is ensured that there is only one imaginary frequency. All the calculated activation energy (E_a) and enthalpy changes (ΔH) of the reaction are modified by zero–point energy.

3. Results and Discussion

3.1. Optimized Geometry of FLO

Figure 1a shows the optimized geometry of FLO. The calculation results indicate that although the benzene ring is symmetrical, the bond lengths are not the same. For example, the C2–C3 bond (1.466 Å) is longer than the C2–C1 bond (1.434 Å), and the C5–C4/C4–C3 bond (1.470 Å / 1.356 Å) is shorter than the C5–C6/C6–C1 bond (1.473 Å / 1.360 Å). Figure 1b shows the electronic absorption spectra of FLO. The calculated maximum absorption wavelength of FLO is 229 nm. Experimental results show that the maximum absorption wavelength of FLO is 224 nm [11]. This indicates that the theoretical calculation results are consistent with the experimental results. Due to absorption by the atmosphere, the shortest wavelength of solar radiation reaching the Earth's surface is 290 nm [43]. In natural water, FLO can partially undergo direct photodegradation to a certain extent.

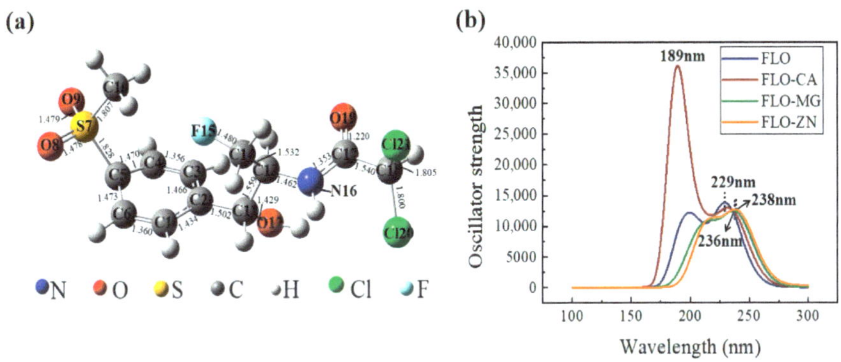

Figure 1. (**a**) Optimized geometries of FLO along with selected bond lengths (Å) and (**b**) calculated electronic absorption spectra of monomer FLO and complexes FLO–Ca^{2+}/FLO–Mg^{2+}/FLO–Zn^{2+} with maximum absorption wavelength (nm).

3.2. Direct Photolysis of FLO in Water

The direct photodegradation pathway of FLO in water with four reaction sites is shown in Figure S1. Figure S2 shows the optimized geometries of Rs, TSs, and Ps for the four reaction sites. The distance between the two sites of the broken bonds in the TSs for the four reaction sites is 1.799–2.100 Å. The ΔH values show that the cleavage of C11–C13/C5–S7 bond (Path 1/Path 4) is exothermic, and the cleavage of C13–N16/C17–C18 bond (Path 2/Path 3) is endothermic. The C11–C13 bond cleavage (Path 1) released the most heat (ΔH = −36.2 kcal/mol), indicating that the products P1a and P1(b) were the most stable.

Pathways 1, 2, 3, and 4 involve C11–C13, C13–N16, C17–C18, and C5–S7 bond cleavage (Path 1, Path 2, Path 3, and Path 4). The energy barrier diagram of the direct photolysis of FLO is shown in Figure 2. The E_a values of C11–C13/C13–N16/C5–S7 bond cleavage (Path 1/Path 2/Path 4) of FLO in T_1 state are 10.0 kcal/mol, 35.0 kcal/mol, and 2.0 kcal/mol, respectively, which were obviously lower than the cleavage of C17–C18 bond

(E_a = 94.8 kcal/mol) (Path 3). The relatively high E_a value (94.8 kcal/mol) indicates that the C17–C18 cleavage reaction (Path 3) does not easily occur during the direct photodegradation of FLO. The E_a value (2.0 kcal/mol) for the C5–S7 bond cleavage (Path 4) is the lowest among all the four paths, indicating that C5–S7 bond cleavage (Path 4) is most likely to occur in direct photodegradation, and this reaction produces P4a and P4b products.

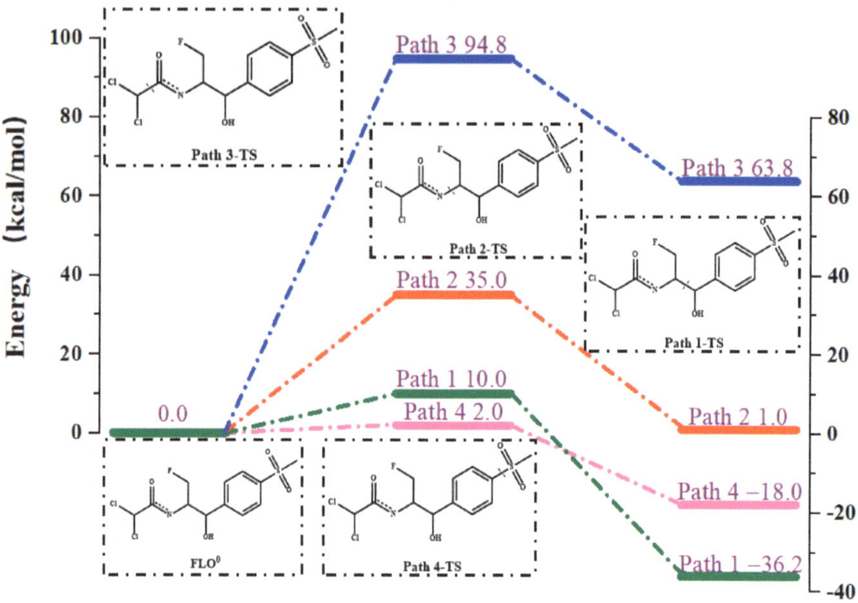

Figure 2. Reaction energy barrier diagram of the direct photolysis of FLO.

3.3. Indirect Photolysis of FLO in Water

3.3.1. Indirect Photolysis Mechanism of FLO and ·OH

Six reaction pathways of FLO with ·OH in water were predicted, and the reaction type of these pathways was ·OH addition reaction (Figure 3). By calculation, the reactions at the C1, C2, C3, C4, C5, and C6 sites on the benzene ring were predicted to be likely to occur. Figure S3 shows the optimized geometries of Rs, TSs, and Ps at six reaction sites. The distance between ·OH and the six reaction sites in the TSs of FLO is 2.011–3.006 Å. Figure 3 shows that all the reaction pathways (Path C1, Path C2, Path C3, Path C4, Path C5, and Path C6) are exothermic, and the ΔH values range from −2.7 kcal/mol to −41.3 kcal/mol. The addition reaction (Path C5) at the C5 site is the most exothermic (ΔH = −41.3 kcal/mol), indicating that the generated products C5_Pa and C5_Pb are the most stable.

OH is predicted to attack six C sites on the benzene ring (Path C1, Path C2, Path C3, Path C4, Path C5, and Path C6) to generate C1_P, C2_P, C3_P, C4_P, C5_P, and C6_P. Zhang et al. [17] also detected that phenolic products were formed after the electrophilic attack of ·OH at the C site of benzene ring in FLO photodegradation experiments. The reaction of ·OH attacking at the C2 site (Path C2) releases the least heat (ΔH = −2.7 kcal/mol). E_a values range from 2.7 kcal/mol to 7.8 kcal/mol, and lower E_a values indicated that ·OH promotes the photolysis of FLO, which is consistent with the results of Li et al. [44]. The E_a value for ·OH (Path C1) attacking at the C1 reaction site is the lowest (2.7 kcal/mol), indicating that this reaction is the easiest to occur and most conducive to the formation of C1_P. In addition, since the E_a value (7.8 kcal/mol) for ·OH attacking at C2 (Path C2) is the maximum value of all the reactions of ·OH attacking at FLO, Path C2 is not conducive to occur. A similar reaction was also shown at the C3 site. The E_a value for ·OH attacking at the C3 site (Path C6) is 7.3 kcal/mol. The data with little difference indicate that C2_P and C3_P are likely to be generated simultaneously

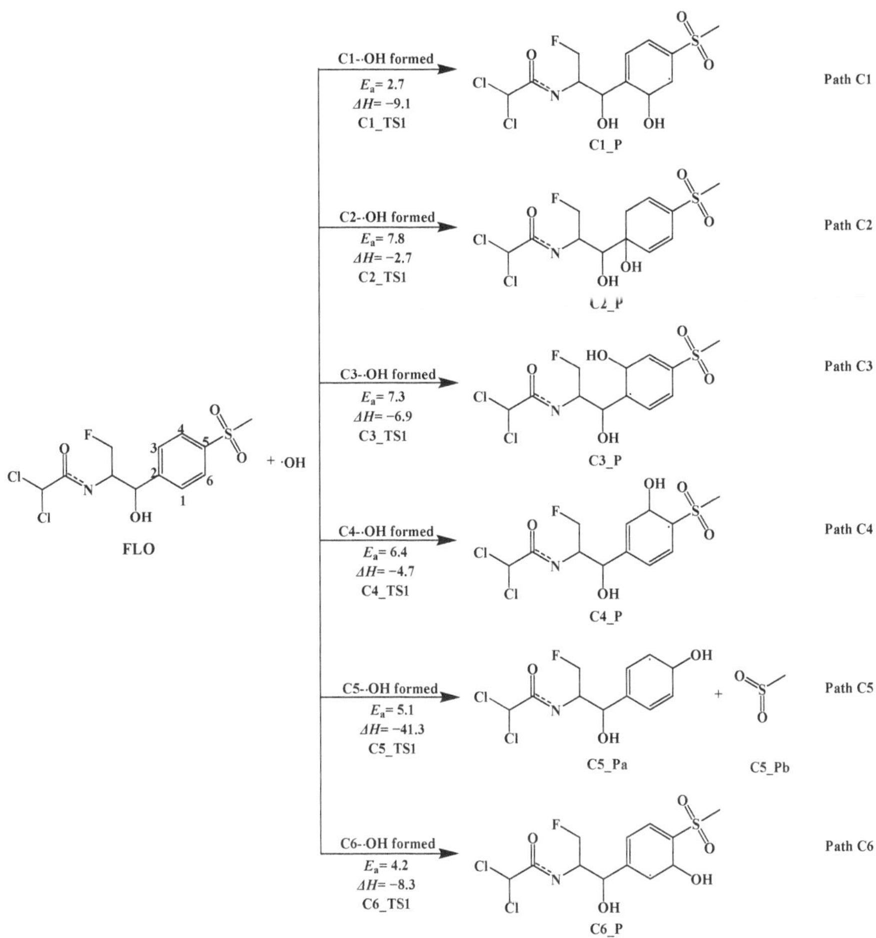

Figure 3. Indirect photodegradation pathways of FLO with ·OH, along with computed activation energies (E_a, kcal/mol) and enthalpy change (ΔH, kcal/mol).

3.3.2. Indirect Photolysis Mechanism of FLO and $\cdot NO_3$

Three possible pathways (Path C1^N, Path C4^N, and Path C6^N) of the indirect photodegradation of FLO by $\cdot NO_3$ in water are shown in Figure 4, and the reaction type of these pathways is an addition reaction. There are three possible reaction sites on the benzene ring: C1, C4, and C6. Figure S4 shows the geometric optimization of Rs, TSs, and Ps for the three reaction sites. The distance between $\cdot NO_3$ and the three reaction sites in the TSs is 1.711–1.866 Å. Figure 4 shows that the three reaction pathways (Path C1^N, Path C4^N, and Path C6^N) are endothermic, with ΔH values ranging from 4.4 kcal/mol to 4.8 kcal/mol.

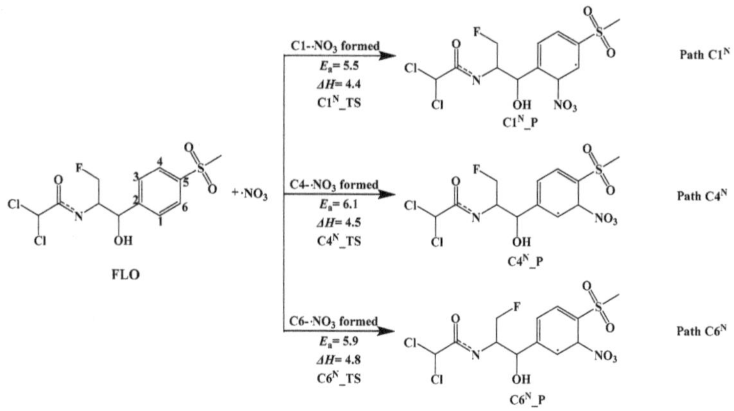

Figure 4. Indirect photodegradation pathways of FLO with $\cdot NO_3$, along with computed activation energies (E_a, kcal/mol) and enthalpy change (ΔH, kcal/mol).

The relatively low E_a value (5.5 kcal/mol) indicates that ·OH is most likely to react at the C1 site (Path C1^N) and is most conducive to the formation of C1^N_Pa. Similar reactions were also shown at the C4 and C6 sites. The E_a values for adding ·OH at the C4 site (Path C4^N) and C6 site (Path C6^N) are 6.1 kcal/mol and 5.9 kcal/mol, respectively. The data with little difference show that sites C1, C4, and C6 are beneficial to react with ·OH.

3.3.3. Indirect Photolysis Mechanism of FLO and $\cdot SO_4^-$

Four possible pathways (Path C1^S, Path C3^S, Path C4^S, and Path C6^S) of the indirect photodegradation of FLO with $\cdot SO_4^-$ in water are shown in Figure 5, and the reaction type is an addition reaction. Figure S5 shows the optimized geometries of Rs, TSs, and Ps for the four reaction sites. The distance between $\cdot SO_4^-$ and the four reaction sites in the TSs is 1.910–1.964 Å. All the pathways (Path C1^S, Path C3^S, Path C4^S, and Path C6^S) are predicted to be endothermic, with ΔH ranging from 4.5 kcal/mol to 8.2 kcal/mol (Figure 5).

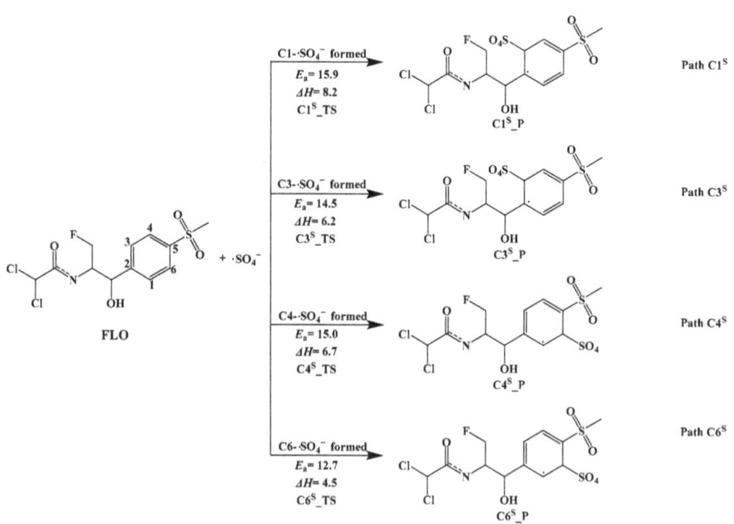

Figure 5. Indirect photodegradation pathways of FLO with $\cdot SO_4^-$, along with computed activation energies (E_a, kcal/mol) and enthalpy change (ΔH, kcal/mol).

The lowest E_a value (12.7 kcal/mol) for Path C6^S among all the four pathways indicates that ·OH is most likely to react at the C6 site and is most conducive to the formation of C6^S_Pa. The E_a value for adding ·SO_4^- to the C1 site was 15.9 kcal/mol (Path C1^S), which is the maximum value of all the four pathways. This indicates that the reaction of ·SO_4^- attacking at the C1 site (Path C1^S) is not conducive to proceed.

3.3.4. Optimized Geometries of Complexes FLO–Ca^{2+}/FLO–Mg^{2+}/FLO–Zn^{2+}

Figure S6 shows the optimized geometries of complexes FLO–Ca^{2+}/FLO–Mg^{2+}/FLO–Zn^{2+}. The calculation results show that the bond lengths of the complexes are different from those of the monomer FLO. For example, the bond lengths of S7–O9/ S7–C5/C2–C11 in complex FLO–Mg^{2+} are longer than those of FLO, conversely, the bond lengths of C11–C13 in complex FLO–Mg^{2+} are shorter than those of FLO (Figures 1a and S9).

Figure 1b shows the electronic absorption spectra of complexes FLO–Ca^{2+}/FLO–Mg^{2+}/FLO–Zn^{2+} in water at the TDDFT/B3LYP/6-311+G(d,p) level of theory. The bond length of Ca–O8 in the complex FLO–Ca^{2+} is the longest (2.329 Å) and the maximum absorption wavelength is the smallest (189 nm) among all the three metal ions complexes. The bond length of Zn–O8 in the complex FLO–Zn^{2+} is the shortest (1.926 Å), and the maximum absorption wavelength is the largest (238 nm).

Compared with the monomer, there is a blue shift in the maximum electron absorbance peak position of complex FLO–Ca^{2+}, while there is a red shift in those of the complexes FLO–Mg^{2+}/FLO–Zn^{2+}. Therefore, the metal ions Ca^{2+}/Mg^{2+}/Zn^{2+} have an effect on the structure and UV absorption spectrum of FLO and may also have an effect on the photodegradation of FLO.

3.3.5. Indirect Photolysis of Complexes FLO–Ca^{2+}/FLO–Mg^{2+}/FLO–Zn^{2+} with ·OH

Mechanism of Indirect Photolysis of Complex FLO–Ca^{2+} with ·OH in Water

Figure S7 shows the optimized geometries of complex FLO–Ca^{2+}. Results show that the complex FLO–Ca^{2+} has four possible geometries. By calculating the minimum value of the potential energy surface, the most stable geometric structure (FLO–Ca^{2+}–OPT3) with the lowest single-point energy is obtained among the four structures (Figure S8). All of the following calculations are based on the FLO–Ca^{2+}–OPT3 structure.

Two possible pathways (Path C3^{Ca} and Path C4^{Ca}) of the indirect photolysis of complex FLO–Ca^{2+} with ·OH are shown in Figure 6, and the reaction type is an addition reaction. There are two possible reaction sites on the benzene ring: C3 and C4. Figure S7 shows the optimized geometries of Rs, TSs, and Ps for the two reaction sites. The distance between ·OH and the two reaction sites in the TSs is 2.028 and 2.033 Å, and the difference is very small. Both the two pathways (Path C3^{Ca} and Path C4^{Ca}) shown in Figure S7 are exothermic, with ΔH values ranging from −8.8 to −9.3 kcal/mol.

The lower E_a value (3.8 kcal/mol) indicates that ·OH is more likely to react at the C3 site (Path C3^{Ca}) than that at the C4 site. The E_a value for adding ·OH to the C4 site (Path C4^{Ca}) is 4.3 kcal/mol. For the complex FLO–Ca^{2+}, the calculated E_a values for the two pathways by ·OH (Path C3^{Ca} and Path C4^{Ca}) are lower than that for the indirect photolysis of monomer FLO. This indicates that Ca^{2+} can promote the ·OH addition reaction at C3 and C4 sites. In addition, as shown in Figure 3, ·OH is more prone to attack the C4 site than the C3 site for the indirect photolysis of monomer FLO. However, in the presence of Ca^{2+}, the E_a value for adding ·OH to the C3 site (Path C3^{Ca}, E_a = 3.8 kcal/mol) is lower than that for the C4 site (Path C4^{Ca}, E_a = 4.3 kcal/mol). This indicates that Ca^{2+} can change the main indirect photolysis pathway of FLO.

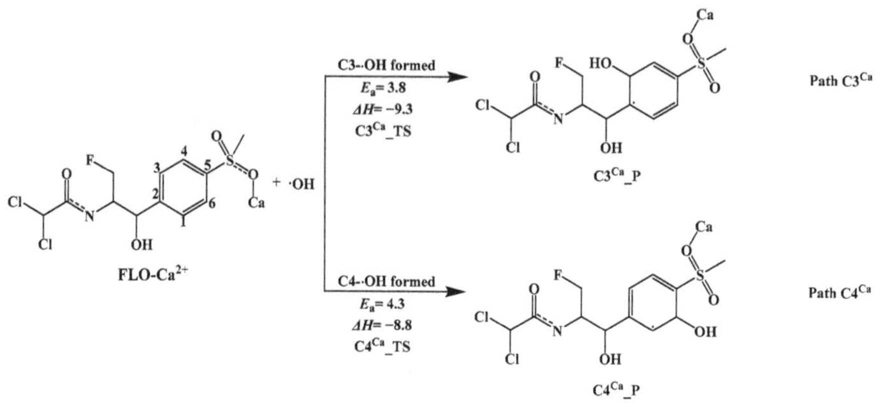

Figure 6. Indirect photodegradation pathways of complex FLO–Ca^{2+} with ·OH, along with computed activation energies (E_a, kcal/mol) and enthalpy change (ΔH, kcal/mol).

Indirect Photodegradation Mechanism of Complex FLO–Mg^{2+} and ·OH in Water

The optimized geometries of complex FLO–Mg^{2+} are shown in Figure S9. The results show that complex FLO–Mg^{2+} has five possible geometries. The most stable geometric structure (FLO–Mg^{2+}–OPT4 structure) with the lowest single-point energy is obtained among the five structures (Figure S9). All calculations are based on the FLO–Mg^{2+}–OPT4 structure.

Four possible pathways (Path $C3^{Mg}$, Path $C4^{Mg}$, Path $C5^{Mg}$, and Path $C6^{Mg}$) of the indirect photodegradation of complex FLO–Mg^{2+} with ·OH are shown in Figure 7, and the reaction type of these pathways is an addition reaction. Four reaction sites (C3, C4, C5, and C6) on the benzene ring were predicted to undergo reactions. Figure S10 shows the optimized geometries of Rs, TSs, and Ps for these four reaction sites. The distance between ·OH and the four reaction sites in the TSs is 1.998–2.049 Å. Figure 7 shows that the ΔH values for all the pathways (Path $C3^{Mg}$, Path $C4^{Mg}$, Path $C5^{Mg}$, and Path $C6^{Mg}$) range from −6.4 kcal/mol to −36.4 kcal/mol, and the four pathways are exothermic. The ΔH value (−36.4 kcal/mol) for ·OH attacking at the C5 site is the lowest among the four pathways, indicating that the products $C5^{Mg}$_Pa and $C5^{Mg}$_Pb are the most stable.

The E_a value (2.9 kcal/mol) for Path $C4^{Mg}$ of all the four pathways is the lowest and indicates that ·OH is most likely to react at the C4 site. A similar reaction was also obtained at the C6 site. The E_a value for adding ·OH at the C6 site is 4.3 kcal/mol. In addition, the E_a value (6.8 kcal/mol) for ·OH attacking at the C3 site is the highest of all the four pathways, so Path $C3^{Mg}$ is the most difficult for undergoing photolysis reactions with ·OH. For C3 and C4 sites, the calculated E_a values of the indirect photolysis of the complex FLO–Mg^{2+} with ·OH (Path $C3^{Mg}$ and Path $C4^{Mg}$) are lower than those of monomer FLO with ·OH(Path C3 and Path C4). In addition, E_a values for the Path $C5^{Mg}$ and Path $C6^{Mg}$ of complex FLO–Mg^{2+} are higher than those of the reaction of monomer FLO with ·OH at C5 and C6 sites(Path C5 and Path C6). This indicates that Mg^{2+} has a dual role in the photolysis of FLO. Mg^{2+} can promote ·OH attacking at the C3 and C4 sites (Path $C3^{Mg}$ and Path $C4^{Mg}$) and inhibit ·OH attacking at the C5 and C6 sites (Path $C5^{Mg}$ and Path $C6^{Mg}$). In addition, Mg^{2+} can also change the main photolysis pathway of FLO with ·OH. In the presence of Mg^{2+}, the order of difficulty in the indirect photolysis with ·OH is C3 > C5 > C6 > C4, and for monomer FLO, the order is C3 > C4 > C5 > C6.

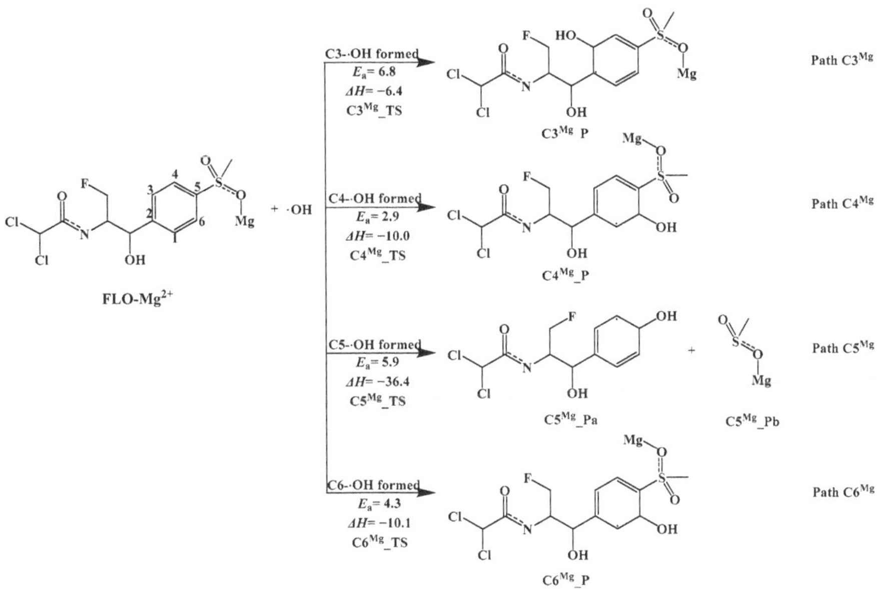

Figure 7. Indirect photodegradation pathways of complex FLO–Mg^{2+} with ·OH, along with computed activation energies (E_a, kcal/mol) and enthalpy change (ΔH, kcal/mol).

Mechanism of Indirect Photodegradation of Complex FLO–Zn^{2+} with ·OH in Water

The optimized geometries of complex FLO–Zn^{2+} are shown in Figure S11. The results show that the complex FLO–Zn^{2+} has four possible geometries, and the most stable geometric structure with the lowest single-point energy is FLO–Zn^{2+}–OPT1 (Figure S11). Thus, all calculations are based on the FLO–Zn^{2+}–OPT1 structure.

Four possible pathways (Path C1^{Zn}, Path C3^{Zn}, Path C4^{Zn}, and Path C6^{Zn}) of the indirect photodegradation of the complex FLO–Zn^{2+} with ·OH are shown in Figure 8, and the reaction type is an addition reaction. Figure S12 shows the optimized geometries of Rs, TSs, and Ps for the four reaction sites (C1, C3, C4, and C6). The distance between ·OH and the four reaction sites is 2.013–2.036 Å. Figure 8 shows that ΔH values for all the four pathways (Path C1^{Zn}, Path C3^{Zn}, Path C4^{Zn}, and Path C6^{Zn}) range from −9.4 kcal/mol to −11.6 kcal/mol, and these pathways are exothermic. The ΔH value (−11.6 kcal/mol) for ·OH attacking at the C6 site is the lowest among all the four pathways, indicating that product C6^{Zn}_P is the most stable.

The E_a values for ·OH attacking at the C1, C3, and C6 sites are all 3.1 kcal/mol, which are lower than that for the C4 cite (4.5 kcal/mol). This indicates that ·OH is more likely to react at the C1, C3, and C6 sites (Path C1^{Zn}, Path C3^{Zn}, and Path C6^{Zn}). In addition, the calculated E_a values for Path C3^{Zn}, Path C4^{Zn}, and Path C6^{Zn} of the complex FLO–Zn^{2+} are lower than those for the indirect photolysis of monomer FLO with ·OH, and the E_a value for Path C1^{Zn} of complex FLO–Zn^{2+} is higher than that for the indirect photolysis of monomer FLO with ·OH. This indicates that Zn^{2+} has a dual role in the indirect photolysis of FLO with ·OH, i.e., Zn^{2+} can promote ·OH attacking at the C3, C4, and C6 sites and inhibit ·OH attacking at the C1 site. In addition, Zn^{2+} can also change the main photodegradation pathway of FLO with ·OH, i.e., in the presence of Zn^{2+}, the order of difficulty of ·OH attacking at the different reaction sites in the indirect photodegradation of FLO with ·OH is C4 > C6 = C1 = C3, which is different from that of monomer FLO (C3 > C4 > C6 > C1).

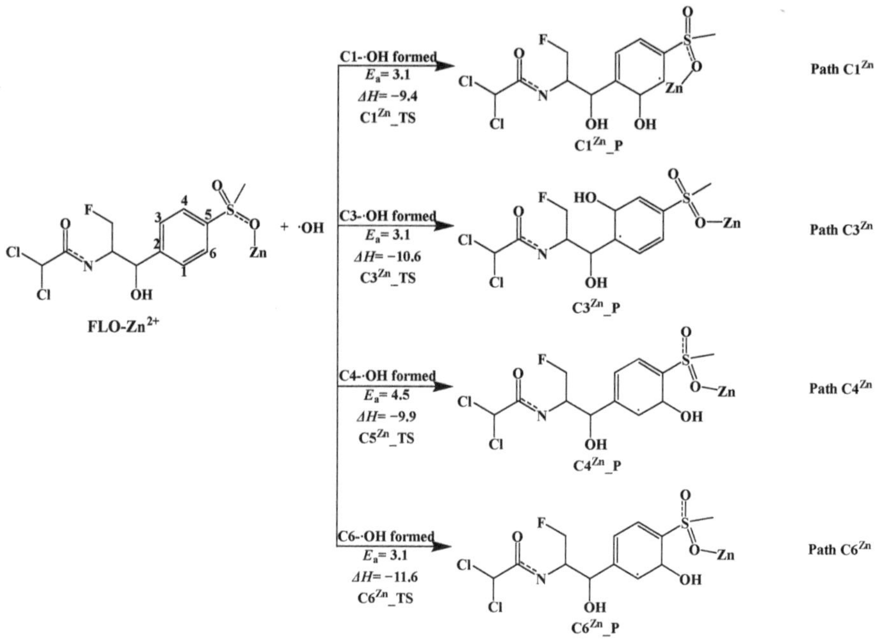

Figure 8. Indirect photodegradation pathways of complex FLO–Zn^{2+} with ·OH, along with computed activation energies (E_a, kcal/mol) and enthalpy change (ΔH, kcal/mol).

4. Conclusions

This study provides a basis for further understanding the photodegradation mechanism of FLO in water. These mechanisms involve direct photodegradation, indirect photodegradation with free radicals (·OH, ·NO_3, and ·SO_4^-), and the effect of metal ions (Ca^{2+}/Mg^{2+}/Zn^{2+}) on indirect photodegradation with ·OH. The calculation results show that the direct photodegradation of FLO involves three reaction types (C–C/C–N/C–S bond cleavage). The relatively high E_a value (94.8 kcal/mol) indicates that C17–C18 cleavage cannot easily occur during the direct photodegradation of FLO. The lower E_a value (2.0 kcal/mol) indicates that C5–S7 cleavage is the main pathway of the direct photodegradation process. The indirect photodegradation of FLO by free radicals is an addition reaction. The ΔH (−41.3 kcal/mol) for adding ·OH at the C5 site is higher than that for other sites on the benzene ring, indicating that the products such as 2,2-dichloro-N-(3-fluoro-1-hydroxy-1-(4-hydroxyphenyl) propan-2-yl) acetamide are the most stable. The indirect photodegradation of FLO by ·NO_3 and ·SO_4^- is endothermic. The NO_3-addition reaction is the most likely to occur at the C1 site, while the ·SO_4-addition reaction is the least likely to occur at the C1 site. The metal ion Ca^{2+} can promote the indirect photodegradation with ·OH, and the metal ion Mg^{2+}/ Zn^{2+} has a dual effect on the indirect photodegradation with ·OH. Mg^{2+} can not only promote the ·OH addition reaction at C3 and C4 sites but also inhibit the ·OH addition reaction at C5 and C6 sites. Zn^{2+} can not only promote the ·OH addition reaction at C3, C4, and C6 sites but also inhibit the ·OH addition reaction at C1 site. In addition, in the presence of metal ion Mg^{2+}, OH-addition is the most likely to occur at the C4 site, but it is the least likely to occur at the C3 site. These results indicate that the quantum chemical calculation method can be used to study the mechanism of the water-based photochemical transformation of antibacterial drugs and to guide their safe use and risk avoidance.

Supplementary Materials: The following supporting information can be downloaded at: https://www.mdpi.com/article/10.3390/toxics12020127/s1, Figure S1: Direct photodegradation pathways of FLO, along with computed activation energies (Ea, kcal/mol) and enthalpy change (ΔH, kcal/mol). The numbers in parentheses are NBO charge and electron spin density (NBO charge, electron spin density); Figure S2: Optimized geometries of reactants, transition states, and products in direct photolysis pathways (Path 1, Path 2, Path 3, and Path 4) of FLO, along with selected bond lengths (Å); Figure S3: Optimized geometries of reactants, transition states, and products in indirect photolysis pathways (Path C1, Path C2, Path C3, and Path C4) of FLO with ·OH, along with selected bond lengths (Å); Figure S4: Optimized geometries of reactants, transition states, and products in indirect photolysis pathways (Path C1N, Path C4N, and Path C6N) of FLO with ·NO3, along with selected bond lengths (Å); Figure S5: Optimized geometries of reactants, transition states, and products in indirect photolysis pathways (Path C1S, Path C3S, Path C4S, and Path C6S) of FLO with ·SO4-, along with selected bond lengths (Å); Figure S5: Optimized geometries of complexes FLO–Ca2+/FLO–Mg2+/FLO–Zn2+, along with selected bond lengths (Å); Figure S7: Four optimized geometries (OPT1, OPT2, OPT3, and OPT4) of complex FLO–Ca2+ along with selected bond length (Å). The energies of geometries are relative to that of the most stable geometry FLO–Ca2+–OPT3; Figure S8: Optimized geometries of reactants, transition states, and products in indirect photolysis pathways (Path C3Ca and Path C4Ca) of complex FLO–Ca2+ with ·OH, along with selected bond lengths (Å); Figure S9: Five optimized geometries (OPT1, OPT2, OPT3, OPT4, and OPT5) of complex FLO–Mg2+ along with selected bond length (Å). The energies of geometries are relative to that of the most stable geometry FLO–Mg2+–OPT4; Figure S10: Optimized geometries of reactants, transition states, and products in indirect photolysis pathways (Path C3Mg, Path C4Mg, Path C5Mg, and Path C6Mg) of complex FLO–Mg2+ with ·OH, along with selected bond lengths (Å); Figure S11: Four optimized geometries (OPT1, OPT2, OPT3, and OPT4) of complex FLO–Zn2+ along with selected bond length (Å). The energies of geometries are relative to that of the most stable geometry FLO–Zn2+–OPT1; Figure S12: Optimized geometries of reactants, transition states, and products in indirect photolysis pathways (Path C1Zn, Path C3Zn, Path C4Zn, and Path C6Zn) of complex FLO–Zn2+ with ·OH, along with selected bond lengths (Å).

Author Contributions: Y.K. performed the theoretical calculations and wrote the manuscript; Y.L. performed the data analyses and manuscript preparation; S.W. contributed to the conception of the study and wrote the manuscript. All authors have read and agreed to the published version of the manuscript.

Funding: This research was funded by the National Natural Science Foundation of China (41601519) and the Jiangsu Students' Platform for innovation and entrepreneurship training program (202210300099Y). Thank the NUIST Students' Platform for Innovation and Entrepreneurship Training Program for financial support (XJDC202210300157).

Institutional Review Board Statement: Not applicable.

Informed Consent Statement: Not applicable.

Data Availability Statement: Data will be made available on request. The dataset used and/or analyzed in this study is available from the corresponding author on reasonable request.

Conflicts of Interest: The authors declare that they have no known competing financial interests or personal relationships that could have appeared to influence the work reported in this paper.

References

1. Schunicht, O.C.; Booker, C.W.; Guichon, P.T.; Jim, G.K.; Wildman, B.K.; Pittman, T.J.; Perrett, T. An evaluation of the relative efficacy of tulathromycin for the treatment of undifferentiated fever in feedlot calves in Nebraska. *Can. Vet. J.-Rev. Vet. Can.* **2007**, *48*, 600–606.
2. Leng, N.N.; Ju, M.J.; Jiang, Y.T.; Guan, D.; Liu, J.H.; Chen, W.; Algharib, S.A.; Dawood, A.; Luo, W.H. The therapeutic effect of florfenicol-loaded carboxymethyl chitosan-gelatin shell nanogels against Escherichia coli infection in mice Running title: Therapeutic effect of florfenicol shell nanogels. *J. Mol. Struct.* **2022**, *1269*, 7. [CrossRef]
3. Li, Z.; Yang, Y.J.; Qin, Z.; Li, S.H.; Bai, L.X.; Li, J.Y.; Liu, X.W. Florfenicol-Polyarginine Conjugates Exhibit Promising Antibacterial Activity Against Resistant Strains. *Front. Chem.* **2022**, *10*, 13. [CrossRef] [PubMed]
4. Liu, J.H.; Ju, M.J.; Wu, Y.F.; Leng, N.N.; Algharib, S.A.; Luo, W.H. Antibacterial activity of florfenicol composite nanogels against Staphylococcus aureus small colony variants. *J. Vet. Sci.* **2022**, *23*, 13. [CrossRef]

5. Park, B.-K.; Lim, J.-H.; Kim, M.-S.; Hwang, Y.-H.; Yun, H.-I. Pharmacokinetics of florfenicol and its metabolite, florfenicol amine, in dogs. *Res. Vet. Sci.* **2008**, *84*, 85–89. [CrossRef]
6. Barreto, F.M.; da Silva, M.R.; Braga, P.A.C.; Bragotto, A.P.A.; Hisano, H.; Reyes, F.G.R. Evaluation of the leaching of florfenicol from coated medicated fish feed into water. *Environ. Pollut.* **2018**, *242*, 1245–1252. [CrossRef] [PubMed]
7. Pouliquen, H.; Delpépée, R.; Larhantec-Verdier, M.; Morvan, M.L.; Le Bris, H. Comparative hydrolysis and photolysis of four antibacterial agents (oxytetracycline oxolinic acid, flumequine and florfenicol) in deionised water, freshwater and seawater under abiotic conditions. *Aquaculture* **2007**, *262*, 23–28. [CrossRef]
8. Bu, Q.W.; Wang, B.; Huang, J.; Deng, S.B.; Yu, G. Pharmaceuticals and personal care products in the aquatic environment in China: A review. *J. Hazard. Mater.* **2013**, *262*, 189–211. [CrossRef]
9. Wei, R.C.; Ge, F.; Chen, M.; Wang, R. Occurrence of Ciprofloxacin, Enrofloxacin, and Florfenicol in Animal Wastewater and Water Resources. *J. Environ. Qual.* **2012**, *41*, 1481–1486. [CrossRef]
10. Martinez, J.L. Environmental pollution by antibiotics and by antibiotic resistance determinants. *Environ. Pollut.* **2009**, *157*, 2893–2902. [CrossRef]
11. Martins, Y.A.; De Oliveira, C. Development and Validation of an UV Spectrophotometric Method for Determination of Thiamphenicol in Dosage Form. *J. Appl. Spectrosc.* **2019**, *86*, 629–635. [CrossRef]
12. Sun, Z.J.; Zhang, L.W.; Dong, D.M.; Guo, Z.Y. Optimizing the multimedia fate model for characterizing environmental risks of florfenicol in seasonally ice-covered reservoirs. *Environ. Pollut.* **2023**, *323*, 10. [CrossRef]
13. Yévenes, K.; Pokrant, E.; Pérez, F.; Riquelme, R.; Avello, C.; Maddaleno, A.; San Martín, B.; Cornejo, J. Assessment of Three Antimicrobial Residue Concentrations in Broiler Chicken Droppings as a Potential Risk Factor for Public Health and Environment. *Int. J. Environ. Res. Public Health.* **2018**, *16*, 24. [CrossRef] [PubMed]
14. Ding, Y.C.; Liang, B.; Jiang, W.L.; Han, J.L.; Guadie, A.; Yun, H.; Cheng, H.Y.; Yang, R.J.; Liu, S.J.; Wang, A.J.; et al. Effect of preferential UV photolysis on the source control of antibiotic resistome during subsequent biological treatment systems. *J. Hazard. Mater.* **2021**, *414*, 10. [CrossRef] [PubMed]
15. Song, C.F.; Wei, Y.L.; Qiu, Y.T.; Qi, Y.; Li, Y.; Kitamura, Y. Biodegradability and mechanism of florfenicol via Chlorella sp UTEX1602 and L38: Experimental study. *Bioresour. Technol.* **2019**, *272*, 529–534. [CrossRef] [PubMed]
16. Zhang, H.M.; Wei, X.X.; Song, X.D.; Shah, S.; Chen, J.W.; Liu, J.H.; Hao, C.; Chen, Z.F. Photophysical and photochemical insights into the photodegradation of sulfapyridine in water: A joint experimental and theoretical study. *Chemosphere* **2018**, *191*, 1021–1027. [CrossRef] [PubMed]
17. Zhang, Y.; Li, J.H.; Zhou, L.; Wang, G.Q.; Feng, Y.H.; Wang, Z.Y.; Yang, X. Aqueous photodegradation of antibiotic florfenicol: Kinetics and degradation pathway studies. *Environ. Sci. Pollut. Res.* **2016**, *23*, 6982–6989. [CrossRef] [PubMed]
18. Liu, N.; Sijak, S.; Zheng, M.; Tang, L.; Xu, G.; Wu, M.H. Aquatic photolysis of florfenicol and thiamphenicol under direct UV irradiation, UV/H_2O_2 and UV/Fe(II) processes. *Chem. Eng. J.* **2015**, *260*, 826–834. [CrossRef]
19. Gao, Y.Q.; Gao, N.Y.; Deng, Y.; Yin, D.Q.; Zhang, Y.S. Degradation of florfenicol in water by UV/$Na_2S_2O_8$ process. *Environ. Sci. Pollut. Res.* **2015**, *22*, 8693–8701. [CrossRef]
20. Xu, M.F.; Qian, M.R.; Zhang, H.; Ma, J.W.; Wang, J.M.; Wu, H.Z. Simultaneous determination of florfenicol with its metabolite based on modified quick, easy, cheap, effective, rugged, and safe sample pretreatment and evaluation of their degradation behavior in agricultural soils. *J. Sep. Sci.* **2015**, *38*, 211–217. [CrossRef]
21. Silva, C.; Louros, V.; Silva, V.; Otero, M.; Lima, D. Antibiotics in Aquaculture Wastewater: Is It Feasible to Use a Photodegradation-Based Treatment for Their Removal? *Toxics* **2021**, *9*, 194. [CrossRef] [PubMed]
22. Tian, J.; Li, B.B.; Qu, R.J.; Zhou, D.M.; Sun, C.; Wang, Z.Y.; Zhu, F. Influence of anions on ozonation of bisphenol AF: Kinetics, reaction pathways, and toxicity assessment. *Chemosphere* **2022**, *286*, 131864. [CrossRef] [PubMed]
23. Malvaldi, M.; Bruzzone, S.; Chiappe, C.; Gusarov, S.; Kovalenko, A. Ab Initio Study of Ionic Liquids by KS-DFT/3D-RISM-KH Theory. *J. Phys. Chem. B* **2009**, *113*, 3536–3542. [CrossRef] [PubMed]
24. Ge, L.; Chen, J.W.; Lin, J.; Cai, X.Y. Light-Source-Dependent Effects of Main Water Constituents on Photodegradation of Phenicol Antibiotics: Mechanism and Kinetics. *Environ. Sci. Technol.* **2009**, *43*, 3101–3107. [CrossRef]
25. Jiang, W.L.; Ding, Y.C.; Haider, M.R.; Han, J.L.; Liang, B.; Xia, X.; Yang, L.M.; Wang, H.C.; Peng, Y.Z.; Wang, A.J. A novel TiO_2/graphite felt photoanode assisted electro-Fenton catalytic membrane process for sequential degradation of antibiotic florfenicol and elimination of its antibacterial activity. *Chem. Eng. J.* **2020**, *391*, 12. [CrossRef]
26. Liu, C.G.; Guan, X.H.; Su, Z.M. TDDFT study on the second-order nonlinear optical properties of a series of mono- and di-nuclear 60 fullerene complexes. *Comput. Theor. Chem.* **2011**, *963*, 98–103. [CrossRef]
27. Minaev, B.F.; Baryshnikov, G.V.; Korop, A.A.; Minaeva, V.A.; Kaplunov, M.G. Theoretical investigation of the structure and electronic absorption spectrum of a complex zinc bis- 8-(3,5-difluorophenylsulfanylamino)quinolinate. *Opt. Spectrosc.* **2012**, *113*, 298–304. [CrossRef]
28. Berland, K.; Londero, E.; Schröder, E.; Hyldgaard, P. Harris-type van der Waals density functional scheme. *Phys. Rev. B* **2013**, *88*, 10. [CrossRef]
29. Wolny, J.A.; Paulsen, H.; Winkler, H.; Trautwein, A.X.; Tuchagues, J.P. DFT calculations as a tool to analyse quadrupole splittings of spin crossover Fe(II) complexes. *Hyperfine Interact.* **2005**, *166*, 495–498. [CrossRef]

30. Shah, T.A.; Alam, U.; Alam, M.; Park, S.; Muneer, M. Single crystal X-ray structure, spectroscopic and DFT studies of Imidazo [2,1-b]thiazole: 2-(3-hydroxy-3-phenylimidazo[2,1-b]thiazol-2(3H)-ylidene)-1-phenylethanone. *J. Mol. Struct.* **2018**, *1157*, 638–653. [CrossRef]
31. Benzon, K.B.; Mary, Y.S.; Varghese, H.T.; Panicker, C.Y.; Armakovic, S.; Armakovic, S.J.; Pradhan, K.; Nanda, A.K.; Van Alsenoy, C. Spectroscopic, DFT, molecular dynamics and molecular docking study of 1-butyl-2-(4-hydroxypheny1)-4,5-dimethyl-imidazole 3-oxide. *J. Mol. Struct.* **2017**, *1134*, 330–344. [CrossRef]
32. Wilson, P.B.; Williams, I.H. Critical evaluation of anharmonic corrections to the equilibrium isotope effect for methyl cation transfer from vacuum to dielectric continuum. *Mol. Phys.* **2015**, *113*, 1704–1711. [CrossRef]
33. Arrigoni, F.; Zampella, G.; Gioia, L.D.; Greco, C.; Bertini, L.J.I. The Photochemistry of $Fe_2(S_2C_3H_6)(CO)_6$(-CO) and Its Oxidized Form, Two Simple [FeFe]-Hydrogenase CO-Inhibited Models. A DFT and TDDFT Investigation. *Inorganics.* **2021**, *9*, 16. [CrossRef]
34. Ni, Z.; Chen, Z.; Zhang, X.; Yang, X.; Zhou, L. Photolysis of fungicide triadimefon: A combined experimental and theoretical investigation on homolytic C-O and C-N bonds dissociation mechanisms. *J. Photochem. Photobiol. A Chem.* **2023**, *436*, 114402. [CrossRef]
35. Polska, K.; Zielonka, J.; Chomicz, L.; Czerwicka, M.; Stepnowski, P.; Guzow, K.; Wiczk, W.; Smużyńska, M.; Kasprzykowski, F.; Zylicz-Stachula, A.; et al. Unexpected photoproduct generated via the acetone-sensitized photolysis of 5-bromo-2′-deoxyuridine in a water/isopropanol solution: Experimental and computational studies. *J. Phys. Chem. B* **2010**, *114*, 16902–16907. [CrossRef]
36. Sun, Y.; Li, M.; Hadizadeh, M.H.; Liu, L.; Xu, F. Theoretical insights into the degradation mechanisms, kinetics and eco-toxicity of oxcarbazepine initiated by OH radicals in aqueous environments. *J. Environ. Sci.* **2023**, *129*, 189–201. [CrossRef]
37. Noh, S.H.; Kwak, D.H.; Seo, M.H.; Ohsaka, T.; Han, B. First principles study of oxygen reduction reaction mechanisms on N-doped graphene with a transition metal support. *Electrochim. Acta* **2014**, *140*, 225–231. [CrossRef]
38. Becke, A.D. Density-functional thermochemistry. III. The role of exact exchange. *J. Chem. Phys.* **1993**, *98*, 5648–5652. [CrossRef]
39. Tomasi, J.; Mennucci, B.; Cammi, R.J.C.R. Quantum mechanical continuum salvation models. *Chem. Rev.* **2005**, *105*, 2999–3094. [CrossRef]
40. Yamashita, K.; Yamabe, T.; Fukui, K. IRC approach to chemical dynamics: Toward mode-selective chemical reactions. *Chem. Phys. Lett.* **1981**, *84*, 123–126. [CrossRef]
41. Zhao, G.J.; Han, K.L. Excited State Electronic Structures and Photochemistry of Heterocyclic Annulated Perylene (HAP) Materials Tuned by Heteroatoms: S, Se, N, O, C, Si, and B. *J. Phys. Chem. A* **2009**, *113*, 4788–4794. [CrossRef]
42. Marques, M.A.L.; Gross, E.K.U. Time-Dependent Density Functional Theory. *Annu. Rev. Phys. Chem.* **2008**, *55*, 427–455. [CrossRef] [PubMed]
43. Hölzle, E.; Hönigsmann, H. Ultraviolette Strahlung—Quellen, Spektren, Umwelteinflüsse: UV-radiation—Sources, Wavelength, Environment. *JDDG J. Der Dtsch. Dermatol. Ges.* **2010**, *3*, S3–S10. [CrossRef] [PubMed]
44. Li, K.; Zhang, P.; Ge, L.K.; Ren, H.L.; Yu, C.Y.; Chen, X.Y.; Zhao, Y.F. Concentration-dependent photodegradation kinetics and hydroxyl-radical oxidation of phenicol antibiotics. *Chemosphere* **2014**, *111*, 278–282. [CrossRef] [PubMed]

toxics

Article

Z-Type Heterojunction MnO_2@g-C_3N_4 Photocatalyst-Activated Peroxymonosulfate for the Removal of Tetracycline Hydrochloride in Water

Guanglu Lu [1], Xinjuan Li [1], Peng Lu [1], He Guo [2], Zimo Wang [3], Qian Zhang [1], Yuchao Li [4], Wenbo Sun [1], Jiutao An [1,*] and Zijian Zhang [1,*]

1 College of Resources and Environment Engineering, Shandong University of Technology, Zibo 255000, China; lgl8899168@163.com (G.L.); 15908037096@163.com (X.L.); 15834482173@163.com (P.L.); zq@sdut.edu.cn (Q.Z.); wenbosun@sdut.edu.cn (W.S.)
2 Department of Environmental Engineering, College of Biology and the Environment, Nanjing Forestry University, Nanjing 210037, China; heguo@njfu.edu.cn
3 Department of Marine Engineering, Jimei University, Xiamen 361021, China; wangzimo@163.com
4 Research Institute of Clean Chemical Technology, School of Chemistry and Chemical Engineering, Shandong University of Technology, Zibo 255049, China; cyulee@126.com
* Correspondence: anjiutao@163.com (J.A.); zhangzj@sdut.edu.cn (Z.Z.)

Abstract: A Z-type heterojunction MnO_2@g-C_3N_4 photocatalyst with excellent performance was synthesized by an easy high-temperature thermal polymerization approach and combined with peroxymonosulfate (PMS) oxidation technology for highly efficient degrading of tetracycline hydrochloride (TC). Analysis of the morphological structural and photoelectric properties of the catalysts was achieved through different characterization approaches, showing that the addition of MnO_2 heightened visible light absorption by g-C_3N_4. The Mn_1-CN_1/PMS system showed the best degradation of TC wastewater, with a TC degradation efficiency of 96.97% following 180 min of treatment. This was an approximate 38.65% increase over the g-C_3N_4/PMS system. Additionally, the Mn_1-CN_1 catalyst exhibited excellent stability and reusability. The active species trapping experiment indicated •OH and $SO_4^{\bullet-}$ remained the primary active species to degrade TC in the combined system. TC degradation pathways and intermediate products were determined. The Three-Dimensional Excitation-Emission Matrix (3DEEM) was employed for analyzing changes in the molecular structure in TC photocatalytic degradation. The biological toxicity of TC and its degradation intermediates were investigated via the Toxicity Estimation Software Test (T.E.S.T.). The research offers fresh thinking for water environment pollution treatment.

Keywords: MnO_2@g-C_3N_4; peroxymonosulfate; photocatalytic; degradation mechanism; TC

Citation: Lu, G.; Li, X.; Lu, P.; Guo, H.; Wang, Z.; Zhang, Q.; Li, Y.; Sun, W.; An, J.; Zhang, Z. Z-Type Heterojunction MnO_2@g-C_3N_4 Photocatalyst-Activated Peroxymonosulfate for the Removal of Tetracycline Hydrochloride in Water. *Toxics* **2024**, *12*, 70. https://doi.org/10.3390/toxics12010070

Academic Editor: Roberto Rosal

Received: 10 November 2023
Revised: 10 January 2024
Accepted: 10 January 2024
Published: 14 January 2024

1. Introduction

Tetracycline antibiotics are among the most frequently adopted antibiotics in livestock, aquaculture, and healthcare [1]. Tetracycline hydrochloride (TC), a tetracycline antibiotic, cannot be completely broken down in the environment because of its stable chemical properties [2]. The misuse and release of antibiotics in the ecosystem can promote the development of various drug-resistant bacteria and the enhancement of resistance genes [3], causing severe risk to global health and the ecosystem [4,5]. As a result, it is important to find a green, effective, and economical way to degrade TC effluent.

Currently, the commonly used remediation methods for environmental pollutants include adsorption [6], flocculation precipitation [7], ion exchange [8], and biochemical methods [9,10]. However, the above methods have the disadvantages of complicated procedures, high costs, and low removal efficiency. Advanced oxidation processes (AOPs) are considered a hopeful degradation technology because of their ability to generate highly reactive radicals, including hydroxyl radicals (•OH) and sulfate radicals ($SO_4^{\bullet-}$) [11,12].

Interestingly, peroxymonosulfate (PMS)-based AOPs (PMS-AOPs) can produce $SO_4^{\bullet-}$ with higher redox potential, better selectivity, and a more comprehensive pH selection range [13]. The PMS could be excited by photocatalysis, carbon materials, and transition metal oxides to generate $SO_4^{\bullet-}$ [13–17]. Photocatalysis activation of PMS is one of the most widely used techniques because photocatalytic methods are inexpensive, simple, and efficient [18]. Many scholars have reported the effective degradation of pollutants by combining photocatalysts with PMS activation [15]. Liu et al. [19] found that the combination of graphitic carbon nitride (g-C_3N_4) nanosheets and PMS enhanced the removal of Bisphenol A under visible light, with complete removal in 90 min. Guan et al. [20] investigated whether the introduction of persulfate in the Co_3O_4/CeO_2 system improved the removal efficacy of TC. Therefore, the combination of PMS and photocatalytic systems can effectively degrade antibiotic wastewater.

Recently, with good stability, reaction to light in the visible range, and unique properties of affordability, g-C_3N_4 has gained much attention as a non-metallic semiconductor material [21]. Meanwhile, research has now demonstrated that g-C_3N_4 can activate PMS to degrade organic pollutants [19]. Unfortunately, g-C_3N_4 still has significant limitations, including weak conductivity and quick recombination of photogenerated electron-holes (e^--h^+), which restricts its practical applicability in environmental purification [22]. Hence, numerous techniques have been developed for improving the photocatalytic activity of g-C_3N_4, involving metal or non-metal doping [23,24], construction nanostructures [25], and combinations of different semiconductors [26]. Among all modification methods, the combination of other semiconductors and g-C_3N_4 remains an excellent method for constructing Z-type heterojunction materials [27,28].

Transition metal oxide semiconductors can serve as catalysts and PMS activators to produce strongly oxidizing $SO_4^{\bullet-}$. As a typical transition metal oxide, manganese dioxide (MnO_2) has the benefits of low cost, environmental friendliness, and good conductivity [29]. MnO_2 is an effective PMS activator due to the abundant oxygen mobility in the MnO_2 lattice and the electron transfer between Mn species [30,31]. Therefore, the addition of MnO_2 to g-C_3N_4 for forming a Z-type heterojunction is a viable approach for improving photocatalytic performance of the catalyst [32,33].

In our research, a MnO_2@g-C_3N_4 heterojunction with a Z-type structure was prepared for efficient TC photocatalytic degradation with PMS and visible light. Various characterization approaches were employed for analyzing the morphology, chemical compositions, and optoelectronic properties of samples. The photocatalytic degradation performance of TC was studied under diverse reaction environment parameters. Meanwhile, the intermediates of TC and possible reaction pathways were determined. The Toxicity Estimation Software Test (T.E.S.T.) was employed for assessing TC toxicity and its intermediates. The average TC removal rate observed under the current procedures at the wastewater treatment plants was 96.97%. This research presents a safe, green, and environmentally friendly treatment method for purifying and treating antibiotic wastewater.

2. Experimental Section

2.1. Materials

The chemical reagents used in our research are all listed in Supplementary file Text 1.

2.2. Photocatalyst Sample Preparation

2.2.1. Preparation of g-C_3N_4

The g-C_3N_4 preparation had two steps. In the first step, bulk g-C_3N_4 was produced through thermal polycondensation [34]. A total of 5 g of melamine was laid in a muffle furnace, with the temperature heated to 550 °C at 5 °C min^{-1} and held for 4 h. After cessation of heating, the yellowish solid was ground for 30 min to form a fine powder. To pyrolytically peel the prepared bulk g-C_3N_4 into thinner g-C_3N_4, in the second step, the yellowish g-C_3N_4 powder was divided into several portions, and the g-C_3N_4 was laid in crucible and placed in a muffle furnace, with the temperature heated to 500 °C at 10 °C min^{-1} within 2 h to obtain g-C_3N_4.

2.2.2. Preparation of MnO_2

MnO_2 was synthesized by redox coprecipitation using $MnCl_2$-$4H_2O$ and $KMnO_4$ (molar ratio 3:2). The brown precipitate was separated by adding $KMnO_4$ solution (2 mol L^{-1} 100 mL) to $MnCl_2$ solution (3 mol L^{-1} 100 mL) dropwise at 5 mL min^{-1} during magnetic stirring and drying for 4 h at 90 °C. The solution was stirred again to redissolve the dried powder in water. The precipitate was washed properly with alcohol by filtration and dried for 12 h at 90 °C to obtain MnO_2.

2.2.3. Preparation of MnO_2@g-C_3N_4

Three mass ratios of MnO_2 and g-C_3N_4 (MnO_2:g-C_3N_4 = 1:2, 1:1, and 2:1) were weighed and were recorded as Mn_1-CN_2, Mn_1-CN_1, and Mn_2-CN_1, respectively. The mixture was ground for 30 min to form a fine powder and then laid in crucible in a muffle furnace, with the temperature heated to 400 °C at 5 °C min^{-1} within 4 h and chilled to normal temperature to obtain Mn-CN photocatalytic composite materials with different ratios.

2.3. Photocatalyst Characterization

The morphology and dimensions of the photocatalysts were identified using scanning electron microscopy (SEM) techniques. X-ray diffraction (XRD) patterns were obtained using a Brucker D8 Advance diffraction apparatus. The surface chemistry of the samples was characterized by X-ray photoelectron spectroscopy (XPS). Solid-state UV-Vis diffuse reflectance spectroscopy (UV-Vis-DRS) was used to study the optical properties and separation of photogenerated electrons and holes. Electrochemical impedance spectroscopy (EIS) was measured using a CHI 660B electrochemical system. The degradation intermediates of TC were analyzed by HPLC-MS. The molecular structural changes during TC degradation were analyzed using Three-Dimensional Excitation-Emission Matrix (3DEEM). The parameters used for all characterizations are given in Supplementary file Text 2.

2.4. Photocatalytic Degradation of TC by Mn-CN

In the experiments, the photocatalytic degradation performance of the catalysts was tested via a dominant wavelength of 400 nm and a 300 W solar simulator xenon lamp (xenon lamp model and manufacturer's declared spectrum as well as photocatalytic device diagram are shown in Supplementary Figure S1). The photocatalyst (0.03 g) was dispersed into 50 mL of 20 mg L^{-1} TC solution for each experiment. The solution was stirred for 30 min without light until adsorption equilibrium was reached. Then, 0.6 mM PMS was added, and the lamp was switched on. Samples were taken every 30 min during the photocatalytic process. Supernatant measurement was performed with UV-Vis spectrophotometer (UV-5100B) at 357 nm after filtration through a 0.22 μm microfilter membrane.

3. Results and Discussion

3.1. Properties of the Material

Figure 1 displays the SEM and TEM pictures of the samples. The g-C_3N_4 was composed of many irregular lamellar and layered structures [35], with a large specific surface area, as shown in Figure 1a [36]. MnO_2 exhibited regular small particle-size nanospheres with a rough surface but a relatively uniform particle size distribution (Figure 1b). Moreover, MnO_2 was dispersed on the g-C_3N_4 layered structure in Mn_1-CN_1 (Figure 1c). Figure 1d–f shows the TEM images of the sample. Many MnO_2 particles were found on the g-C_3N_4 surface, indicating that the composites of MnO_2 and g-C_3N_4 were well prepared [37]. As depicted in Figure 1g–j, C, N, O, and Mn were evenly scattered on the Mn_1-CN_1 surface. The percentages of C, N, O and Mn in Mn_1-CN_1 were 50.36%, 16.42%, 6.05%, and 27.17%, respectively (Figure 1k). Figure 1l shows the HRTEM image of Mn_1-CN_1; two different types of lattice stripes can be found from the (002) and (111) facets of the crystal structure at crystallographic plane distances of approximately 0.34 and 0.24 nm, respectively.

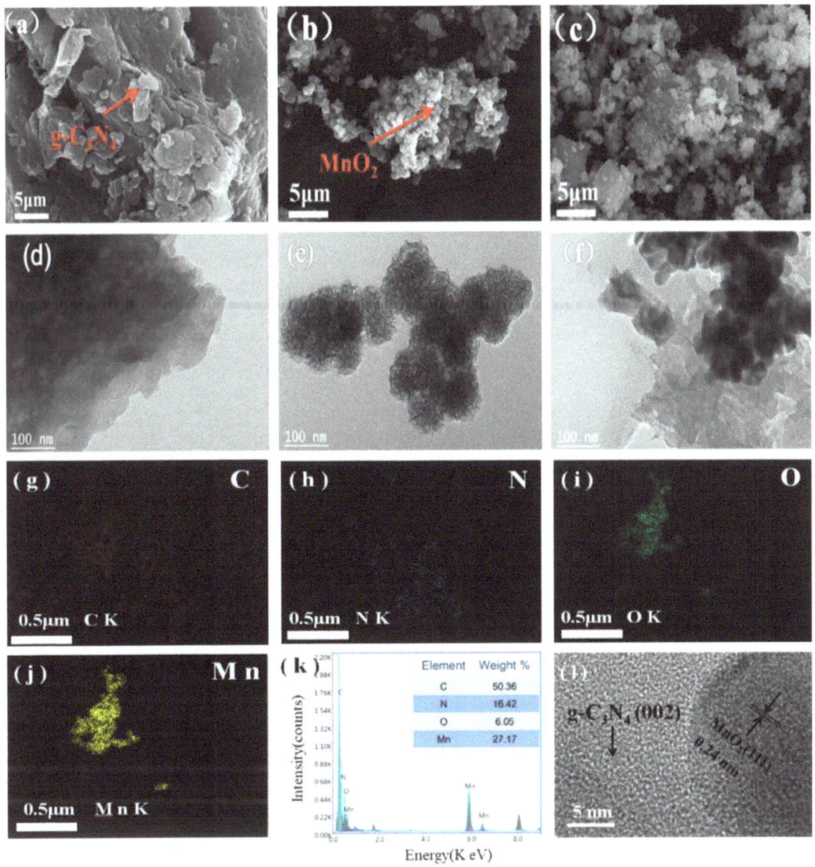

Figure 1. SEM and TEM pictures of (**a**,**d**) g-C_3N_4, (**b**,**e**) MnO_2, (**c**,**f**) Mn_1-CN_1; (**g**–**j**) elemental mapping images of C, N, O, and Mn in Mn_1-CN_1; (**k**) EDS spectrum of Mn_1-CN_1; (**l**) HRTEM image of Mn_1-CN_1.

Figure 2a displays the XRD patterns of g-C_3N_4, MnO_2, and Mn_1-CN_1. The g-C_3N_4 and Mn_1-CN_1 have the same peaks located at 13.1° and 27.4° [38]. The diffraction peak at 13.1° was weak and belonged to the (100) crystal plane of g-C_3N_4, which was made up of repeating structural units of the heptazine ring. The diffraction peak at 27.4° was strong and belonged to the g-C_3N_4 (002) crystal plane. This indicates that the sample was a lamellar stacking of graphite-like material [39]. The location and intensity of MnO_2 diffraction peaks were relatively uniform and smooth, with the diffraction peaks at 2θ = 12.7°, 25.2°, 37.1°, and 66.0° for lattice planes (001), (002), (111), and (311), respectively [40]. The relatively weak intensity and broad peak pattern of these diffraction peaks indicated that the product was a compact growth of amorphous and weakly crystalline MnO_2. Furthermore, all prominent peaks of g-C_3N_4 and MnO_2 could be seen in the Mn_1-CN_1 spectra, which indicates that MnO_2 was compounded into g-C_3N_4 to form Mn_1-CN_1.

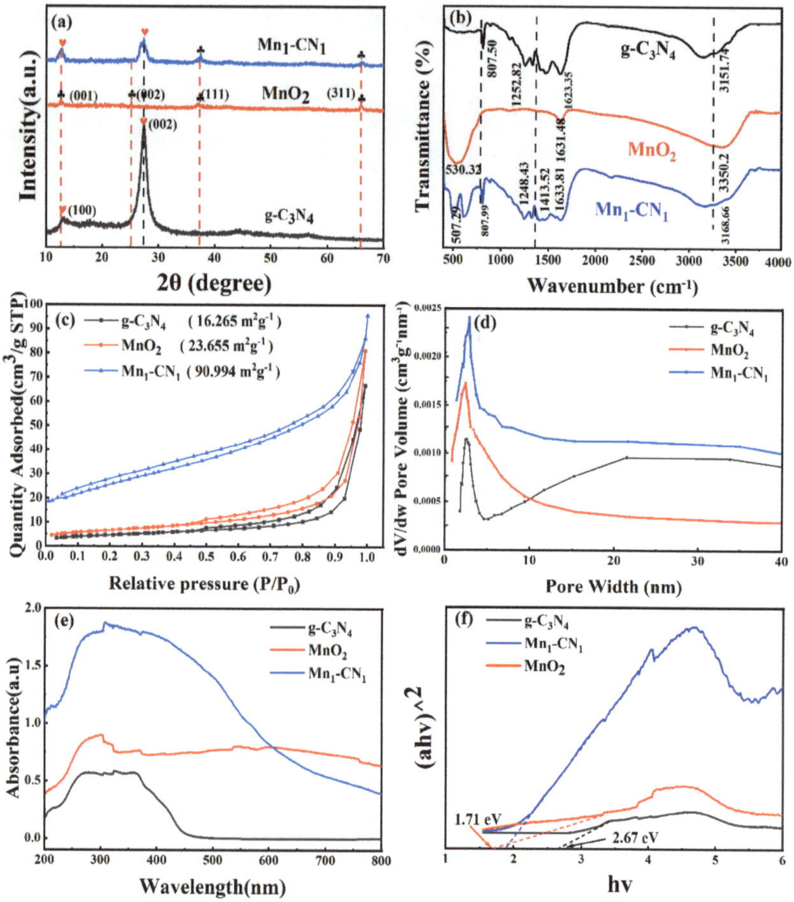

Figure 2. (**a**) XRD patterns (The hearts and clubs in 2a are the locations of the g−C_3N_4 and MnO_2 characteristic peaks, respectively), (**b**) FT-IR spectra, (**c**) N_2 adsorption-desorption isotherm curve, (**d**) pore size distribution of g-C_3N_4, MnO_2, and Mn_1-CN_1, (**e**) UV-vis DRS spectra of the samples, (**f**) diffuse reflectance spectra after Tauc fitting for the different photocatalysts.

Figure 2b displays the FT-IR of the sample. The peak of g-C_3N_4 at around 807.50 cm^{-1} can be related to the triazinic moiety [41]. Vibrational absorption peaks at 1252.82 and 1623.35 cm^{-1} correspond to the vibrational stretch of the carbon and nitrogen bonds (C-N and C=N). The stretching vibrations of -NH_2 and -NH groups and of the -OH bond correspond to the peak at 3151.74 cm^{-1} [42]. The characteristic peak of MnO_2 was mainly at 530.32 cm^{-1}. The positions of the Mn_1-CN_1 absorption peaks were essentially the same as those of MnO_2 and g-C_3N_4 absorption peaks, illustrating that the MnO_2 had been loaded onto g-C_3N_4.

Samples' N_2 adsorption-desorption isotherms and pore-size profiles were measured, with the results shown in Figure 2c,d. The specific surface area of Mn_1-CN_1 was 90.99 m^2g^{-1}, which was 5.6 and 3.8 times higher relative to g-C_3N_4 and MnO_2, respectively. Due to the large number of MnO_2 particles tightly dispersed on the surface of g-C_3N_4, the specific surface area of the composites was enlarged relative to single g-C_3N_4 and MnO_2, which contributed to TC degradation. Under the International Union of Pure and Applied Chemistry (IUPAC) classification, all three samples displayed the Type IV isotherm of the H_3 type, indicating that narrow pores existed, unique to mesoporous materials, in the prepared samples [43]. From Figure 2d, the pore size distribution curves of all photocatalysts, g-C_3N_4, MnO_2, and Mn_1-CN_1, indicate a mesoporous structure. Moreover, the pore volume of Mn_1-CN_1 was also greater relative to g-C_3N_4 and MnO_2. The larger specific surface area and higher porosity of the catalyst could increase the contact area of active molecules and pollutants, thus enhancing photocatalytic performance [44].

The UV-Vis DRS spectrum shows the optical properties of samples (Figure 2e). The G-C_3N_4 sample absorbed light within the range of 450 nm to UV. Compared with g-C_3N_4, MnO_2 exhibited strong absorption in the wavelength range of 300–800 nm. Notably, the introduction of MnO_2 resulted in a significant red-shift of Mn_1-CN_1, indicating that the MnO_2 was able to extend the absorption range of Mn_1-CN_1 for visible light [45]. The bandgap energy of the samples was calculated via the Kubelka–Munk function [46], as shown below:

$$(\alpha h v)^{1/2} = K(hv - E_g) \tag{1}$$

where α, v, h, K, and E_g are absorbance, photon frequency, Planck's constant, proportional constant, and bandgap energy, respectively. This can be derived from the diffuse reflectance spectra after Tauc fitting with E_g values of 2.67 eV and 1.71 eV for g-C_3N_4 and MnO_2 samples, respectively (Figure 2f). Valence bands (VB) of g-C_3N_4 and MnO_2 can be estimated from the XPS valence band spectra (Supplementary Figure S2), with positions of 1.54 and 2.34 eV, respectively. The conduction band (CB) position was calculated via Equation (2), and the CB potentials of g-C_3N_4 and MnO_2 were −1.13 and 0.63 eV, respectively.

$$E_{CB} = E_{VB} - E_g \tag{2}$$

Figure 3a displays the total XPS spectra of the photocatalysts. The XPS spectrum of Mn_1-CN_1 contained peaks of C 1s and N 1s elements in g-C_3N_4 and peaks of Mn 2p and O 1s elements in MnO_2. In addition, no peaks of other elements were found in the Mn_1-CN_1 catalyst, indicating that the composite was not doped with impurities of other elements. Figure 3b shows the three peaks obtained by fitting the C 1s high resolution spectrum with the XPSPEAK software. The three peaks at 284.2, 285.2, and 287.4 eV resulted from the C-C bond, C=N bond, and N=C-N_2 group, respectively [47]. In Figure 3c, the four characteristic peaks of N 1s spectra at 397.8, 398.7, 400.2, and 403.8 eV originated from the C=N-C bond of sp^2-hybridized aromatic nitrogen, $(C)_3$-N_2 bond of the sp^2-hybridized bridged nitrogen atom, C-N-H of the sp^3-hybridized terminal functional group [48,49], and the charge effect of the π-excited states [50], respectively. These results prove the existence of the graphite phase g-C_3N_4. The two characteristic peaks in Mn 2p spectra (Figure 3d) at 642.0 and 654.0 eV corresponded to the binding energy spin-orbit double peaks of Mn $2p^{3/2}$ and Mn $2p^{1/2}$, respectively, which is consistent with the nature of MnO_2. In addition, the change in the valence of Mn was crucial for the photocatalytic reaction. The presence of characteristic peaks at 652.9 and 641.4 eV indicates that both Mn^{3+} and Mn^{4+} were present in MnO_2. As shown in the O 1s spectra (Figure 3e), the three peaks at 529.1, 530, and 531.1 eV were due to the atomic lattice oxygen of MnO_2, hydroxyl groups attached to the Mn surface, and adsorbed oxygen [51].

In addition, the interfacial charge transfer of the samples was studied using the EIS technique. Photoelectric performance tests were performed on a CHI 660B electrochemical system. The system used a standard three-electrode system with a photocatalyst as the working electrode, a Pt plate counter electrode and an Ag/AgCl reference electrode, and Na_2SO_4 as the electrolyte solution. The smaller arc radius of the EIS curve indicates lower resistance of the sample and a better electron-hole pair separation [52]. From Figure 3f, it can be seen that the electrochemical impedance arc radius of Mn_1-CN_1 was significantly smaller relative to g-C_3N_4 and MnO_2, indicating that Mn_1-CN_1 possessed the smallest charge transfer resistance and quicker interfacial charge transfer efficiency. The above results show that the formation of heterojunctions could speed up the transfer of photo-excited electron-hole pairs, which improved photocatalytic degradation [28].

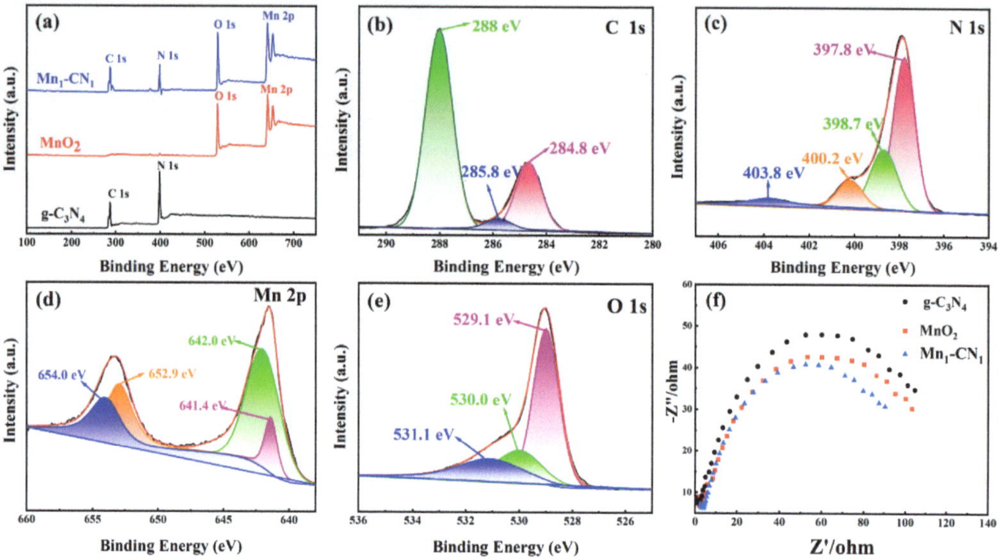

Figure 3. Extended range XPS spectra of g-C_3N_4, MnO_2, and Mn_1-CN_1 (**a**); high-resolution XPS patterns of C 1s (**b**), N 1s (**c**), Mn 2p (**d**), and O 1s (**e**); electrochemical impedance spectra (**f**).

3.2. Photocatalytic Efficiency of the Samples toward TC Degradation

From Figure 4a, it can be seen that the Mn-CN/PMS system exhibited excellent degradation performance, with a degradation efficiency of 85.06%. When MnO_2 was introduced, g-C_3N_4/MnO_2 samples displayed excellent photocatalytic degradation efficiency for TC, which resulted from the formation of the g-C_3N_4/MnO_2 heterojunction. However, degradation efficiency of TC for the sole MnO_2 system was less than 7.0%. Moreover, the TC degradation efficiency for the sole g-C_3N_4 or PMS system was also relatively low, with only 15.5% and 5.4% removal, respectively. When both Mn-CN and PMS were present, the TC degradation efficiency reached 85.06%, suggesting that PMS was activated by the Mn-CN activator.

The photocatalytic degradation performance of different Mn-CN catalysts is shown in Figure 4b. Compared with other samples, the Mn_1-CN_1 catalysts exhibited better degradation performance. TC degradation efficiency achieved 85.06% when PMS existed after 180 min. However, when the MnO_2 ratio increased, the TC degradation efficiency did not significantly change because the excessive MnO_2 increased the absorption and scattering of photons, which affected light utilization by g-C_3N_4 and reduced photocatalytic efficiency. This is consistent with previous reports [23].

3.3. The Effect of Catalyst Dosage

TC degradation efficiency was enhanced when the amount of the Mn_1-CN_1 catalyst increased (Figure 4c). When the Mn_1-CN_1 catalyst was dosed from 0.2 to 0.6 g L^{-1}, TC degradation efficiency improved from 54.3% to 85.06%. With the increase in the amount of the Mn_1-CN_1 catalyst, the number of active sites increased, which was helpful for improving TC photocatalytic degradation efficiency [26]. However, TC degradation efficiency was not significantly improved by more than 1% when the Mn_1-CN_1 dose reached 0.8 g L^{-1}. When the amount of the Mn_1-CN_1 catalyst was too high, the excessive Mn_1-CN_1 impeded the passage of light, further affecting the number of photogenerated electron-hole pairs generated by the photocatalyst and causing the degradation efficiency of TC solution to decrease.

3.4. The Effect of initial TC Concentration

As Figure 4d displays, as the TC concentration was augmented from 10 to 25 mg L^{-1}, the TC degradative efficiency diminished from 99.1% to 77.9% in 180 min photoreaction. When the catalyst and the PMS dosage were constant, the electron-hole pairs produced were almost constant, and the number of active molecules produced by the photocatalytic system tended to be constant. The photocatalytic system was able to provide sufficient active substances while the initial TC concentration was 10 mg L^{-1}. Therefore, each TC molecule was more easily oxidized, improving the utilization rate of the active species. The photocatalytic system provided insufficient active species, and TC degradation efficiency diminished, with gradual augmentation of the initial TC concentration. In addition, the amount of intermediate products produced in TC degradation also increased with the initial TC concentration. This could result in occupation of other active sites [28].

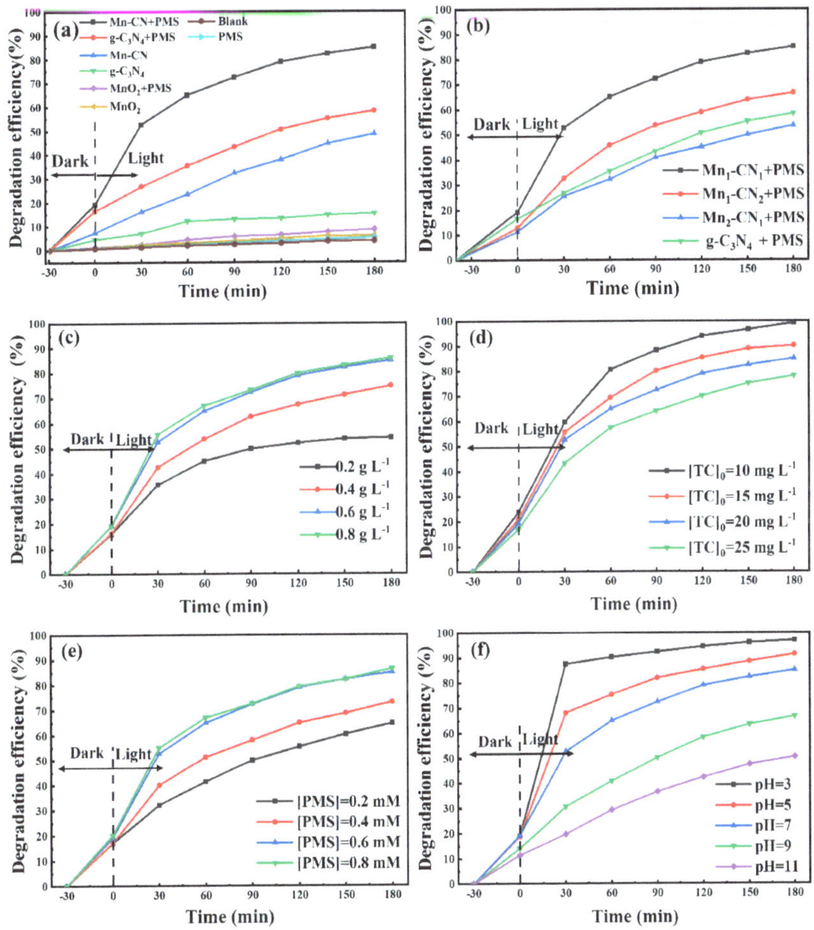

Figure 4. Photocatalytic degradation efficiency of (**a**) the different photocatalysts, (**b**) Mn-CN in different proportions, (**c**) amounts of the photocatalyst, (**d**) initial TC concentration, (**e**) PMS dosage, and (**f**) initial pH. The reaction conditions: [TC] = 20 mg L^{-1}, [photocatalyst] = 0.6 g L^{-1}, [PMS] = 0.6 mM, T = 25 °C, and pH = 7.

3.5. The Effect of PMS Dosage on TC Degradation

The influence of different PMS concentrations on TC degradation is detailed in Figure 4e. As the PMS dosage was augmented, the TC degradation efficiency was gradually enhanced. The photocatalytic degradation efficiency of TC increased from 64.7% to 85.06% when the concentration of PMS was increased from 0.2 to 0.6 mM, which was attributed to the production of more reactive oxygen species by Mn_1-CN_1 [53]. However, the TC degradation capacity was not significantly improved when PMS was dosed to 0.8 mM, with no great improvement in TC degradation efficiency. Excess PMS could react with $SO_4^{\bullet-}$, forming persulfate radicals ($S_2O_8^{2-}$), which have a much lower oxidation capacity than $SO_4^{\bullet-}$ (see Equations (3) and (4)) [54].

$$SO_4^{\bullet-} + S_2O_8^{2-} \rightarrow S_2O_8^{\bullet-} + SO_4^{2-} \tag{3}$$

$$SO_4^{\bullet-} + SO_4^{\bullet-} \rightarrow S_2O_8^{2-} \tag{4}$$

3.6. The Effect of Initial Solution pH

The influence of diverse initial solution pH levels on TC degradation is detailed in Figure 4f. In acidic conditions, the combined system showed better degradation performance, whereas in alkaline conditions, a lower TC degradation efficiency remained. The TC degradation efficiency obtained was 96.97% at a pH of 3, while the lowest degradation efficiency was recorded when the pH was 11 (50.6%). When the solution pH was below the equipotential point of Mn_1-CN_1 in the acidic environment, the Mn_1-CN_1 catalyst was positively charged [55]. This could promote the migration of photoexcited electrons to the catalyst surface and effectively lessen the compounding of e^- and h^+, thus enhancing the catalytic efficiency of the catalyst. The Mn_1-CN_1 was negatively charged in an alkaline environment, which did not assist the transfer of photoexcited electrons to the catalyst surface, thus affecting photocatalytic efficiency.

Supplementary Figure S3 exhibits the UV-visible degradation spectra for each of the influencing factors. The intensity of the typical absorption peaks decreased as the photocatalytic reaction progressed. This indicated that the Mn_1-CN_1/PMS system produced effective active species. The reaction kinetic rate constant for TC degradation can be calculated from the first-order kinetic equation (Equation (5)), where k and t are the degradation rate constant and reaction time (min), respectively, C_t is the TC concentration at t min (mg L^{-1}), and C_0 is the initial TC concentration (mg L^{-1}).

$$\ln(C_t/C_0) = -kt \tag{5}$$

The photocatalytic process of MnO_2/g-C_3N_4 was compared with other g-C_3N_4-based photocatalysts reported in the literature (Table 1). The findings proved that MnO_2/g-C_3N_4 exhibited relatively better photocatalytic activity for organic pollutant removal. In addition, the kinetic diagrams at the TC level for each of the above influences are detailed as Supplementary Figure S4a–f, showing first-order reaction kinetics well explained the TC degradation rate, and the Mn_1-CN_1/PMS system had a high degradation rate constant (Supplementary Table S1).

The TOC analyzer (Analytickjena, multi N/C 2100s) was used to determine the change in total organic carbon in the TC solution. As shown in Figure 5, the removal of TOC increased with increasing of the photocatalytic treatment time. When the time was 180 min, the removal rate of TOC was 81.3%. This indicates that the Mn_1-CN_1/PMS system has significant photocatalytic treatment ability under visible light, which can oxidize industrial recalcitrant pollutants such as TC.

Table 1. Comparison of TC degradation efficiencies with other C_3N_4-based photocatalysts.

Photocatalyst (g L^{-1})	TC Concentration (mg L^{-1})	Light Source Type	Volumes Tested	Time (min)	Removal (%)	References
$AgCl/ZnO/g\text{-}C_3N_4$ (1.0)	20 mg L^{-1}	400–700 nm	40 ml	80 min	92.7%	[56]
$Ag\text{-}C_3N_4/SnS_2$ (0.4)	15 mg L^{-1}	500 W Xe lamp	50 ml	150 min	94.9%	[57]
$g\text{-}C_3N_4/BiPO_4$ (1.0)	20 mg L^{-1}	1000 W Xe lamp	50 ml	180 min	97%	[58]
C_3N_4/WO_3 (0.4)	10 mg L^{-1}	500 W Xe lamp (320–780nm)	100 ml	180 min	79.8%	[59]
$Co_3O_4/g\text{-}C_3N_4$ (0.4)	15 mg L^{-1}	500 W Xe lamp (>420 nm)	100 ml	150 min	92.6%	[60]
$g\text{-}C_3N_4/MnO_2/GO$ (0.5)	10 mg L^{-1}	300 W Xe lamp (420 nm)	100 ml	90 min	91.4%	[61]
$Ag_3PO_4/AgBr/g\text{-}C_3N_4$(0.5)	40 mg L^{-1}	300 W Xe lamp (>420 nm)	100 ml	25 min	80.2%	[28]
$g\text{-}C_3N_4/Bi_2WO_6/AgI$ (0.6)	20 mg L^{-1}	>420 nm	50 ml	60 min	91.1%	[62]
$Mn_1\text{-}CN_1$ (0.6)	20 mg L^{-1}	300W xenon lamp	50 ml	180 min	96.97%	This study

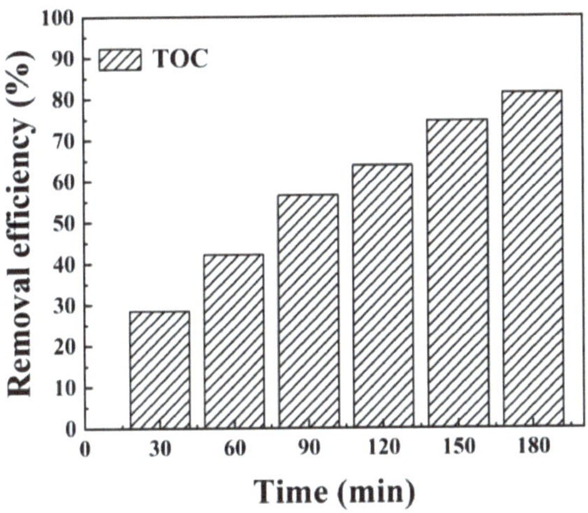

Figure 5. TOC removal by $Mn_1\text{-}CN_1$/PMS system for TC photocatalytic degradation. Reaction conditions: TC = 20 mg L^{-1}; photocatalyst = 0.6 g L^{-1}; PMS = 0.6 mM; T = 25 °C; and pH = 7.

3.7. Stability of Photocatalytic Materials

As illustrated in Figure 6a, there was no appreciable difference between the first used catalyst and the five recycled ones, and TC degradation efficiency only declined from 96.5% to 91.7%. The decrease in TC degradation efficiency could be explained by the accumulation of intermediates on the $Mn_1\text{-}CN_1$ catalyst, which competed with the photocatalyst active site, and the loss of a small amount of photocatalyst during the cycling process experiments [63]. Meanwhile, to further confirm the stability, $Mn_1\text{-}CN_1$ was analyzed by SEM/TEM, XPS, and XRD after five cyclic experiments. The SEM/TEM image (Supplementary Figure S5) shows no great changes in $Mn_1\text{-}CN_1$ morphology. The XRD pattern (Supplementary Figure S6) shows that the diffraction peaks were significantly unchanged after the reaction. Furthermore, the XPS spectra of the recovered $Mn_1\text{-}CN_1$ were similar to those of the original catalyst, indicating that the surface structure of the sample did not change after five cycles of reaction (Supplementary Figure S7). The $Mn_1\text{-}CN_1$ composite photocatalyst was prepared with high stability, which is very useful for practical applications in wastewater treatment.

3.8. Visible Light Photocatalytic Degradation Mechanism

The active species trapping experiment was performed to investigate the active species involved in TC degradation [64,65]. Reagents IPA, BQ, EDTA-2Na, and MA were employed for capturing •OH, superoxide radical (•O_2^-), h^+, and $SO_4^{\bullet-}$, respectively. Supplementary Figure S8 shows the addition of IPA and MA significantly inhibited TC degradation efficiency, which diminished from 97% to 41.08% and 54.9%, correspondingly, revealing that •OH and $SO_4^{\bullet-}$ were primary active species. Additionally, the presence of BQ also inhibited TC degradation efficiency, suggesting that •O_2^- was also involved in the photocatalytic reaction. However, EDTA-2Na only slightly inhibited TC photodegradation, revealing that h^+ was not the major active substance.

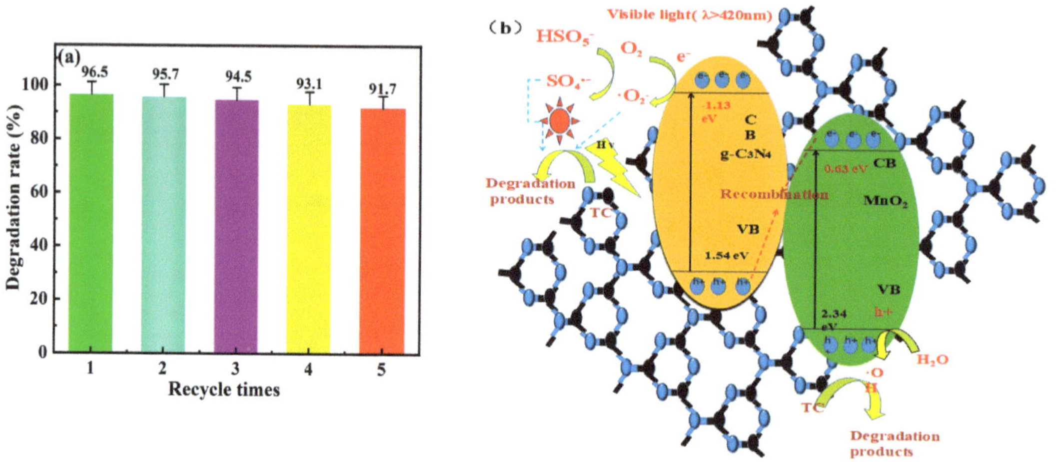

Figure 6. (**a**) Recyclability of Mn_1-CN_1 catalyst; (**b**) photocatalytic mechanism scheme of Mn_1-CN_1 composites.

The possible mechanism of the degrading of TC pollutants by the Mn_1-CN_1/PMS system under visible light was investigated (Figure 6b). Unlike the type II configuration, the Z-type MnO_2/g-C_3N_4 photocatalyst combines a semiconductor of high energy in the valence band (MnO_2) with another semiconductor of lower bandgap and stronger reduction ability in the conduction band (g-C_3N_4). The electrons photogenerated in the MnO_2 are cross-combined with the holes of the g-C_3N_4. This process yields the accumulation of separated charge carriers in the respective phases. The MnO_2's holes keep their strong oxidation ability, while the g-C_3N_4's electrons retain their reduction potential. In Z-scheme heterojunction photocatalysts, the Mn_1-CN_1 generated photo-induced e^- and h^+ (Equation (6)) were excited under visible light irradiation. Recombination reactions are conducted by photogenerated e^- in the CB of MnO_2 and h^+ in the VB of g-C_3N_4. The H_2O or OH^- could be oxidized by h^+ on the VB of MnO_2 to produce •OH (Equations (7) and (8)). Photogenerated e^- on the CB of g-C_3N_4 reacted with dissolved oxygen, forming •O_2^- (Equation (9)). Meanwhile, the PMS received photogenerated e^-, which can be activated for producing $SO_4^{\bullet-}$ (Equation (10)). The $SO_4^{\bullet-}$ was able to react rapidly with OH^- for producing •OH (Equation (11)). The $SO_4^{\bullet-}$ and •OH were vital for TC degradation (Equation (12)). Overall, ideal matching of the energy band structure between g-C_3N_4 and MnO_2 can increase the amount of photogenerated e^- and h^+ [66].

$$MnO_2@g\text{-}C_3N_4 + h\upsilon \rightarrow e^- + h^+ \tag{6}$$

$$H_2O + h^+ \rightarrow \bullet OH + H^+ \tag{7}$$

$$OH^- + h^+ \rightarrow \bullet OH \tag{8}$$

$$O_2 + e^- \rightarrow \bullet O_2^- \quad (9)$$

$$HSO_5^- + e^- \rightarrow OH^- + SO_4^{\bullet-} \quad (10)$$

$$SO_4^{\bullet-} + OH^- \rightarrow \bullet OH + SO_4^{2-} \quad (11)$$

$$\bullet OH + SO_4^{\bullet-} + TC \rightarrow \text{Degradation products} \quad (12)$$

Intermediates of TC degradation were identified by HPLC/MS for future investigation of the TC degradation mechanism. More information on the nine main intermediates can be found in Supplementary Table S2. Three possible transformation reactions, hydroxylation, carboxylation, and N-C bond cleavage, are shown in Figure 7 [67–69]. Since -C=C- was conjugated to the oxygen on the adjacent -OH, P1 (m/z = 461) and P2 (m/z = 476) were derived from the hydroxyl addition of two enol groups. The -OH served as an electron-donating group, weakening the -C=C- linkage, and was attacked by •OH, generating ketone and epoxide group transformations. Meanwhile, P1 was susceptible to attack by reactive radicals and further formation of polyhydroxylated P2 (m/z = 476) because of low electron density of the C(1)=C(2) bond of P1. The C(3)-N bond of P2 was then oxidized for producing P3 (m/z = 447). Finally, P4 (m/z = 495) was obtained through ring-opening the aromatic ring on P3. In degradation pathway 2, the TC molecule underwent deamidation under the attack of reactive radicals for producing P5 (m/z = 401). P6 (m/z = 417) was formed by -OH addition to the deamidated C1. P7 (m/z = 360) was an intermediate product formed when -OH and $-CH_3$ on C5 of P6 were attacked by h^+ and $\bullet O_2^-$ reactive oxide species, which were followed by dimethylamino and ring opening. In degradation pathway 3, the carbocyclic branched chain of P5 was destroyed by strong oxidizing radicals to produce P8 (m/z = 306). Finally, P8 underwent deacylation and demethylation dehydroxylation reactions to give P9 (m/z = 227). These intermediates were unstable and quickly oxidized to CO_2 and H_2O. A comparison of the by-products produced with the study of Yu et al. [28] revealed that the present study produced lesser quantities of by-products. This indicates that the Mn_1-CN_1/PMS system has significant photocatalytic treatment ability under visible light.

Supplementary Figure S9 shows Three-Dimensional Excitation-Emission Matrix (3DEEM) of TC at different photocatalytic times. The fluorescence peak of TC mainly occurred in the wavelength range of Ex/Em= (290–340 nm)/(380–460 nm), suggesting that TC was an aromatic protein-like compound. The intensity of the fluorescence peaks decreased with prolonged photocatalytic reaction time, suggesting that the aromatic benzene ring structure was damaged by the attack of the reactive species. The prominence peak almost disappeared after 180 min, suggesting that the simple aryl ring of TC was almost entirely disrupted (Supplementary Figure S8d).

3.9. TC Degradation Toxicity Analysis

To examine the environmental toxicity of TC, toxicity evaluation of the degradation products was used for contaminant status assessment. The biological toxicity of TC and its degradation intermediates were investigated via T.E.S.T. according to the quantitative structure–activity relationship (QSAR). Figure 8a,b shows the predicted values of daphnia magna LC_{50} and the developmental toxicity of the intermediates. The LC_{50}-48 h values of most degradation intermediates were higher than that of TC, suggesting that TC was slowly transformed into less toxic intermediates. The products P2, P4, and P8 can be regarded as non-toxicants in Figure 8b. In addition, the results on the basis of bioaccumulation factors and mutagenicity illustrated that the photocatalytic reaction reduced the ecosystem hazard of TC and its intermediates [70]. Therefore, TC oxidation by Mn_1-CN_1/PMS system could reduce the toxicity of TC.

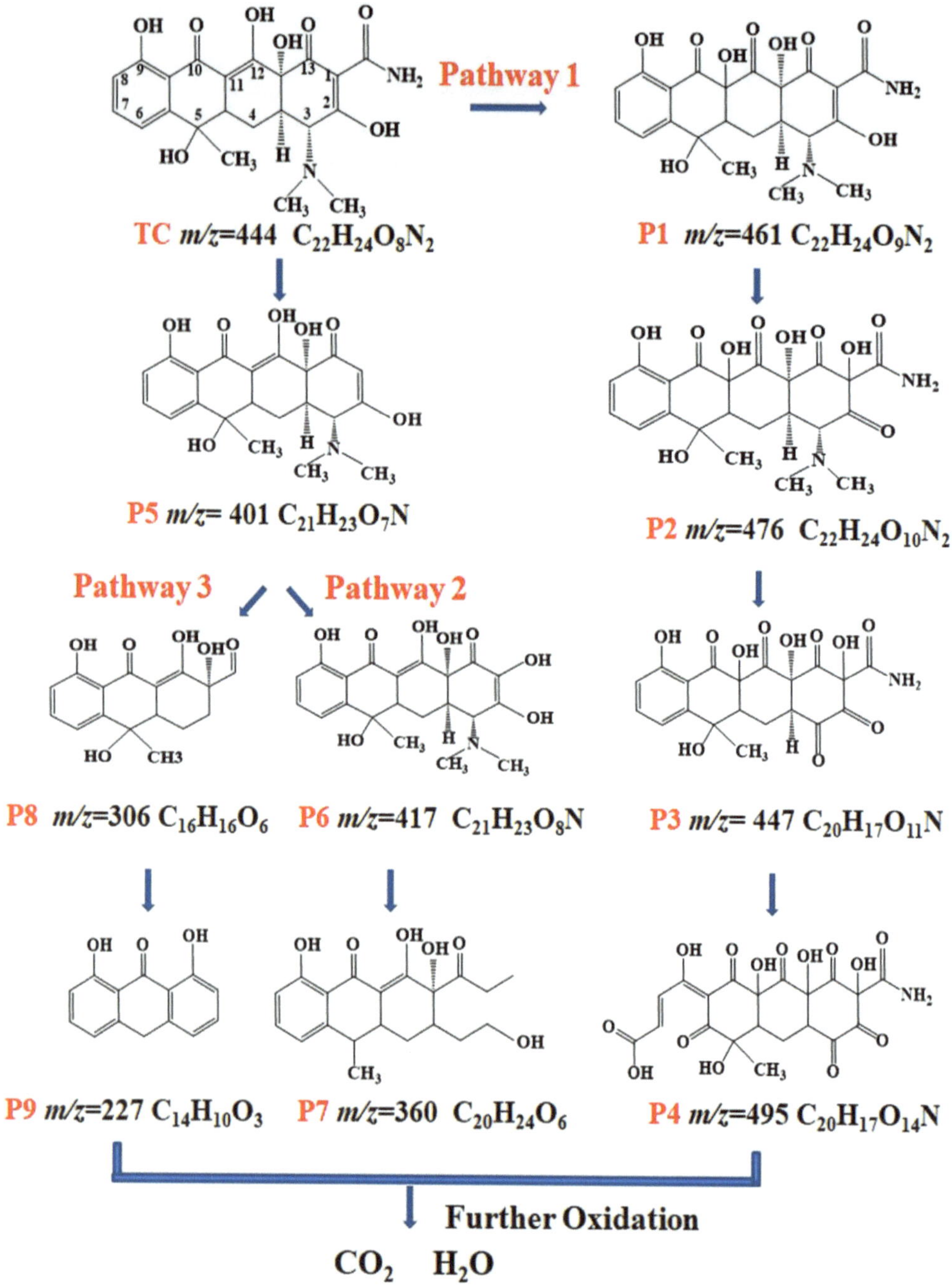

Figure 7. Three possible degradation pathways of TC in the Mn_1-CN_1/PMS system.

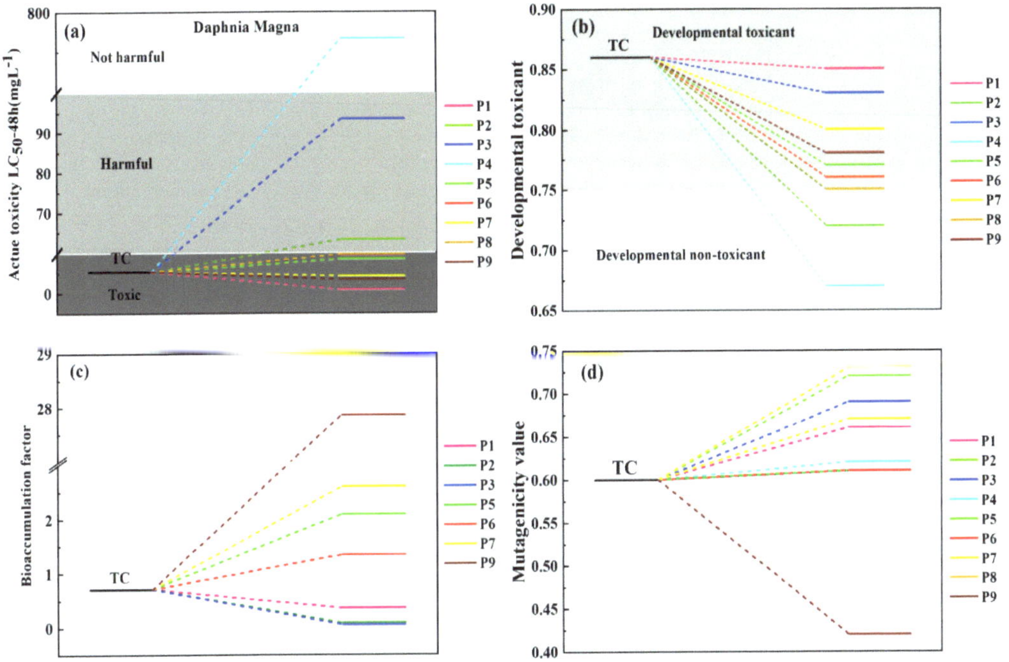

Figure 8. Toxicity analysis of reaction products: (**a**) The Daphnia magna LC_{50}-48 h, (**b**) Developmental toxicity, (**c**) Bioaccumulation factors, (**d**) Mutagenicity.

4. Conclusions

In this study, a Z-type Mn-CN heterojunction composite photocatalyst with excellent performance was successfully prepared by coupling the g-C_3N_4 with MnO_2 via high-temperature thermal polymerization. The XRD, SEM, and TEM analysis showed synthesized Mn_1-CN_1 had good purity and a crystalline state. The catalytic experiments indicated that the Mn_1-CN_1/PMS system was the most effective, with the degradation efficiency of TC reaching 96.97% after 180 min of treatment, which was 38.65% higher relative to the g-C_3N_4/PMS system. The photocatalyst displayed marvelous stability and reusability after five cycle experiments. In addition, the PMS played an essential role in removing TC, while Mn_1-CN_1 was also an efficient activator for PMS under visible light irradiation. The active species trapping experiment showed •OH and $SO_4^{\bullet-}$ were major active species. Nine primary degradation intermediates and three possible degradation pathways were put forward. Toxicological evaluation of TC and its intermediates showed that the Mn_1-CN_1/PMS system could diminish TC toxicity. The findings showed the Mn_1-CN_1/PMS system was advantageous for the photocatalytic degradation of TC, which can offer potential application for the treatment of antibiotics in wastewater.

Supplementary Materials: The following supporting information can be downloaded at: https://www.mdpi.com/article/10.3390/toxics12010070/s1, Figure S1: (a) Spectrogram of the xenon lamp light source BBZM−III, (b) Diagram of the photo-catalytic device; Figure S2: the XPS valence band spectra; Figure S3: The UV absorption spectra of TC degradation under optimal reaction conditions for each influencing factor. (a) Mn1−CN1 photocatalyst, the amount of Mn1−CN1 photocatalyst dosed was 0.6 g L^{-1}, the dosage of PMS was 0.6 mM; (b) The initial concentration of TC was 10 mg L^{-1}; (c) The initial pH was 3; Figure S4: (a) Catalyst type on the reaction kinetics of TC; (b) Reaction kinetics of different ratios of Mn−CN catalysts for the degradation of TC; (c) Reaction kinetics of catalyst dosing for degradation of TC; (d) Reaction ki-netics of TC degradation at initial TC concentrations; (e) Kinetics of PMS dosage in response to TC; (f) Initial pH on the reaction kinetics of TC; Figure S5: SEM (a) and

TEM (b) images of the Mn1−CN1 cycle after reaction; Figure S6: The XRD patterns of Mn1−CN1 before and after reaction; Figure S7: XPS spectra of Mn1−CN1 after reaction: full spectrum (a), C 1s (b), N 1s (c), Mn 2p (d), O1s (e); Figure S8: Active species trapping experiments of as−prepared Mn1−CN1 for TC degra-dation under visible light irradiation. [IPA, BQ, EDTA−2Na] = 1 mM L^{-1}, [MA] = 10 mM L^{-1}. The experimental conditions: [photocatalyst] = 0.6 g L^{-1}, [TC] = 20 mg L^{-1}, [PMS] = 0.6 mM, pH = 7, time=180 min, and T = 25 °C; Figure S9: 3DEEM spectra of TC solutions after different treatment times: (a) 0 min; (b) 60 min; (c) 120 min; (d) 180 min; Table S1: Reaction rate constants for each influencing factor.; Table S2: The TC degradation intermediates detected by HPLC−MS.

Author Contributions: G.L.: conceptualization, investigation, data curation, formal analysis, writing—original draft, visualization. X.L.: data curation, resources. P.L.: methodology, conceptualization. H.G.: investigation. Z.W.: methodology, investigation. Q.Z.: writing—review and editing, validation. Y.L.: formal analysis. W.S.: supervision. J.A.: writing—review and editing, funding acquisition. Z.Z.: writing—review and editing. All authors have read and agreed to the published version of the manuscript.

Funding: This work was supported by Natural Science Foundation of China (21706149 and 22006069); Natural Science Foundation of Shandong Province (ZR2023MB052).

Institutional Review Board Statement: Not applicable.

Informed Consent Statement: Not applicable.

Data Availability Statement: Data will be made available on request.

Acknowledgments: We would like to express our gratitude to the members of the Analysis and Testing Center of Shandong University of Technology and the School of Chemistry and Chemical Engineering for their technical assistance and suggestions.

Conflicts of Interest: The authors declare that they have no known competing financial interests or personal relationships that could have appeared to influence the work reported in this paper.

References

1. Wang, H.B.; Hu, C.; Liu, L.Z.; Xing, X.C. Interaction of ciprofloxacin chlorination products with bacteria in drinking water distribution systems. *J. Hazard. Mater.* **2017**, *339*, 174–181. [CrossRef] [PubMed]
2. Wang, L.P.; Yang, G.P.; Wang, D.; Lu, C.Y.; Guan, W.S.; Li, Y.L.; Deng, J.; Crittenden, J. Fabrication of the flower-flake-like $CuBi_2O_4/Bi_2WO_6$ heterostructure as efficient visible-light driven photocatalysts: Performance, kinetics and mechanism insight. *Appl. Surf. Sci.* **2019**, *495*, 143521. [CrossRef]
3. Li, J.N.; Cheng, W.X.; Xu, L.K.; Jiao, Y.N.; Baig, S.A.; Chen, H. Occurrence and removal of antibiotics and the corresponding resistance genes in wastewater treatment plants: Effluents' influence to downstream water environment. *Environ. Sci. Pollut. Res.* **2016**, *23*, 6826–6835. [CrossRef] [PubMed]
4. Moradi, S.; Sobhgol, S.A.; Hayati, F.; Isari, A.A.; Kakavandi, B.; Bashardoust, P.; Anvaripour, B. Performance and reaction mechanism of MgO/ZnO/Graphene ternary nanocomposite in coupling with LED and ultrasound waves for the degradation of sulfamethoxazole and pharmaceutical wastewater. *Sep. Purif. Technol.* **2020**, *251*, 117373. [CrossRef]
5. Mirzaei, A.; Yerushalmi, L.; Chen, Z.; Haghighat, F. Photocatalytic degradation of sulfamethoxazole by hierarchical magnetic ZnO@g-C_3N_4: RSM optimization, kinetic study, reaction pathway and toxicity evaluation. *J. Hazard. Mater.* **2018**, *359*, 516–526. [CrossRef] [PubMed]
6. Yang, G.R.; Gao, Q.Z.; Yang, S.Y.; Yin, S.H.; Cai, X.; Yu, X.Y.; Zhang, S.S.; Fang, Y.P. Strong adsorption of tetracycline hydrochloride on magnetic carbon-coated cobalt oxide nanoparticles. *Chemosphere* **2020**, *239*, 124831. [CrossRef] [PubMed]
7. Yang, Z.; Jia, S.Y.; Zhuo, N.; Yang, W.B.; Wang, Y.P. Flocculation of copper(II) and tetracycline from water using a novel pH- and temperature-responsive flocculants. *Chemosphere* **2015**, *141*, 112–119. [CrossRef]
8. Meng, X.; Liu, Z.M.; Deng, C.; Zhu, M.F.; Wang, D.Y.; Li, K.; Deng, Y.; Jiang, M.M. Microporous nano-MgO/diatomite ceramic membrane with high positive surface charge for tetracycline removal. *J. Hazard. Mater.* **2016**, *320*, 495–503. [CrossRef]
9. Shao, S.C.; Wu, X.W. Microbial degradation of tetracycline in the aquatic environment: A review. *Crit. Rev. Biotechnol.* **2020**, *40*, 1010–1018. [CrossRef]
10. Zhang, X.L.; Chen, J.X.; Li, J. The removal of microplastics in the wastewater treatment process and their potential impact on anaerobic digestion due to pollutants association. *Chemosphere* **2020**, *251*, 126360. [CrossRef]
11. Khan, A.; Zhang, K.K.; Taraqqi-A-Kamal, A.; Wang, X.G.; Chen, Y.; Zhang, Y.R. Degradation of antibiotics in aqueous media using manganese nanocatalyst-activated peroxymonosulfate. *J. Colloid Interface Sci.* **2021**, *599*, 805–818. [CrossRef] [PubMed]

12. Nguyen, T.B.; Huang, C.P.; Doong, R.A.; Wang, M.H.; Chen, C.W.; Dong, C.D. Manipulating the morphology of 3D flower-like $CoMn_2O_4$ bimetallic catalyst for enhancing the activation of peroxymonosulfate toward the degradation of selected persistent pharmaceuticals in water. *Chem. Eng. J.* **2022**, *436*, 135244. [CrossRef]
13. Gao, Y.; Wang, Q.; Ji, G.Z.; Li, A.M. Degradation of antibiotic pollutants by persulfate activated with various carbon materials. *Chem. Eng. J.* **2022**, *429*, 132387. [CrossRef]
14. Zhang, Y.; Zhou, J.B.; Chen, X.; Wang, L.; Cai, W.Q. Coupling of heterogeneous advanced oxidation processes and photocatalysis in efficient degradation of tetracycline hydrochloride by Fe-based MOFs: Synergistic effect and degradation pathway. *Chem. Eng. J.* **2019**, *369*, 745–757. [CrossRef]
15. Chen, G.Y.; Yu, Y.; Liang, L.; Duan, X.G.; Li, R.; Lu, X.K.; Yan, B.B.; Li, N.; Wang, S.B. Remediation of antibiotic wastewater by coupled photocatalytic and persulfate oxidation system: A critical review. *J. Hazard. Mater.* **2021**, *408*, 124461. [CrossRef]
16. Adil, S.; Maryam, B.; Kim, E.J.; Dulova, N. Individual and simultaneous degradation of sulfamethoxazole and trimethoprim by ozone, ozone/hydrogen peroxide and ozone/persulfate processes: A comparative study. *Environ. Res.* **2020**, *189*, 109889. [CrossRef]
17. Ji, Y.F.; Dong, C.X.; Kong, D.Y.; Lu, J.H.; Zhou, Q.S. Heat-activated persulfate oxidation of atrazine: Implications for remediation of groundwater contaminated by herbicides. *Chem. Eng. J.* **2015**, *263*, 45–54. [CrossRef]
18. Li, L.; Yuan, X.P.; Zhou, Z.P.; Tang, R.D.; Deng, Y.C.; Huang, Y.; Xiong, S.; Su, L.; Zhao, J.; Gong, D.X. Research progress of photocatalytic activated persulfate removal of environmental organic pollutants by metal and nonmetal based photocatalysts. *J. Clean. Prod.* **2022**, *372*, 133420. [CrossRef]
19. Liu, B.C.; Qiao, M.; Wang, Y.B.; Wang, L.J.; Gong, Y.; Guo, T.; Zhao, X. Persulfate enhanced photocatalytic degradation of bisphenol A by g-C_3N_4 nanosheets under visible light irradiation. *Chemosphere* **2017**, *189*, 115–122. [CrossRef]
20. Guan, R.P.; Yuan, X.Z.; Wu, Z.B.; Jiang, L.B.; Zhang, J.; Li, Y.F.; Zeng, G.M.; Mo, D. Efficient degradation of tetracycline by heterogeneous cobalt oxide/cerium oxide composites mediated with persulfate. *Sep. Purif. Technol.* **2019**, *212*, 223–232. [CrossRef]
21. Jia, J.; Zhang, Q.Q.; Li, K.K.; Zhang, Y.T.; Liu, E.Z.; Li, X. Recent advances on g-C_3N_4-based Z-scheme photocatalysts: Structural design and photocatalytic applications. *Int. J. Hydrogen Energy* **2022**, *48*, 196–231. [CrossRef]
22. Jiang, L.B.; Yuan, X.Z.; Zeng, G.M.; Liang, J.; Chen, X.H.; Yu, H.B.; Wang, H.; Wu, Z.B.; Zhang, J.; Xiong, T. In-situ synthesis of direct solid-state dual Z-scheme WO_3/g-C_3N_4/Bi_2O_3 photocatalyst for the degradation of refractory pollutant. *Appl. Catal. B Environ.* **2018**, *227*, 376–385. [CrossRef]
23. Song, Y.L.; Qi, J.Y.; Tian, J.Y.; Gao, S.S.; Cui, F.Y. Construction of Ag/g-C_3N_4 photocatalysts with visible-light photocatalytic activity for sulfamethoxazole degradation. *Chem. Eng. J.* **2018**, *341*, 547–555. [CrossRef]
24. Guo, F.; Li, M.Y.; Ren, H.J.; Huang, X.L.; Shu, K.K.; Shi, W.L.; Lu, C.Y. Facile bottom-up preparation of Cl-doped porous g-C_3N_4 nanosheets for enhanced photocatalytic degradation of tetracycline under visible light. *Sep. Purif. Technol.* **2019**, *228*, 115770. [CrossRef]
25. Gao, X.X.; Yang, B.Z.; Yao, W.Q.; Wang, Y.J.; Zong, R.L.; Wang, J.; Li, X.C.; Jin, W.J.; Tao, D.P. Enhanced photocatalytic activity of ZnO/g-C_3N_4 composites by regulating stacked thickness of g-C_3N_4 nanosheets. *Environ. Pollut.* **2020**, *257*, 113577. [CrossRef]
26. Wang, J.J.; Tang, L.; Zeng, G.M.; Deng, Y.C.; Liu, Y.N.; Wang, L.G.; Zhou, Y.Y.; Guo, Z.; Wang, J.J.; Zhang, C. Atomic scale g-C_3N_4/Bi_2WO_6 2D/2D heterojunction with enhanced photocatalytic degradation of ibuprofen under visible light irradiation. *Appl. Catal. B Environ.* **2017**, *209*, 285–294. [CrossRef]
27. Xue, S.Y.; Wu, C.Z.; Pu, S.Y.; Hou, Y.Q.; Tong, T.; Yang, G.; Qin, Z.J.; Wang, Z.M.; Bao, J.M. Direct Z-Scheme charge transfer in heterostructured MoO_3/g-C_3N_4 photocatalysts and the generation of active radicals in photocatalytic dye degradations. *Environ. Pollut.* **2019**, *250*, 338–345. [CrossRef]
28. Yu, H.B.; Wang, D.Y.; Zhao, B.; Lu, Y.; Wang, X.H.; Zhu, S.Y.; Qin, W.C.; Huo, M.X. Enhanced photocatalytic degradation of tetracycline under visible light by using a ternary photocatalyst of Ag_3PO_4/AgBr/g-C_3N_4 with dual Z-scheme heterojunction. *Sep. Purif. Technol.* **2020**, *237*, 116365. [CrossRef]
29. Eslami, A.; Hashemi, M.; Ghanbari, F. Degradation of 4-chlorophenol using catalyzed peroxymonosulfate with nano-MnO_2/UV irradiation: Toxicity assessment and evaluation for industrial wastewater treatment. *J. Clean. Prod.* **2018**, *195*, 1389–1397. [CrossRef]
30. Jawad, A.; Zhan, K.; Wang, H.B.; Shahzad, A.; Zeng, Z.H.; Wang, J.; Zhou, X.Q.; Ullah, H.; Chen, Z.L.; Chen, Z.Q. Tuning of Persulfate Activation from a Free Radical to a Nonradical Pathway through the Incorporation of Non-Redox Magnesium Oxide. *Environ. Sci. Technol.* **2020**, *54*, 2476–2488. [CrossRef]
31. Li, J.N.; Li, X.Y.; Wang, X.; Zeng, L.B.; Chen, X.; Mu, J.C.; Chen, G.H. Multiple regulations of Mn-based oxides in boosting peroxymonosulfate activation for norfloxacin removal. *Appl. Catal. A Gen.* **2019**, *584*, 117170. [CrossRef]
32. Zhang, Q.; Peng, Y.; Deng, F.; Wang, M.; Chen, D.Z. Porous Z-scheme MnO_2/Mn-modified alkalinized g-C_3N_4 heterojunction with excellent Fenton-like photocatalytic activity for efficient degradation of pharmaceutical pollutants. *Sep. Purif. Technol.* **2020**, *246*, 116890. [CrossRef]
33. Nguyen, M.D.; Nguyen, T.B.; Tran, L.H.; Nguyen, T.G.; Fatimah, I.; Kuncoro, E.P.; Doong, R.A. Z-scheme S, B co-doped g-C_3N_4 nanotube@MnO_2 heterojunction with visible-light-responsive for enhanced photodegradation of diclofenac by peroxymonosulfate activation. *Chem. Eng. J.* **2023**, *452*, 139249. [CrossRef]

34. Dong, F.; Zhao, Z.W.; Xiong, T.; Ni, Z.L.; Zhang, W.D.; Sun, Y.J.; Ho, W.K. In Situ Construction of g-C_3N_4/g-C_3N_4 Metal-Free Heterojunction for Enhanced Visible-Light Photocatalysis. *ACS Appl. Mater. Interfaces* **2013**, *145*, 11392–11401. [CrossRef] [PubMed]
35. Wang, Y.Z.; Yu, J.Y.; Peng, W.D.; Tian, J.; Yang, C. Novel multilayer TiO_2 heterojunction decorated by low g-C_3N_4 content and its enhanced photocatalytic activity under UV, visible and solar light irradiation. *Sci. Rep.* **2019**, *9*, 5932. [CrossRef]
36. Du, X.R.; Zou, G.J.; Wang, Z.H.; Wang, X.L. A scalable chemical route to soluble acidified graphitic carbon nitride: An ideal precursor for isolated ultrathin g-C_3N_4 nanosheets. *Nanoscale* **2015**, *7*, 8701–8706. [CrossRef]
37. Panimalar, S.; Uthrakumar, R.; Selvi, E.T.; Gomathy, P.; Inmozhi, C.; Kaviyarasu, K.; Kennedy, J. Studies of MnO_2/g-C_3N_4 hetrostructure efficient of visible light photocatalyst for pollutants degradation by sol-gel technique. *Surf. Interfaces* **2020**, *20*, 100512. [CrossRef]
38. Bai, X.J.; Yan, S.C.; Wang, J.J.; Li, W.; Jiang, W.J.; Wu, S.L.; Sun, C.P.; Zhu, Y.F. A simple and efficient strategy for the synthesis of a chemically tailored g-C_3N_4 material. *J. Mater. Chem. A* **2014**, *2*, 17521–17529. [CrossRef]
39. Liu, X.M.; Liu, Y.; Zhang, W.K.; Zhong, Q.Y.; Ma, X.Y. In situ self-assembly of 3D hierarchical 2D/2D CdS/g-C_3N_4 hereojunction with excellent photocatalytic performance. *Mat. Sci. Semicon. Proc.* **2020**, *105*, 104734. [CrossRef]
40. Rivera-Lugo, Y.Y.; Félix-Navarro, R.M.; Trujillo-Navarrete, B.; Reynoso-Soto, E.A.; Silva-Carrillo, C.; Cruz-Gutiérrez, C.A.; Quiroga-González, E.; Calva-Yáez, J.C. Flower-like δ-MnO_2 as cathode material of Li-ion batteries of high charge-discharge rates. *Fuel* **2021**, *287*, 119463. [CrossRef]
41. Shen, G.D.; Pu, Y.P.; Cui, Y.F.; Jing, P.P. Easy synthesis of TiO_2/g-C_3N_4 Heterostructure Photocatalyst with Large Surface Area and Excellent Photocatalytic Activity. *Ceram. Int.* **2017**, *43*, S664–S670. [CrossRef]
42. Dai, X.; Xie, M.L.; Meng, S.G.; Fu, X.L.; Chen, S.F. Coupled systems for selective oxidation of aromatic alcohols to aldehydes and reduction of nitrobenzene into aniline using CdS/g-C_3N_4 photocatalyst under visible light irradiation. *Appl. Catal. B Environ.* **2014**, *158*, 382–390. [CrossRef]
43. Yu, J.G.; Wang, S.H.; Cheng, B.; Zhang, L.; Huang, F. Noble metal-free $Ni(OH)_2$-g-C_3N_4 composite photocatalyst with enhanced visible-light photocatalytic H_2-production activity. *Catal. Sci. Technol.* **2013**, *3*, 1782–1789. [CrossRef]
44. Zhang, Q.K.; Han, F.X.; Yan, Y.H.; Dai, Q.L.; Proctor, G.; Cheah, P.; Avijit, P.; Chandra, R.P.; Kang, N.; Hu, M.G.; et al. Preparation and properties of visible light responsive RGO/In_2 TiO_5 nanobelts for photocatalytic degradation of organic pollutants. *Appl. Surf. Sci.* **2019**, *485*, 547–553. [CrossRef]
45. Nanda, B.; Pradhan, A.C.; Parida, K.M. A comparative study on adsorption and photocatalytic dye degradation under visible light irradiation by mesoporous MnO_2 modified MCM-41 nanocomposite. *Micropor. Mesopor. Mat.* **2016**, *226*, 229–242. [CrossRef]
46. Zhang, Y.G.; Wu, M.Y.; Kwok, Y.H.; Wang, Y.F.; Zhao, W.; Zhao, X.L.; Huang, H.B.; Leung, D.Y.C. In-situ synthesis of heterojunction TiO_2/MnO_2 nanostructure with excellent performance in vacuum ultraviolet photocatalytic oxidation of toluene. *Appl. Catal. B Environ.* **2019**, *259*, 118034. [CrossRef]
47. Zhu, B.C.; Xia, P.F.; Ho, W.K.; Yu, J.G. Isoelectric point and adsorption activity of porous g-C_3N_4. *Appl. Surf. Sci.* **2015**, *344*, 188–195. [CrossRef]
48. Wu, W.T.; Zhang, J.Q.; Fan, W.Y.; Li, Z.T.; Wang, L.Z.; Li, X.M.; Wang, Y.; Wang, R.Q.; Zheng, J.T.; Wu, M.B.; et al. Remedying Defects in Carbon Nitride to Improve both Photooxidation and H_2 Generation Efficiencies. *ACS Catal.* **2016**, *6*, 3365–3371. [CrossRef]
49. Lin, L.H.; Ou, H.H.; Zhang, Y.F.; Wang, X.C. Tri-s-triazine-Based Crystalline Graphitic Carbon Nitrides for Highly Efficient Hydrogen Evolution Photocatalysis. *ACS Catal.* **2016**, *6*, 3921–3931. [CrossRef]
50. Qiu, P.X.; Chen, H.; Xu, C.M.; Zhou, N.; Jiang, F.; Wang, X.; Fu, Y.S. Fabrication of an exfoliated graphitic carbon nitride as a highly active visible light photocatalyst. *J. Mater. Chem. A* **2015**, *3*, 24237–24244. [CrossRef]
51. Zhao, J.H.; Nan, J.; Zhao, Z.W.; Li, N.; Liu, J.; Cui, F.Y. Energy-efficient fabrication of a novel multivalence Mn_3O_4-MnO_2 heterojunction for dye degradation under visible light irradiation. *Appl. Catal. B Environ.* **2017**, *202*, 509–517. [CrossRef]
52. Li, B.X.; Shao, X.K.; Liu, T.X.; Shao, L.Z.; Zhang, B.S. Construction of metal/$WO_{2.72}$/rGO ternary nanocomposites with optimized adsorption, photocatalytic and photoelectrochemical properties. *Appl. Catal. B Environ.* **2016**, *198*, 325–333. [CrossRef]
53. Wang, Y.W.; Xiang, L.R.; Li, Z.; Han, J.G.; Guo, H. Sulfite activation by water film dielectric barrier discharge plasma for ibuprofen degradation: Efficiency, comparison of persulfate, mechanism, active substances dominant to pathway, and toxicity evaluation. *Sep. Purif. Technol.* **2024**, *330*, 125531. [CrossRef]
54. Forouzesh, M.; Ebadi, A.; Aghaeinejad-Meybodi, A. Degradation of metronidazole antibiotic in aqueous medium using activated carbon as a persulfate activator. *Sep. Purif. Technol.* **2019**, *210*, 145–151. [CrossRef]
55. Huang, D.L.; Li, J.; Zeng, G.M.; Xue, W.J.; Chen, S.; Li, Z.H.; Deng, R.; Yang, Y.; Cheng, M. Facile construction of hierarchical flower-like Z-scheme AgBr/Bi_2WO_6 photocatalysts for effective removal of tetracycline: Degradation pathways and mechanism. *Chem. Eng. J.* **2019**, *375*, 121991. [CrossRef]
56. Ding, C.M.; Zhu, Q.R.; Yang, B.; Petropoulos, E.; Xue, L.H.; Feng, Y.F.; He, S.Y.; Yang, L.Z. Efficient photocatalysis of tetracycline hydrochloride (TC-HCl) from pharmaceutical wastewater using AgCl/ZnO/g-C_3N_4 composite under visible light: Process and mechanisms. *J. Environ. Sci.* **2023**, *126*, 249–262. [CrossRef] [PubMed]
57. Zhao, W.; Li, Y.J.; Zhao, P.S.; Zhang, L.L.; Dai, B.L.; Xu, J.M.; Huang, H.B.; He, Y.L.; Dennis, Y.C. Novel Z-scheme Ag-C_3N_4/SnS_2 plasmonic heterojunction photocatalyst for degradation of tetracycline and H_2 production. *Chem. Eng. J.* **2021**, *405*, 126555. [CrossRef]

58. Zhu, X.; Wang, Y.T.; Guo, Y.; Wan, J.Z.; Yan, Y.; Zhou, Y.X.; Sun, C. Environmental-friendly synthesis of heterojunction photocatalysts g-C_3N_4/$BiPO_4$ with enhanced photocatalytic performance. *Appl. Surf. Sci.* **2021**, *544*, 148872. [CrossRef]
59. Jing, H.C.; Ou, R.P.; Yu, H.B.; Zhao, Y.H.; Lu, Y.; Huo, M.X.; Huo, H.L.; Wang, X.Z. Engineering of g-C_3N_4 nanoparticles/WO_3 hollow microspheres photocatalyst with Z-scheme heterostructure for boosting tetracycline hydrochloride degradation. *Sep. Purif. Technol.* **2021**, *255*, 117646. [CrossRef]
60. Zhao, W.; Li, J.; She, T.T.; Ma, S.S.; Cheng, Z.P.; Wang, G.X.; Zhao, P.S.; Wei, W.; Xia, D.H.; Dennis, Y.C. Study on the Photocatalysis Mechanism of the Z-Scheme Cobalt Oxide Nanocubes/Carbon Nitride Nanosheets Heterojunction Photocatalyst with High Photocatalytic Performances. *J. Hazard. Mater.* **2021**, *402*, 123839. [CrossRef]
61. Du, C.Y.; Zhang, Z.; Tan, S.Y.; Yu, G.L.; Chen, H.; Zhou, L.; Yu, L.; Su, Y.H.; Zhang, Y.; Deng, F.F.; et al. Construction of Z-scheme g-C_3N_4 / MnO_2 /GO ternary photocatalyst with enhanced photodegradation ability of tetracycline hydrochloride under visible light radiation. *Environ. Res.* **2021**, *200*, 111427. [CrossRef] [PubMed]
62. Xue, W.J.; Huang, D.L.; Li, J.; Zeng, G.M.; Deng, R.; Yang, Y.; Chen, S.; Li, Z.H.; Gong, X.M.; Li, B. Assembly of AgI nanoparticles and ultrathin g-C_3N_4 nanosheets codecorated Bi_2WO_6 direct dual Z-scheme photocatalyst: An efficient, sustainable and heterogeneous catalyst with enhanced photocatalytic performance. *Chem. Eng. J.* **2019**, *373*, 1144–1157. [CrossRef]
63. Nguyen, M.D.; Nguyen, T.B.; Thamilselvan, A.; Nguyen, T.G.; Kuncoro, E.P.; Doong, R.A. Fabrication of visible-light-driven tubular F, P-codoped graphitic carbon nitride for enhanced photocatalytic degradation of tetracycline. *J. Environ. Chem. Eng.* **2021**, *10*, 106905. [CrossRef]
64. Su, Y.Y.; Yang, Y.X.; Jiang, W.X.; Han, J.G.; Guo, H. A novel strategy of peracetic acid activation by dielectric barrier discharge plasma for bisphenol a degradation: Feasibility, mechanism and active species dominant to degradation pathway. *Chem. Eng. J.* **2023**, *476*, 146469. [CrossRef]
65. Jiang, W.X.; Han, J.G.; Guo, H. Highlight the plasma-generated reactive oxygen species (ROSs) dominant to degradation of emerging contaminants based on experiment and density functional theory. *Sep. Purif. Technol.* **2024**, *330*, 125309. [CrossRef]
66. Kaviyarasu, K.; Magdalane, C.M.; Manikandan, E.; Jayachandran, M.; Maaza, M. Well-Aligned Graphene Oxide Nanosheets Decorated with Zinc Oxide Nanocrystals for High Performance Photocatalytic Application. *Int. J. Nanosci.* **2015**, *14*, 8064. [CrossRef]
67. Khan, M.H.; Bae, H.K.; Jung, J.Y. Tetracycline degradation by ozonation in the aqueous phase: Proposed degradation intermediates and pathway. *J. Hazard. Mater.* **2010**, *181*, 659–665. [CrossRef]
68. Song, C.; Guo, B.B.; Sun, X.F.; Wang, S.G.; Li, Y.T. Enrichment and degradation of tetracycline using three-dimensional graphene/MnO_2 composites. *Chem. Eng. J.* **2019**, *358*, 1139–1146. [CrossRef]
69. Cheng, J.; Xie, Y.R.; Wei, Y.; Xie, D.R.; Sun, W.B.; Zhang, Y.; Li, M.H.; An, J.T. Degradation of tetracycline hydrochloride in aqueous via combined dielectric barrier discharge plasma and Fe–Mn doped AC. *Chemosphere* **2022**, *286*, 131841. [CrossRef]
70. Ren, J.Y.; Song, H.Z.; Guo, H.; Yao, Z.Z.; Wei, Q.; Jiao, K.Q.; Li, Z.Y.; Zhong, C.C.; Wang, J.; Zhen, Y.Z. Removal of chloramphenicol in water by an improved water falling film dielectric barrier discharge reactor: Performance, mechanism, degradation pathway and toxicity evaluation. *J. Clean. Prod.* **2021**, *325*, 129332. [CrossRef]

toxics

MDPI

Article

Degradation of Butylated Hydroxyanisole by the Combined Use of Peroxymonosulfate and Ferrate(VI): Reaction Kinetics, Mechanism and Toxicity Evaluation

Peiduan Shi [1,†], Xin Yue [1,†], Xiaolei Teng [1,*], Ruijuan Qu [1], Ahmed Rady [2], Saleh Maodaa [2], Ahmed A. Allam [3], Zunyao Wang [1] and Zongli Huo [4,*]

1 State Key Laboratory of Pollution Control and Resources Reuse, School of the Environment, Nanjing University, Nanjing 210023, China; peiduan_shi@163.com (P.S.); point_xinyue@163.com (X.Y.); quruijuan0404@nju.edu.cn (R.Q.); wangzy@nju.edu.cn (Z.W.)

2 Department of Zoology, College of Science, King Saud University, P.O. Box 2455, Riyadh 11451, Saudi Arabia; jajarem@ksu.edu.sa (A.R.); maodaa_28@yahoo.com (S.M.)

3 Department of Zoology, Faculty of Science, Beni-Suef University, Beni-Suef 65211, Egypt; allam1081981@yahoo.com

4 Jiangsu Provincial Center for Disease Control and Prevention, No. 172 Jiangsu Road, Nanjing 210009, China

* Correspondence: tengxiaolei@smail.nju.edu.cn (X.T.); huozl@jscdc.cn (Z.H.)

† These authors contributed equally to this work.

Abstract: Butylated hydroxyanisole (BHA), a synthetic phenolic antioxidant (SPA), is now widely present in natural waters. To improve the degradation efficiency of BHA and reduce product toxicity, a combination of peroxymonosulfate (PMS) and Ferrate(VI) (Fe(VI)) was used in this study. We systematically investigated the reaction kinetics, mechanism and product toxicity in the degradation of BHA through the combined use of PMS and Fe(VI). The results showed that PMS and Fe(VI) have synergistic effects on the degradation of BHA. The effects of operational factors, including PMS dosage, pH and coexisting ions (Cl^-, SO_4^{2-}, HCO_3^-, K^+, NH_4^+ and Mg^{2+}), and different water matrices were investigated through a series of kinetic experiments. When T = 25 °C, the initial pH was 8.0, the initial BHA concentration was 100 μM, the initial concentration ratio of $[PMS]_0$:$[Fe(VI)]_0$:$[BHA]_0$ was 100:1:1 and the degradation rate could reach 92.4% within 30 min. Through liquid chromatography time-of-flight mass spectrometry (LC-TOF-MS) identification, it was determined that the oxidation pathway of BHA caused by PMS/Fe(VI) mainly includes hydroxylation, ring-opening and coupling reactions. Density functional theory (DFT) calculations indicated that •OH was most likely to attack BHA and generate hydroxylated products. The comprehensive comparison of product toxicity results showed that the PMS/Fe(VI) system can effectively reduce the environmental risk of a reaction. This study contributes to the development of PMS/Fe(VI) for water treatment applications.

Keywords: ferrate; peroxymonosulfate; butylated hydroxyanisole; reaction pathways; toxicity evaluation

Citation: Shi, P.; Yue, X.; Teng, X.; Qu, R.; Rady, A.; Maodaa, S.; Allam, A.A.; Wang, Z.; Huo, Z. Degradation of Butylated Hydroxyanisole by the Combined Use of Peroxymonosulfate and Ferrate(VI): Reaction Kinetics, Mechanism and Toxicity Evaluation. *Toxics* **2024**, *12*, 54. https://doi.org/10.3390/toxics12010054

Academic Editor: Mark Taggart

Received: 30 November 2023
Revised: 7 January 2024
Accepted: 9 January 2024
Published: 10 January 2024

1. Introduction

Butylated hydroxyanisole (BHA), a synthetic phenolic antioxidant (SPA), is used as an additive and preservative in a wide range of food, packaging, pharmaceutical, cosmetic and other product categories [1]. Due to the broad application of BHA, it has been widely detected in water, with the highest detected concentrations reaching 17.4 ng/g and 135 ng/L in wastewater treatment plants in China and Spain, respectively [2,3]. The widespread presence of BHA has led to growing concerns about its impact on environmental health [4–6]. Toxicological studies have indicated that BHA may have adverse effects on developmental and reproductive systems, can exert estrogenic or antiestrogenic activities that interfere with the endocrine system, and can induce adipogenesis [7,8]. BHA is acutely and developmentally toxic to zebrafish, leading to organogenesis retardation in the early embryonic development of zebrafish [9]. Dietary exposure to BHA can also lead to pregastric papilloma and squamous cell carcinoma and induce pregastric tumors in rodents [10]. In addition, Some BHA metabolites may pose a health risk as they were reported to generate peroxides in mice and rats and may induce cellular DNA damage or have the capacity to be cancer initiators [2].

Current methods of BHA removal from water are mainly focused on photolysis, ozonation and chlorination. However, chlorination and ozonation are prone to produce toxic disinfection byproducts, and photoreaction devices are complex and expensive [11–13]. Advanced oxidation technology based on peroxymonosulfate (PMS), as a new oxidation technology, has attracted attention in the field of organic pollutant treatment due to its high redox potential, long half-life and wide application range [14–16]. The key to PMS-based oxidation is the formation of highly reactive species. The general methods for activating PMS include the use of external energy (such as heat, ultraviolet light radiation, ultrasound and microwave), chemical activators and transition metals or metal oxides [17–21]. However, $SO_4^{\bullet-}$ is also susceptible to the influence of aqueous components (e.g., halides and natural organic matter), leading to the generation of some potentially toxic byproducts and unwanted oxidant consumption [22,23]. PMS-based advanced oxidation processes have been reported to produce more organochlorine acids, thereby increasing the toxicity of wastewater to mammalian cells [24].

In recent years, Fe(VI), as a multifunctional oxidant, has been used in water treatment [25]. It has been reported that Fe(VI)-based approaches have also been shown to be highly cost-effective for water remediation because they can be used more than once in the treatment procedure [26]. And the coupling of Fe(VI) and other oxidants (e.g., hydrogen peroxide (H_2O_2) and PMS) can be used to treat a wide range of micropollutants [27,28]. The combination of PMS and Fe(VI) has been reported to synergistically improve the removal of fluoroquinolones [28,29]. Wu et al. used Fe(VI) and PMS to degrade atrazine and showed that there was indeed a synergistic effect between Fe(VI) and PMS [30,31]. Therefore, the combination of PMS and Fe(VI) may be a valuable water treatment process. However, the degradation kinetics and potential mechanisms of PMS/Fe(VI) for the treatment of pollutants are not clear, especially regarding the toxicity changes in the intermediates during the reaction process.

Therefore, this study presents a systematic investigation regarding the feasibility of BHA degradation through the combined use of PMS and Fe(VI). The BHA degradation efficiencies of PMS alone, Fe(VI) alone and PMS/Fe(VI) were compared. In addition, the effects of solution pH, common inorganic ions and different water matrices on the removal rate were examined. The main active species of the reaction system were identified by EPR tests, and this result was correlated with theoretical calculations. Thermodynamic calculations demonstrated a possible synergistic reaction between PMS and Fe(VI). Detailed comparative analyses of the type, content and toxicity of the degradation products in the PMS/Fe(VI) and single-PMS systems were carried out with liquid chromatography–tandem mass spectrometry (LCMS/MS). This study provides new insights into the degradation of organic pollutants by PMS in combination with Fe(VI).

2. Materials and Methods

2.1. Chemicals and Water Samples

BHA, containing 90% 3-BHA and 10% 2-BHA, was obtained from Hebei Jijie Biotechnology Co., Ltd. (Hebei, China). Solid potassium ferrate (K_2FeO_4, >98%) was chemically synthesized by the wet method [32]. 5,5-dimethyl-1-pyrroline-N-oxide (DMPO, purity >99%) was purchased from Aladdin Chemical Co., Ltd. (Shanghai, China). Other chemicals used in the experiments to determine the influence of water chemistry, including PMS (>99%), NaCl (>99%), Na_2SO_4 (>98%), $Na_2S_2O_3$ (>99%), $NaHCO_3$ (>99%), K_2SO_4 (>99%), $(NH_4)_2SO_4$ (>99%) and $MgSO_4$ (>99%), were obtained from Aladdin Chemistry Co., Ltd. (Shanghai, China) and used directly without further purification.

Tap water (TW), synthetic wastewater (SW), secondary clarifier effluent (SCE) and river water (RW) were used to examine the effect of water matrices on PMS/Fe(VI). Synthetic wastewater was prepared from moderately hard freshwater according to the composition of local industrial wastewater, which contains NaCl (1000 mg L^{-1}), citric acid (50 mg L^{-1}), ascorbic acid (30 mg L^{-1}), D(+)-sucrose (100 mg L^{-1}) and Na_2HPO_4 (230 mg L^{-1}). Secondary clarifier effluent and river water were collected from the Wulongkou sewage treatment plant (Zhengzhou, China) and Jiuxiang River (Nanjing, China), respectively. The main physicochemical parameters of the four water samples are listed in Table S1.

2.2. Experimental Procedures

The oxidative degradation experiments were conducted in 50 mL conical flasks. Then, a 100 μM BHA solution and borate buffer of a certain pH were added. Next, the conical flasks were placed on magnetic stirrers, undergoing rapid stirring to achieve a uniform reaction. The reaction was started when a certain amount of oxidant was added. In total, 1 mL of the reaction solution was removed and put into a 2 mL plastic centrifuge tube preadded with 200 μL of methanol and 100 μL of 0.2 M $Na_2S_2O_3$ to terminate the reaction. The reaction temperature was 25 °C, and all experiments were performed in triplicate.

2.3. Analytical Methods

The BHA content was detected by means of a PerkinElmer liquid chromatograph (USA), the Flexar model, under a UV detector. The detailed HPLC setups can be found in Text S1. Possible intermediates in the reaction system were detected using an Agilent 1260 infinity HPLC system tandem. The detailed LC-MS/MS setup can be found in Text S2. The reactive oxygen species were detected by an electron paramagnetic resonance spectrometer (EPR; MiniScope MS 5000, Freiberg Instruments, Germany).

2.4. Theory Calculations

Density functional theory calculations were run on Gaussian 09 software. And the theory level used for the calculation was B3LYP/6-311G**. The geometric optimization and frequency calculations of reaction types, including reactants (REs), transition states (TSs) and products (PRs), were carried out. The Gibbs free energy (G) was obtained from the Gaussian output file [33–36]. The transition state had only one imaginary frequency, which was further verified by the intrinsic reaction coordinate (IRC). By contrast, a fully optimized configuration of the reactants and products had no imaginary frequency. To study the aqueous phase reaction, the Solvation Density Model was used to consider the solvent effect.

2.5. Toxicity Evaluation

The acute toxicity of BHA and its reaction intermediates to fish, Daphnia and green algae was evaluated using ECOSAR software [37]. The software has been approved by the Organization for Economic Cooperation and Development (OECD), the Environmental Protection Agency of the United States (US EPA) and the European Union (EU).

3. Results and Discussion

3.1. Degradation of BHA

The experiment of oxidative degradation of BHA by PMS alone was carried out under the conditions of $[BHA]_0$ = 100 μM, pH = 8.0 and T = 25 °C. As shown in Figure 1a, when the proportion of PMS was constantly increased, the improvement in degradation rates was not significant enough. Within 30 min, when the ratio of $[PMS]_0$:$[BHA]_0$ was 10:1, 50:1, 100:1, 150:1 and 200:1, the degradation rates of BHA were 5.9%, 9.3%, 15.1%, 29.7% and 39.8%, respectively; this was because PMS did not readily generate more $SO_4^{\bullet-}$ without the influence of external energy, which was consistent with the results from previous studies [38].

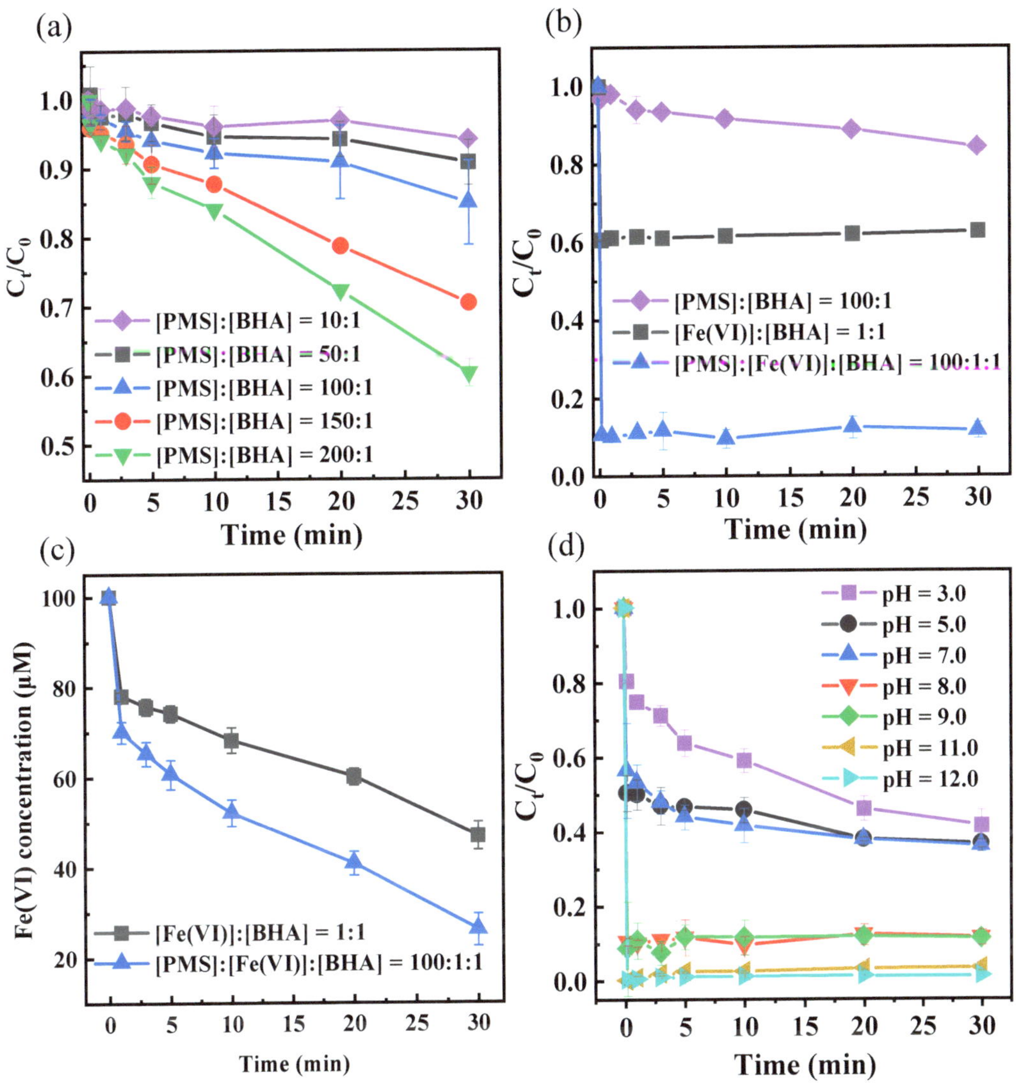

Figure 1. (**a**) Effect of PMS dosage on the degradation of BHA. Experimental conditions: $[BHA]_0$ = 100 μM, pH = 8.0 and T = 25 °C. (**b**) Degradation of BHA by different processes. Experimental conditions: $[BHA]_0$ = 100 μM, pH = 8.0 BS T = 25 °C. (**c**) Change in Fe(VI) concentration during degradation of BHA by Fe(VI) and PMS/Fe(VI). Experimental conditions: $[BHA]_0$ = 100 μM, pH = 8.0 and T = 25 °C. (**d**) Effect of different pH values on degradation of BHA in PMS/Fe(VI). Experimental conditions: $[BHA]_0$ = 100 μM, $[PMS]_0$:$[Fe(VI)]_0$:$[BHA]_0$ = 100:1:1 and T = 25 °C.

As shown in Figure 1b, at $[PMS]_0$:$[BHA]_0$ = 100:1, the degradation rate of BHA was only 15.1% in 30 min. In contrast, while the Fe(VI) process alone rapidly degraded 37.4% of the BHA in 30 s, the BHA was barely oxidized for the next 30 min. The effect on the degradation of BHA when PMS and Fe(VI) were involved in the reaction simultaneously was investigated. When $[PMS]_0$:$[Fe(VI)]_0$:$[BHA]_0$ = 100:1:1, the degradation trend was essentially the same as for the Fe(VI) system alone, but the degradation rate was signifi cantly faster, reaching 92.4%. The degradation efficiency of BHA using the PMS/Fe(VI) process was improved by 29.3% compared to that when the PMS system and Fe(VI) system were combined. Changes in the concentration of Fe(VI) were measured to explain why the addition of PMS to Fe(VI) increased the degradation of BHA, as shown in Figure 1c. It is clear that the presence of PMS accelerates the reduction in Fe(VI), suggesting that the addition of PMS promotes the rapid reaction of Fe(VI) with BHA. These results confirmed that the combined process of Fe(VI) and PMS had a stronger synergistic influence on the degradation of BHA.

3.2. Effect of pH on the Reaction

The pH value plays an essential role in the oxidation and degradation process in pollutants [39]. Thus, the initial pH of the reaction solutions was varied from 3.0 to 12.0 to examine its effect on the degradation of BHA. The experimental results are shown in Figure 1d. The degradation rate of BHA increased with an increasing pH value. The reaction degradation rate reached 100% when the pH value was greater than 11.0.

This result was a combination of multiple factors. First, the pK_a value of BHA in aqueous solution is 8.8. When the pH value is greater than 8.8, BHA presents in an ion state, which has a higher activity than when in a molecular state [40]. Second, Fe(VI) has three pK_a values in aqueous solution, 1.6, 3.5 and 7.3, corresponding to H_2FeO_4, $HFeO_4^-$, and FeO_4^{2-}, respectively (Equations (1)–(3)) [41]. With an increasing solution pH, Fe(VI) gradually changed from $HFeO_4^-$ under acidic conditions to FeO_4^{2-} under alkaline conditions. Although the redox potential of Fe(VI) decreased from 2.2 V to 0.7 V, the stability increased and the self-decomposition rate decreased; then, the reaction of Fe(VI) with pollutants was more effective under alkaline conditions [42,43]. In addition, Fe^{2+} (generated through Fe(VI) decomposition) can react with PMS to produce more $SO_4^{\bullet-}$ ((Equation (4)) [44]. However, under acidic conditions, excess H^+ combines with HSO_5^- to form hydrogen bonds, preventing this reaction from proceeding [30]. Due to the combined effects of the above reasons, the degradation rate of BHA showed a trend of pH-related variation.

$$H_3FeO_4^+ \leftrightharpoons H^+ + H_2FeO_4 \ (pK_a = 1.6) \tag{1}$$

$$H_2FeO_4 \leftrightharpoons H^+ + HFeO_4^- \ (pK_a = 3.5) \tag{2}$$

$$HFeO_4^- \leftrightharpoons H^+ + FeO_4^{2-} \ (pK_a = 7.3) \tag{3}$$

$$HSO_5^- + Fe^{2+} \rightarrow SO_4^{\bullet-} + Fe^{3+} + OH^- \tag{4}$$

3.3. Effect of Water Constituents

It has been reported that the degradation efficiency of pollutants may be affected by the presence of various inorganic ions in natural waters [45]. The effect of several common inorganic anion ions and cations in natural water on the degradation of BHA was investigated. Three anions (Cl^-, SO_4^{2-} and HCO_3^-) and three cations (K^+, NH_4^+ and Mg^{2+}) were included in the experiment (Figure 2a,b).

In previous studies, Fe(VI) degraded micropollutants, such as chlorophene, 4-tert-butylphenol and chlorpyrifos, and Cl^- had no significant effect on the Fe(VI) degradation of pollutants [43,45]. However, Cl^- can compete with contaminants for $^\bullet OH$ and $SO_4^{\bullet-}$ in PMS/Fe(VI) systems [46,47]. Therefore, as the concentration of Cl^- increases, the BHA degradation rate of the reaction decreases. Compared with Cl^-, the inhibition was more pronounced with an increasing HCO_3^- concentration. This is due to the ability of HCO_3^- to quench $^\bullet OH$ and $SO_4^{\bullet-}$ [48]. And HCO_3^- has a certain inhibitory effect on the oxidation of Fe (VI) [49]. SO_4^{2-}, K^+, NH_4^+ and Mg^{2+} had almost no influence on the degradation rate of BHA. Similar results were also reported in previous studies [28,50]. This study also tested the effects of humic acid at different concentrations on the degradation of BHA by PMS/Fe(VI). As shown in Figure 2c, the addition of HA inhibited the efficiency of the PMS/Fe(VI) degradation of BHA, and the inhibition effect was more obvious at high concentrations of HA. This is due to the ability of HA to competitively consume the oxidant, resulting in a decrease in the removal of the target pollutant [51].

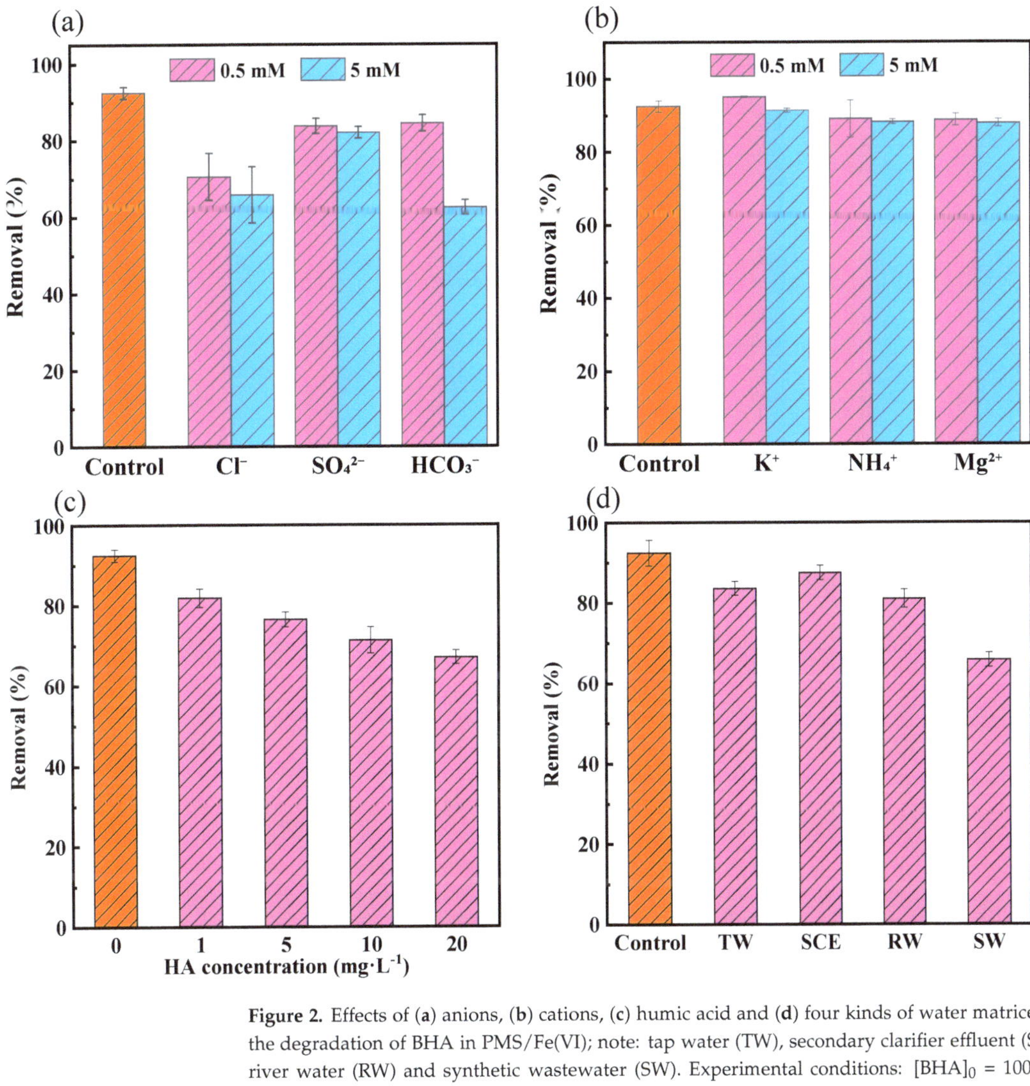

Figure 2. Effects of (**a**) anions, (**b**) cations, (**c**) humic acid and (**d**) four kinds of water matrices on the degradation of BHA in PMS/Fe(VI); note: tap water (TW), secondary clarifier effluent (SCE), river water (RW) and synthetic wastewater (SW). Experimental conditions: $[BHA]_0$ = 100 μM, $[PMS]_0$:$[Fe(VI)]_0$:$[BHA]_0$ = 100:1:1, pH = 8.0, T = 25 °C and reaction time = 30 min.

3.4. Effect of Water Matrices

In addition to ions and HA, the composition of natural waters are very complex. It is crucial to evaluate the feasibility of using PMS/Fe(VI) for BHA removal in real natural waters. The degradation of BHA by PMS/Fe(VI) in four different water matrices (tap water (TW), secondary clarifier effluent (SCE), river water (RW) and synthetic wastewater (SW)) is shown in Figure 2d. TW, SCE and RW had a slight inhibitory effect on the degradation of BHA in the PMS/Fe (VI) system. The lowest degradation rate of BHA was observed in

SW. This is due to the fact that the highest content of Cl^- and TOC in SW will compete with the target compounds to consume the oxidant, which is consistent with the findings of previous research studies [35]. Therefore, the amount of PMS/Fe(VI) needs to be increased in organic-rich wastewater. Overall, PMS/Fe(VI) can be used as an effective technique to degrade coexisting BHA in natural waters.

3.5. Identification of Main Reactive Species

According to the results of previous studies, in the PMS system, $SO_4^{\bullet-}$ and $^{\bullet}OH$ were the main species, while in the Fe(VI) system, $^{\bullet}OH$ was produced by the self-decay of Fe(VI) [52,53]. Potentially active species were identified using EPR techniques. As shown in Figure 3, adding DMPO to the PMS system alone did not result in spikes. In contrast, a distinct quartet spectrum with an intensity ratio of 1:2:2:1 and a splitting constant of $\alpha N = \alpha H = 14.9$ G was present in the PMS/Fe(VI) and Fe(VI) systems, which was considered to be the signaling peak of the hydroxyl radical. In addition, the intensity of the DMPO-$^{\bullet}OH$ peak in the PMS/Fe(VI) system was significantly higher than that in the Fe(VI) system, which also indicated a synergistic effect between PMS and Fe(VI). At the same time, $SO_4^{\bullet-}$ was also produced in the PMS/Fe(VI) oxidation system, with a center of symmetry of g = 2.005 and an intensity ratio of 1:1:1:1:1:1 for the six-line signals [54]. Therefore, the accelerated degradation of BHA in the PMS/Fe(VI) oxidation system was attributed to the synergistic effect between PMS and Fe(VI) and the generation of reactive radicals.

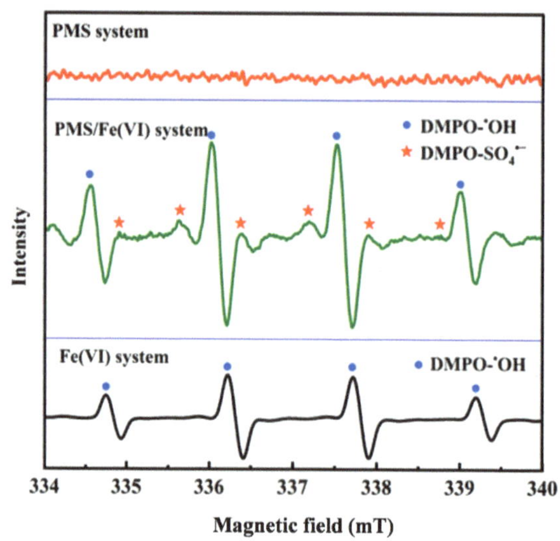

Figure 3. EPR spectra of hydroxyl and sulfate radicals trapped by DMPO in deionized water.

3.6. The Reaction Pathway

By using LC/MS-MS, it was found that the reaction intermediate species of BHA during degradation in the single-PMS system and PMS/Fe(VI) system were the same. A total of 13 products (P1–P13) were identified, and the detailed mass spectral information and MS/MS fragmentation patterns are shown in Table S2 and Figure S1. The differences between the calculated and measured masses never exceeded 5.0 ppm, which confirmed that the proposed molecular formulae were highly reliable. Based on these identified products, as shown in Figure 4, three different transformation pathways for the oxidative degradation of BHA by PMS and PMS/Fe(VI) were proposed, which mainly included hydroxylated, ring-opening and coupling reactions.

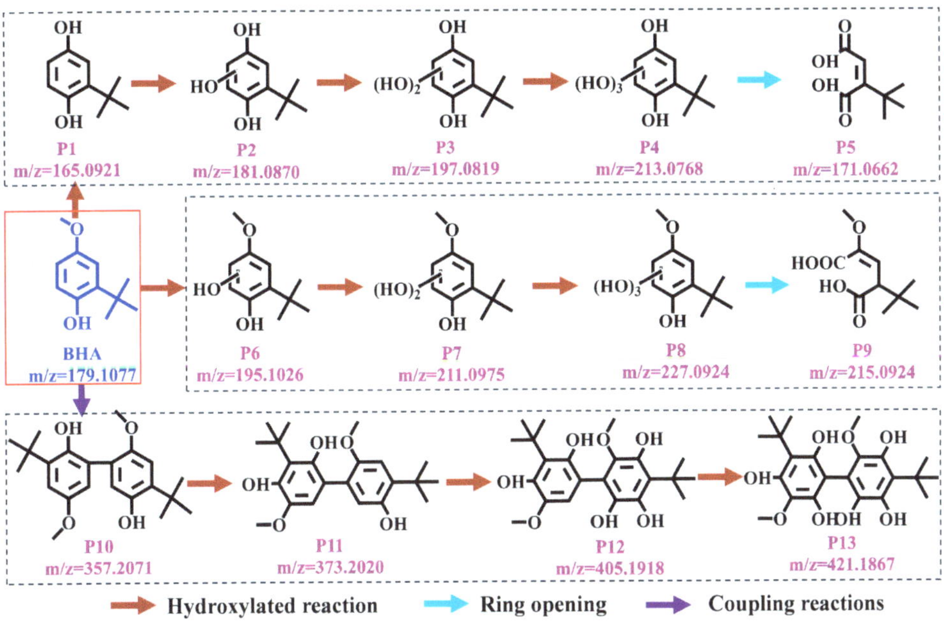

Figure 4. The proposed reaction pathway of BHA oxidation by PMS and PMS/Fe(VI).

In pathway I, $^{\bullet}$OH replaced the methoxy group of BHA to produce the hydroxylated product P1 and methyl alcohol. In pathway II, $^{\bullet}$OH attacked the aromatic C atom of BHA to generate a hydroxylation product P6. P1 and P6 continued the hydroxylation reaction under the attack of $^{\bullet}$OH to form polyhydroxylated products (P2→P3→P4 and P7→P8). Furthermore, in addition to the $^{\bullet}$OH attack, these polyhydroxylated products can also be generated by a mechanism of singlet oxygen transfer via Fe(VI) oxidation [34]. It has been reported that two adjacent or opposite hydroxyl groups on the benzene ring can be easily oxidized to form ketone groups [52]. Thus, the hydroxylation products P1–P4 and P6–P8 were oxidized into quinone intermediates. These intermediates underwent a ring-opening reaction to produce the carboxylic acid intermediates P5 and P9.

Another important pathway for the oxidation of phenol by Fe(VI) and PMS was the coupling reaction. Phenolic compounds are susceptible to the formation of phenoxy radicals by chemical oxidizers through hydrogen extraction or single-electron transfer [55]. Fe(VI) and PMS attack the electron-rich phenolic groups of phenol to generate the corresponding phenoxy radicals [45,50]. These unpaired-electron phenoxy radicals are not stable and tend to couple to each other, eventually forming stabilized products with C-C or C-O-C bonds. The polymerization product P10 was observed during the degradation of BHA by single PMS and PMS/Fe(VI). P10 was attacked by $^{\bullet}$OH to continue the hydroxylation reaction to form polyhydroxylated polymerization products (P11→P12→P13). In addition, DFT theory was used to calculate the potential attack sites of BHA, and a detailed analysis was conducted, the results of which are presented in Text S3 and Figure S2. Compared with Fe(VI) and $SO_4^{\bullet -}$, the activation energy of the $^{\bullet}$OH attack on BHA is lower, and the corresponding hydroxylation products are easily generated. This was consistent with the degradation pathways which showed that the degradation intermediates were predominantly hydroxylated products.

3.7. Toxicity Evaluation

It is critical to give attention to the ecological risks in water bodies after treatment. BHA degrades into the same product types within PMS and PMS/Fe(VI) systems; however, the abundance of the products varies greatly between the two systems. Therefore, a comparison of the changes in product abundance and toxicity in the two systems is extremely necessary. The acute toxicity of the products of each pathway was predicted by the ECOSAR software, as shown in Figure 5. BHA was found to be extremely toxic to fish, Daphnia and green algae. The toxicity of the transformation intermediates showed an overall decreasing trend compared to the toxicity of the parent compounds. In pathways I and II, P1–P4 and P6–P8 remained at toxic levels for fish, Daphnia and algae. However, the aquatic toxicity to fish and Daphnia was generally reduced. The end products of both pathways, P5 and P9, were virtually nontoxic. The molecular weight of the generated polymerization coupling products (P10–P13) is high, and therefore, accurate toxicity predictions cannot be obtained by ECOSAR. According to the available studies, polymerization products of large molecules have low solubility and low bioavailability properties [56]. As a result, they can be separated from water by simple physical means, which facilitated the control of environmental risks.

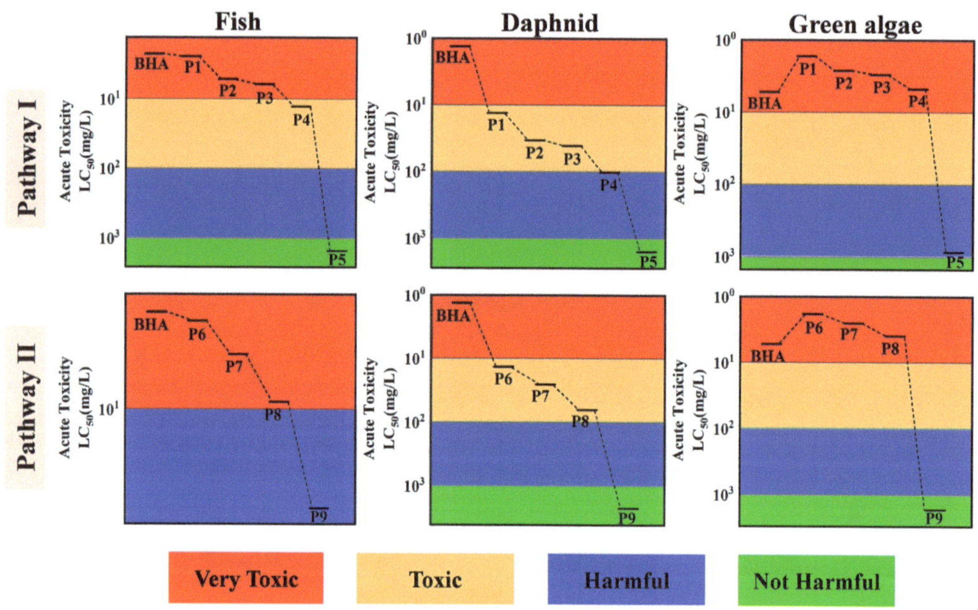

Figure 5. PMS/Fe(VI) oxidation BHA reaction intermediates' toxicity evaluation heatmap.

In the degradation of BHA, the presence of intermediates cannot be ignored. Analyzing the toxicity of a single product is not representative of assessing changes in the toxicity of the entire reaction system. It is more informative to assess the toxicity of the reaction system in terms of changes in the abundance of all products. In pathway I, although the P1–P4 toxicity level was relatively high, the overall concentration decreased with reaction time in the PMS/Fe(VI) system, and the peak area at the end of the reaction was similar to that of the PMS system (Figure 6a). As the end product of pathway I, P5 was nontoxic, and the peak area of the PMS/Fe(VI) system was significantly higher than that of the PMS system (Figure 6b), therefore, the toxicity of PMS/Fe(VI) in pathway I was lower than that of PMS.

In pathway II, the P6–P8 toxicity level remained relatively high, but in the PMS/Fe(VI) system, the peak area decreased with reaction time (Figure 6c). In contrast, as the reaction progressed, the peak areas of P6–P8 in the PMS system showed a tendency to increase and then decrease, and eventually, the peak area of P6–P8 in the PMS/Fe(VI) system was half of that in the PMS system. P9 was the final product of pathway II and was harmless to three aquatic organisms. The PMS/Fe(VI) system produced more P9 during the reaction (approximately 3 times more than the PMS system) (Figure 6d).

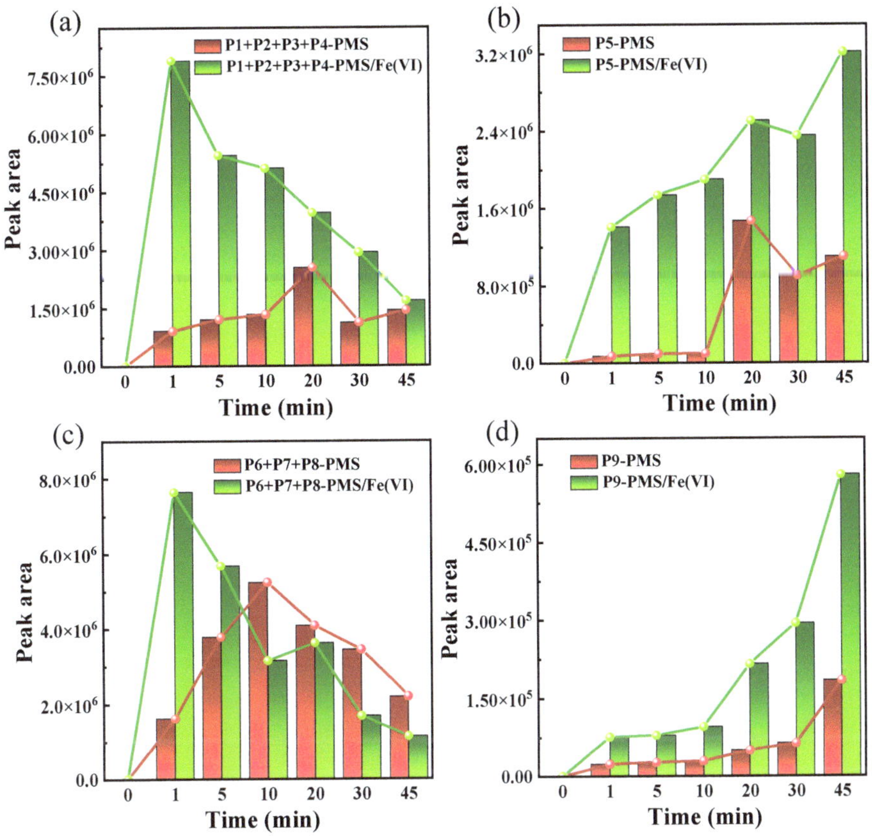

Figure 6. Changes in abundance of (**a**) P1-P4, (**b**) P5, (**c**) P6-P8 and (**d**) P9 of BHA oxidation by PMS/Fe(VI) and PMS alone. PMS/Fe(VI) oxidation experiment: $[BHA]_0$ = 100 μM, $[PMS]_0$:$[Fe(VI)]_0$:$[BHA]_0$ = 100:1:1, pH = 8.0, T = 25 °C. PMS oxidation experiment: $[BHA]_0$ = 100 μM, $[PMS]_0$:$[BHA]_0$ = 200:1, T = 50 °C.

In summary, the toxicity of the combined PMS/Fe(VI) system was lower than that of the single-PMS system in all cases, suggesting that the combined approach can reduce the toxicity during BHA degradation. Overall, this PMS/Fe(VI) technology can be a promising and safe method for treating contaminants.

4. Conclusions

The results showed that PMS/Fe(VI) had a synergistic effect on the degradation of BHA compared to a single-PMS system. When T = 25 °C, the initial pH was 8.0 and the initial BHA concentration was 100 μM, the initial concentration ratio of $[PMS]_0$:$[Fe(VI)]_0$:$[BHA]_0$ was 100:1:1, and the degradation rate could reach 92.4% within 30 min. In addition, the degradation efficiency of BHA increased with increasing solution pH in the PMS/Fe (VI) system. K^+, NH_4^+ and Mg^{2+} had a promoting effect on the reaction. Based on the intermediates identified with LC-QTOF-MS, the oxidation of BHA by PMS/Fe(VI) included hydroxylation, ring-opening and coupling reactions. The main active species of the reaction system were identified with EPR tests, and this result was correlated with theoretical calculations. Based on the results of DFT calculations, it can be seen that BHA was more susceptible to attack by hydroxyl radicals, generating hydroxylation products A comprehensive comparison of the toxicity was performed, and the PMS/Fe(VI) system reduced the environmental risk compared to the single-PMS system. Importantly, the results of these studies may provide useful information for the removal of pollutants in real water bodies and may also deepen the understanding of the behavior of the PMS/Fe oxidation system in transforming pollutants. The above results indicate that the combined treatment of PMS and Fe(VI) is a safe and efficient route for BHA degradation. This study provides more favorable results that support the use of this combined oxidation method for water treatment.

Supplementary Materials: The following supporting information can be downloaded at: https://www.mdpi.com/article/10.3390/toxics12010054/s1, Text S1: Analytical method of HPLC; Text S2: Analytical method of LC-MS; Text S3: Theoretical Analysis; Table S1: Water quality parameters of natural water samples*; Table S2: Accurate mass measurements of the degradation products of BHA during oxidation as determined by LC- MS. Noted that Rt means retention time; Figure S1: Product ion spectra of reaction intermediates during PMS/Fe(VI) oxidation of BHA; Figure S2: Energy barriers for the reaction of (a) Fe(VI), (b) $SO_4^{\bullet-}$ and (c) $^{\bullet}OH$ with BHA calculated at the b3lyp/6-311 g(d,p) level.

Author Contributions: Conceptualization, A.A.A., Z.W. and Z.H.; methodology, P.S. and A.R.; validation, X.Y., A.R. and S.M.; formal analysis, X.T.; investigation, X.T. and X.Y.; data curation, A.R., S.M. and A.A.A.; writing—original draft preparation, X.T. and P.S.; writing—review and editing, R.Q.; visualization, R.Q., S.M. and A.A.A.; supervision, Z.W. and Z.H.; project administration, A.R., S.M., Z.W. and Z.H.; funding acquisition, A.A.A., Z.W. and Z.H. All authors have read and agreed to the published version of the manuscript.

Funding: This research was funded by the National Natural Science Foundation of China (No. 22076076), the Medical Scientific Research Project of Jiangsu Provincial Health Commission (M2021028) the Social Development Projects of Key R&D Plans in Jiangsu Province (BE2021740), the Researchers Supporting Project (number RSPD2024R691), King Saud University, Riyadh, Saudi Arabia and College Students Innovative Entrepreneurial Training Plan Program of Nanjing University.

Institutional Review Board Statement: Not applicable.

Informed Consent Statement: Not applicable.

Data Availability Statement: Data will be made available on request.

Conflicts of Interest: The authors declare no conflicts of interest.

References

1. Zhang, R.; Li, C.; Li, Y.; Cui, X.; Ma, L.Q. Determination of 2,6-di-tert-butyl-hydroxytoluene and its transformation products in indoor dust and sediment by gas chromatography–mass spectrometry coupled with precolumn derivatization. *Sci. Total Environ.* **2018**, *619–620*, 552–558. [CrossRef] [PubMed]
2. Liu, R.; Song, S.; Lin, Y.; Ruan, T.; Jiang, G. Occurrence of synthetic phenolic antioxidants and major metabolites in municipal sewage sludge in China. *Environ. Sci. Technol.* **2015**, *49*, 2073–2080. [CrossRef]
3. Rodil, R.; Quintana, J.B.; Basaglia, G.; Pietrogrande, M.C.; Cela, R. Determination of synthetic phenolic antioxidants and their metabolites in water samples by downscaled solid-phase extraction, silylation and gas chromatography–mass spectrometry. *J. Chromatogr. A* **2010**, *1217*, 6428–6435. [CrossRef]
4. Wang, Y.; Li, X.; Sun, X. The transformation mechanism and eco-toxicity evaluation of butylated hydroxyanisole in environment. *Ecotoxicol. Environ. Saf.* **2022**, *231*, 113179. [CrossRef] [PubMed]
5. Soubra, L.; Sarkis, D.; Hilan, C.; Verger, P. Dietary exposure of children and teenagers to benzoates, sulphites, butylhydroxyanisol (BHA) and butylhydroxytoluen (BHT) in Beirut (Lebanon). *Regul. Toxicol. Pharmacol.* **2007**, *47*, 68–77. [CrossRef] [PubMed]
6. Lee, M.-R.; Lin, C.-Y.; Li, Z.-G.; Tsai, T.-F. Simultaneous analysis of antioxidants and preservatives in cosmetics by supercritical fluid extraction combined with liquid chromatography–mass spectrometry. *J. Chromatogr. A* **2006**, *1120*, 244–251. [CrossRef] [PubMed]
7. Sun, Z.; Yang, X.; Liu, Q.S.; Li, C.; Zhou, Q.; Fiedler, H.; Liao, C.; Zhang, J.; Jiang, G. Butylated hydroxyanisole isomers induce distinct adipogenesis in 3T3-L1 cells. *J. Hazard. Mater.* **2019**, *379*, 120794. [CrossRef]
8. Yang, X.; Song, W.; Liu, N.; Sun, Z.; Liu, R.; Liu, Q.S.; Zhou, Q.; Jiang, G. Synthetic Phenolic Antioxidants Cause Perturbation in Steroidogenesis in Vitro and in Vivo. *Environ. Sci. Technol.* **2018**, *52*, 850–858. [CrossRef]

9. Yang, X.; Sun, Z.; Wang, W.; Zhou, Q.; Shi, G.; Wei, F.; Jiang, G. Developmental toxicity of synthetic phenolic antioxidants to the early life stage of zebrafish. *Sci. Total Environ.* **2018**, *643*, 559–568. [CrossRef]
10. Felter, S.P.; Zhang, X.; Thompson, C. Butylated hydroxyanisole: Carcinogenic food additive to be avoided or harmless antioxidant important to protect food supply? *Regul. Toxicol. Pharmacol.* **2021**, *121*, 104887. [CrossRef]
11. Ye, Z.; Brillas, E.; Centellas, F.; Cabot, P.L.; Sirés, I. Electrochemical treatment of butylated hydroxyanisole: Electrocoagulation versus advanced oxidation. *Sep. Purif. Technol.* **2019**, *208*, 19–26. [CrossRef]
12. Rodil, R.; Quintana, J.B.; Cela, R. Oxidation of synthetic phenolic antioxidants during water chlorination. *J. Hazard. Mater.* **2012**, *199–200*, 73–81. [CrossRef] [PubMed]
13. Chu, W.; Lau, T. Ozonation of endocrine disrupting chemical BHA under the suppression effect by salt additive—With and without H_2O_2. *J. Hazard. Mater.* **2007**, *144*, 249–254. [CrossRef] [PubMed]
14. Xiao, S.; Cheng, M.; Zhong, H.; Liu, Z.; Liu, Y.; Yang, X.; Liang, Q. Iron-mediated activation of persulfate and peroxymonosulfate in both homogeneous and heterogeneous ways: A review. *Chem. Eng. J.* **2020**, *384*, 123265. [CrossRef]
15. Yang, Y.; Banerjee, G.; Brudvig, G.W.; Kim, J.-H.; Pignatello, J.J. Oxidation of Organic Compounds in Water by Unactivated Peroxymonosulfate. *Environ. Sci. Technol.* **2018**, *52*, 5911–5919. [CrossRef]
16. Zhou, Q.; Song, C.; Wang, P.; Zhao, Z.; Li, Y.; Zhan, S. Generating dual-active species by triple-atom sites through peroxymonosulfate activation for treating micropollutants in complex water. *Proc. Natl. Acad. Sci. USA* **2023**, *120*, e2300085120. [CrossRef]
17. Ghanbari, F.; Moradi, M. Application of peroxymonosulfate and its activation methods for degradation of environmental organic pollutants: Review. *Chem. Eng. J.* **2017**, *310*, 41–62. [CrossRef]
18. Antoniou, M.G.; de la Cruz, A.A.; Dionysiou, D.D. Degradation of microcystin-LR using sulfate radicals generated through photolysis, thermolysis and e−transfer mechanisms. *Appl. Catal. B Environ.* **2010**, *96*, 290–298. [CrossRef]
19. Mahamuni, N.N.; Adewuyi, Y.G. Advanced oxidation processes (AOPs) involving ultrasound for waste water treatment: A review with emphasis on cost estimation. *Ultrason. Sonochemistry* **2010**, *17*, 990–1003. [CrossRef]
20. Zhang, B.-T.; Zhang, Y.; Teng, Y.; Fan, M. Sulfate Radical and Its Application in Decontamination Technologies. *Crit. Rev. Environ. Sci. Technol.* **2014**, *45*, 1756–1800. [CrossRef]
21. Huang, W.; Xiao, S.; Zhong, H.; Yan, M.; Yang, X. Activation of persulfates by carbonaceous materials: A review. *Chem. Eng. J.* **2021**, *418*, 129297. [CrossRef]
22. von Gunten, U. Oxidation Processes in Water Treatment: Are We on Track? *Environ. Sci. Technol.* **2018**, *52*, 5062–5075. [CrossRef] [PubMed]
23. Lee, J.; von Gunten, U.; Kim, J.-H. Persulfate-Based Advanced Oxidation: Critical Assessment of Opportunities and Roadblocks. *Environ. Sci. Technol.* **2020**, *54*, 3064–3081. [CrossRef]
24. Du, Y.; Wang, W.-L.; Wang, Z.-W.; Yuan, C.-J.; Ye, M.-Q.; Wu, Q.-Y. Overlooked Cytotoxicity and Genotoxicity to Mammalian Cells Caused by the Oxidant Peroxymonosulfate during Wastewater Treatment Compared with the Sulfate Radical-Based Ultraviolet/Peroxymonosulfate Process. *Environ. Sci. Technol.* **2023**, *57*, 3311–3322. [CrossRef] [PubMed]
25. Teng, X.; Qi, Y.; Qin, C.; Tang, X.; Yan, C.; Wang, Z.; Qu, R. Mixed oxidation of chlorophene and 4-tert-butylphenol by ferrate(VI): Reaction kinetics, cross-coupling products and improved utilization efficiency of ferrate(VI). *Sep. Purif. Technol.* **2022**, *301*, 121902. [CrossRef]
26. Rai, P.K.; Lee, J.; Kailasa, S.K.; Kwon, E.E.; Tsang, Y.F.; Ok, Y.S.; Kim, K.-H. A critical review of ferrate(VI)-based remediation of soil and groundwater. *Environ. Res.* **2018**, *160*, 420–448. [CrossRef]
27. Luo, M.; Zhou, H.; Zhou, P.; Lai, L.; Liu, W.; Ao, Z.; Yao, G.; Zhang, H.; Lai, B. Insights into the role of in-situ and ex-situ hydrogen peroxide for enhanced ferrate(VI) towards oxidation of organic contaminants. *Water Res.* **2021**, *203*, 117548. [CrossRef]
28. Feng, M.; Cizmas, L.; Wang, Z.; Sharma, V.K. Synergistic effect of aqueous removal of fluoroquinolones by a combined use of peroxymonosulfate and ferrate(VI). *Chemosphere* **2017**, *177*, 144–148. [CrossRef]
29. Kralchevska, R.P.; Prucek, R.; Kolařík, J.; Tuček, J.; Machala, L.; Filip, J.; Sharma, V.K.; Zbořil, R. Remarkable efficiency of phosphate removal: Ferrate(VI)-induced in situ sorption on core-shell nanoparticles. *Water Res.* **2016**, *103*, 83–91. [CrossRef]
30. Wu, S.; Li, H.; Li, X.; He, H.; Yang, C. Performances and mechanisms of efficient degradation of atrazine using peroxymonosulfate and ferrate as oxidants. *Chem. Eng. J.* **2018**, *353*, 533–541. [CrossRef]
31. Sharma, V.K. Ferrate(VI) and ferrate(V) oxidation of organic compounds: Kinetics and mechanism. *Co-ord. Chem. Rev.* **2013**, *257*, 495–510. [CrossRef]
32. Feng, M.; Wang, X.; Chen, J.; Qu, R.; Sui, Y.; Cizmas, L.; Wang, Z.; Sharma, V.K. Degradation of fluoroquinolone antibiotics by ferrate(VI): Effects of water constituents and oxidized products. *Water Res.* **2016**, *103*, 48–57. [CrossRef]
33. Gao, Y.; Ji, Y.; Li, G.; An, T. Theoretical investigation on the kinetics and mechanisms of hydroxyl radical-induced transformation of parabens and its consequences for toxicity: Influence of alkyl-chain length. *Water Res.* **2016**, *91*, 77–85. [CrossRef] [PubMed]
34. Liu, M.; Wu, N.; Tian, B.; Zhou, D.; Yan, C.; Huo, Z.; Qu, R. Experimental and theoretical study on the degradation of Benzophenone-1 by Ferrate(VI): New insights into the oxidation mechanism. *J. Hazard. Mater.* **2022**, *425*, 127877. [CrossRef] [PubMed]
35. Tian, B.; Wu, N.; Pan, X.; Wang, Z.; Yan, C.; Sharma, V.K.; Qu, R. Ferrate(VI) oxidation of bisphenol E–Kinetics, removal performance, and dihydroxylation mechanism. *Water Res.* **2022**, *210*, 118025. [CrossRef]

36. Ma, J.; Minakata, D.; O'Shea, K.; Bai, L.; Dionysiou, D.D.; Spinney, R.; Xiao, R.; Wei, Z. Determination and Environmental Implications of Aqueous-Phase Rate Constants in Radical Reactions. *Water Res.* **2021**, *190*, 116746. [CrossRef]
37. Sanderson, H.; Johnson, D.J.; Wilson, C.J.; Brain, R.A.; Solomon, K.R. Probabilistic hazard assessment of environmentally occurring pharmaceuticals toxicity to fish, daphnids and algae by ECOSAR screening. *Toxicol. Lett.* **2003**, *144*, 383–395. [CrossRef]
38. Olmez-Hanci, T.; Arslan-Alaton, I.; Genc, B. Bisphenol A treatment by the hot persulfate process: Oxidation products and acute toxicity. *J. Hazard. Mater.* **2013**, *263*, 283–290. [CrossRef]
39. Teng, X.; Li, J.; Wang, J.; Liu, J.; Ge, X.; Gu, T. Effective degradation of atrazine in wastewater by three-dimensional electrochemical system using fly ash-red mud particle electrode: Mechanism and pathway. *Sep. Purif. Technol.* **2021**, *267*, 118661. [CrossRef]
40. Kaposi, A.D.; Wright, W.W.; Fidy, J.; Stavrov, S.S.; Vanderkooi, J.M.; Rasnik, I. Carbonmonoxy horseradish peroxidase as a function of ph and substrate: Influence of local electric fields on the optical and infrared spectra. *Biochemistry* **2001**, *40*, 3483–3491. [CrossRef]
41. Anquandah, G.A.K.; Sharma, V.K.; Knight, D.A.; Batchu, S.R.; Gardinali, P.R. Oxidation of Trimethoprim by Ferrate(VI): Kinetics, Products, and Antibacterial Activity. *Environ. Sci. Technol.* **2011**, *45*, 10575–10581. [CrossRef]
42. Li, C.; Li, X.; Graham, N. A study of the preparation and reactivity of potassium ferrate. *Chemosphere* **2005**, *61*, 537–543. [CrossRef]
43. Liu, H.; Chen, J.; Wu, N.; Xu, X.; Qi, Y.; Jiang, L.; Wang, X.; Wang, Z. Oxidative degradation of chlorpyrifos using ferrate(VI): Kinetics and reaction mechanism. *Ecotoxicol. Environ. Saf.* **2019**, *170*, 259–266. [CrossRef]
44. Nie, M.; Yan, C.; Xiong, X.; Wen, X.; Yang, X.; Lv, Z.; Dong, W. Degradation of chloramphenicol using a combination system of simulated solar light, Fe^{2+} and persulfate. *Chem. Eng. J.* **2018**, *348*, 455–463. [CrossRef]
45. Li, X.; Liu, M.; Wu, N.; Sharma, V.K.; Qu, R. Enhanced removal of phenolic compounds by ferrate(VI): Unveiling the Bi(III)-Bi(V) valence cycle with in situ formed bismuth hydroxide as catalyst. *Water Res.* **2024**, *248*, 120827. [CrossRef]
46. Liu, H.; Pan, X.; Chen, J.; Qi, Y.; Qu, R.; Wang, Z. Kinetics and mechanism of the oxidative degradation of parathion by Ferrate(VI). *Chem. Eng. J.* **2019**, *365*, 142–152. [CrossRef]
47. Zhang, R.; Yang, Y.; Huang, C.-H.; Li, N.; Liu, H.; Zhao, L.; Sun, P. UV/H_2O_2 and UV/PDS Treatment of Trimethoprim and Sulfamethoxazole in Synthetic Human Urine: Transformation Products and Toxicity. *Environ. Sci. Technol.* **2016**, *50*, 2573–2583. [CrossRef] [PubMed]
48. Nie, M.; Yan, C.; Li, M.; Wang, X.; Bi, W.; Dong, W. Degradation of chloramphenicol by persulfate activated by Fe 2+ and zerovalent iron. *Chem. Eng. J.* **2015**, *279*, 507–515. [CrossRef]
49. Jain, A.; Sharma, V.K.; Mbuya, O.S. Removal of arsenite by Fe(VI), Fe(VI)/Fe(III), and Fe(VI)/Al(III) salts: Effect of pH and anions. *J. Hazard. Mater.* **2009**, *169*, 339–344. [CrossRef]
50. Zheng, Q.; Wu, N.; Qu, R.; Albasher, G.; Cao, W.; Li, B.; Alsultan, N.; Wang, Z. Kinetics and reaction pathways for the transformation of 4-tert-butylphenol by ferrate(VI). *J. Hazard. Mater.* **2021**, *401*, 123405. [CrossRef]
51. Gao, Y.; Jiang, J.; Zhou, Y.; Pang, S.-Y.; Ma, J.; Jiang, C.; Yang, Y.; Huang, Z.-S.; Gu, J.; Guo, Q.; et al. Chlorination of bisphenol S: Kinetics, products, and effect of humic acid. *Water Res.* **2018**, *131*, 208–217. [CrossRef] [PubMed]
52. Chen, J.; Qi, Y.; Pan, X.; Wu, N.; Zuo, J.; Li, C.; Qu, R.; Wang, Z.; Chen, Z. Mechanistic insights into the reactivity of Ferrate(VI) with phenolic compounds and the formation of coupling products. *Water Res.* **2019**, *158*, 338–349. [CrossRef] [PubMed]
53. Yin, L.; Wei, J.; Qi, Y.; Tu, Z.; Qu, R.; Yan, C.; Wang, Z.; Zhu, F. Degradation of pentachlorophenol in peroxymonosulfate/heat system: Kinetics, mechanism, and theoretical calculations. *Chem. Eng. J.* **2022**, *434*, 134736. [CrossRef]
54. Wang, J.; Wang, C.; Guo, H.; Ye, T.; Liu, Y.; Cheng, X.; Li, W.; Yang, B.; Du, E. Crucial roles of oxygen and superoxide radical in bisulfite-activated persulfate oxidation of bisphenol AF: Mechanisms, kinetics and DFT studies. *J. Hazard. Mater.* **2020**, *391*, 122228. [CrossRef]
55. Dar, A.A.; Chen, J.; Shad, A.; Pan, X.; Yao, J.; Bin-Jumah, M.; Allam, A.A.; Huo, Z.; Zhu, F.; Wang, Z. A combined experimental and computational study on the oxidative degradation of bromophenols by Fe(VI) and the formation of self-coupling products. *Environ. Pollut.* **2020**, *258*, 113678. [CrossRef]
56. Zhong, C.; Zhao, H.; Cao, H.; Huang, Q. Polymerization of micropollutants in natural aquatic environments: A review. *Sci. Total Environ.* **2019**, *693*, 133751. [CrossRef]

toxics

MDPI

Article

Safe Disposal of Accident Wastewater in Chemical Industrial Parks Using Non-Thermal Plasma with ZnO-Fe_3O_4 Composites

Aihua Li [1], Chaofei Wang [2], Chengjiang Qian [1], Jinfeng Wen [1,*] and He Guo [2,*]

[1] College of Safety Science and Engineering, Nanjing Tech University, Nanjing 211816, China; liaihua_njtech@163.com (A.L.)
[2] College of Biology and the Environment, Nanjing Forestry University, Nanjing 210037, China; 19852260029@163.com
* Correspondence: wenjinfeng_njtech@163.com (J.W.); heguo@njfu.edu.cn (H.G.)

Abstract: Chemical wastewater has a high concentration of toxic and hazardous antibiotic pollutants, which not only devastates the ecological environment and disrupts the ecological balance, but also endangers human health. This research proposed a non-thermal plasma (NTP) combined with a ZnO-Fe_3O_4 nano-catalyst system to achieve the efficient degradation of ciprofloxacin (CIP) in chemical wastewater. Firstly, ZnO-Fe_3O_4 composite materials were prepared using hydrothermal method and characterized with scanning electron microscopy (SEM), transmission electron microscope (TEM), X-ray diffractometer (XRD), X-ray photoelectron spectroscopy (XPS), etc. With the sole NTP, NTP/ZnO, and NTP/ZnO-Fe_3O_4 systems, the removal efficiency of CIP can reach 80.1%, 88.2%, and 99.6%, respectively. The optimal doping amount of Fe_3O_4 is 14%. Secondly, the capture agent experiment verified that ·OH, $\cdot O_2^-$, and $^1O_2$ all have a certain effect on CIP degradation. Then, liquid chromatography–mass spectrometry (LC-MS) was used to detect the intermediate and speculate its degradation pathway, which mainly included hydroxyl addition, hydroxyl substitution, and piperazine ring destruction. After treatment with the NTP/ZnO-Fe_3O_4 system, the overall toxicity of the product was reduced. Finally, a cyclic experiment was conducted, and it was found that the prepared ZnO-Fe_3O_4 catalyst has good reusability. The NTP/ZnO-Fe_3O_4 was also applied in practical pharmaceutical wastewater treatment and has practical applicability.

Keywords: non-thermal plasma; ZnO; ZnO-Fe_3O_4; ciprofloxacin; chemical wastewater

check for updates

Citation: Li, A.; Wang, C.; Qian, C.; Wen, J.; Guo, H. Safe Disposal of Accident Wastewater in Chemical Industrial Parks Using Non-Thermal Plasma with ZnO-Fe_3O_4 Composites. *Toxics* **2024**, *12*, 40. https://doi.org/10.3390/toxics12010040

Academic Editor: Xin-Gui Li

Received: 24 November 2023
Revised: 22 December 2023
Accepted: 2 January 2024
Published: 4 January 2024

1. Introduction

Chemical industrial parks are the concentration areas of modern chemical industries, playing an important role in promoting economic development. However, due to the frequent involvement of toxic and harmful substances in the production process of chemical parks, especially antibiotic pollutants, once accidents occur, they may cause severe environmental pollution and personnel harm [1–3]. Therefore, it is very important to establish a comprehensive emergency response plan for pollution accidents in chemical industrial parks, which can in a timely manner and effectively respond to accidents and minimize losses.

Advanced oxidation method can generate strong oxidizing free radicals (such as hydroxyl radical (·OH)), which can completely degrade organic matter into CO_2 and H_2O [4]. The commonly used advanced oxidation methods currently include Fenton method, photocatalytic method, electrochemical method, and ozone catalytic oxidation method [5]. However, these methods generally have associated problems such as harsh reaction conditions, complex operations, and secondary pollution [6]. Non-thermal plasma (NTP) is an emerging advanced oxidation technology in recent years [7]. During the discharge process under applied voltage, it dissociates gas phase molecules and water molecules, producing a large number of active free radicals. It also radiates physical effects such as ultraviolet light and shock waves. Due to its advantages of being fast and efficient, causing no secondary pollution, and having a simple operation, NTP has been proposed for the degradation of organic pollutants in water, mainly including dyes, persistent organic pollutants, endocrine disruptors, and antibiotics [8]. In recent decades, research on plasma technology has also been very active, providing new technologies, methods, and processes for synthesizing new substances and materials and for environmental pollution control. However, its low energy utilization has always limited its development due to the insufficient utilization of its optical and chemical effects [9]. Therefore, it is necessary to seek suitable catalysts to combine with NTP for improving the energy utilization of NTP and the removal efficiency of organic pollutants.

ZnO has attracted widespread attention from researchers due to its stable physical and chemical properties, non-toxicity, rich morphology, and low cost [10]. Previous studies have reported that NTP can successfully activate ZnO for photocatalytic reactions, thereby improving the efficiency of plasma energy utilization and organic matter removal [11–14]. However, ZnO is an n-type semiconductor material with a wide band gap, which limits its ability to utilize near-ultraviolet light with shorter wavelengths and higher energy. It is worth noting that NTP generates a wide range of light, including both ultraviolet and visible light [15]. On the other hand, the generated electron–hole pairs in ZnO are more prone to recombination, resulting in low actual utilization efficiency of photons and insufficient photocatalytic activity. In addition, ZnO nanoparticles are difficult to recover in NTP water treatment systems, which are not conducive for secondary use.

Based on the above issues, some studies consider loading Fe_3O_4 onto ZnO nanoparticles [16]. The ZnO-Fe_3O_4 nanocomposite material can not only shorten the band gap and improve the light absorption range but also reduce the recombination of electrons and holes and improve the photocatalytic performance [17]. In addition, ZnO-Fe_3O_4 nanocomposites also have magnetism, which facilitates separation, recycling, and reuse [18]. However, this material requires an external light source, and the long processing time results in insufficient photocatalytic degradation efficiency. Therefore, this study considers constructing an NTP/ZnO-Fe_3O_4 system as a UV-stimulated luminescent catalyst for NTP radiation, which not only enhances its photocatalytic activity but also enhances the energy utilization efficiency of NTP, achieving the goal of short-term and efficient degradation of pollutants. However, as far as we know, there is still a significant gap in the research on NTP assisted ZnO-Fe_3O_4 nanocomposites for wastewater treatment, and the related degradation pathways and mechanisms are not yet thoroughly studied.

This research proposed the NTP/ZnO-Fe_3O_4 nano-catalyst system to achieve the efficient degradation of ciprofloxacin (CIP) in chemical wastewater. Firstly, ZnO-Fe_3O_4 nanocomposites were synthesized using hydrothermal method and characterized with scanning electron microscopy (SEM), transmission electron microscope (TEM), X-ray diffractometer (XRD), and X-ray photoelectron spectroscopy (XPS). Secondly, this study verified the effect of composite catalyst ZnO-Fe_3O_4 and investigated the effects of different doping and addition amounts of composite catalyst on the degradation efficiency of CIP. Then, the generation and action of free radicals were analyzed using trapping agent experiments. Based on the results of three-dimensional fluorescence, liquid chromatography–mass spectrometry (LC-MS) and DFT analysis decipher the degradation pathway of CIP and determine the catalytic and degradation mechanisms. In addition, to verify whether this system can reduce the toxicity of pollutants, this study used the typical toxicity analysis software ECOSAR (Version 2.0) to analyze the toxicity of CIP and its intermediate products. Finally, this study also conducted catalyst recovery and cycling performance tests and verified whether the NTP/ZnO-Fe_3O_4 system can be applied to the treatment of typical pharmaceutical wastewater. This study constructs the NTP/ZnO-Fe_3O_4 system and delves into the corresponding reaction mechanism, hoping to fill the gap in its application in wastewater treatment and provide some reference for subsequent research.

2. Experimental

2.1. Reagents

Zinc acetate, sodium hydroxide (NaOH), hydrogen peroxide, thick sulfuric acid, ethylene glycol, iron chloride, and ferrous sulfate were bought from China National Pharmaceutical Group Chemical Reagent Co., Ltd. (Beijing, China) CIP, isopropyl alcohol, benzoquinone, sodium oxalate, ethanolamine, and sodium acetate were purchased from Aladdin Biotechnology Co., Ltd (Shanghai, China).

2.2. Experimental Setup and Analysis

The experimental system is shown in Figure 1 and mainly includes a power supply, an oscilloscope, and a reactor. The driving power supply is a high-voltage pulse power supply. The oscilloscope (Tektronix, Beaverton, OR, USA, MDO3012) was used to detect the voltage (Tektronix, USA, MDO3012) and current characteristics (Tektronix, USA, P6021) during the discharge process. It was equipped with voltage and current probes, and the voltage and current waveform during the discharge process were displayed in real time on the oscilloscope. The reactor was composed of needle plate electrodes, and the gas flowed through the flow meter through a gas pump to a drum into the reactor, ultimately forming a bubble plasma. During the experiment, the prepared catalyst was placed in the reactor, and the physicochemical effects generated by plasma discharge can activate the catalyst to further undergo catalytic reactions, thereby improving the pollutant removal efficiency. During the experiment, samples were taken every five minutes using a syringe. The concentration of CIP was detected using a high-performance liquid chromatograph (HPLC). This chromatographic method was equipped with a symmetrical C18 column (Elite, 4.6 mm × 250 mm). The detection wavelength was 276 nm. The mobile phase consisted of acetonitrile and 0.01 M phosphoric acid (pH 2.5) in a ratio of 20:80 v/v, with a flow rate of 0.6 mL/min. The injection volume and residence time were 20 mL and 15 min, respectively.

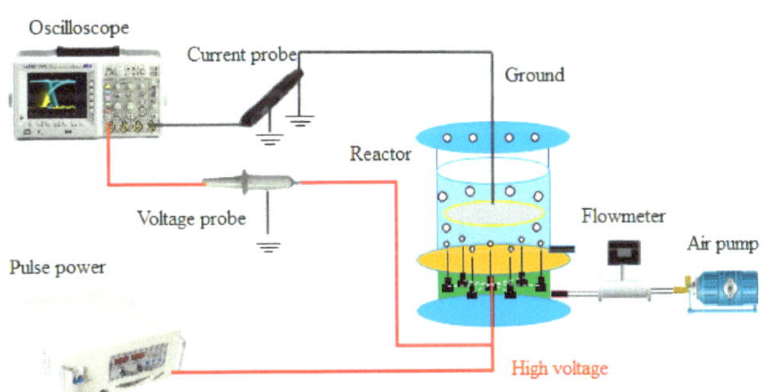

Figure 1. Experimental system diagram.

2.3. ZnO-Fe_3O_4 Preparation and Characterization

ZnO was prepared using hydrothermal method. Firstly, 1 g of zinc acetate was dissolved in 120 mL of ethylene glycol solution, which was then stirred in a 70 °C water bath for 1 h. Then, the mixture solution was dispersed using ultrasonic treatment for 1 h, and NaOH solution was then added to the above solution. Then, it was stirred at 70 °C for 2 h. The solution was placed in a polytetrafluoroethylene reactor and reacted at 160 °C for about 20 h. Finally, the solid particles were centrifuge washed with deionized water and anhydrous ethanol. The obtained ZnO nanoparticles were dried in a vacuum drying furnace (70 °C) for 10 h.

The prepared ZnO was dissolved in 30 mL ethylene glycol solution and was added to 683 mg iron chloride and 584 mg ferrous sulfate. Adjusting the doping amount of Fe_3O_4 can be achieved by proportionally changing the amounts of iron chloride and ferrous sulfate. Then, 0.5 g sodium acetate was added to the suspension after stirring. This solution was further stirred for 0.5 h at 80 °C. To completely mix the colloidal particles, 15 mL ethanolamine was added to the suspension, then heated at 150 °C for 6 h. Finally, the solution was cooled at room temperature and washed with deionized water. The obtained ZnO-Fe_3O_4 composites were dried at 60 °C for 12 h. In the studies, the prepared ratio of ZnO to Fe_3O_4 was 7%, 14%, and 21%.

The morphologies of ZnO-Fe_3O_4 samples were characterized using SEM (FEI, Nova Nano, Lincoln, NE, USA, SEM 450) and TEM (Tecnai F30, FEI, Hillsboro, OR, USA). The structure was tested via XRD (Bruker D8 Focus, Ettlingen, Germany). XPS (Thermo, Waltham, MA, USA, ESCALAB 250X) was adopted for the chemical state of various elements on the sample surface. FT-IR spectra were performed on an FT-IR spectrometer (Bruker MPA FTIR). The ultraviolet–visible diffuse reflectance spectrum (UV-Vis DRS) was determined using a ultraviolet–visible spectrophotometer (Shimadzu, Kyoto, Japan, UV-550) with barium sulfate as the reference material.

2.4. Toxicity Analysis Using ECOSAR

In the toxicity test, the aquatic toxicity of a substance can be predicted according to its structural similarity. To predict the acute and chronic toxicity of CIP and its degradation intermediates, the ECOSAR tool was used [19]. In the calculation process, fish, Daphnia, and green algae were selected as targets.

3. Results and Discussion

3.1. Characterization

The morphology of ZnO-Fe_3O_4 composite material was characterized using SEM and TEM, and the results are shown in Figure 2. From Figure 2A, it can be seen that ZnO exhibits a nanoflower-like structure, while Fe_3O_4 exhibits a granular spherical structure, with direct overlap between the two. Figure 2B shows a high-power scanning electron microscope image, where the lattice spacing of 2.42 Å and 2.53 Å belongs to ZnO and Fe_3O_4 [14,20], respectively. In addition, there is also a phenomenon of interlacing between their lattices. Figure 2C shows the diffraction pattern of ZnO-Fe_3O_4, with its circular structure belonging to the crystal planes of ZnO and Fe_3O_4, indicating that the material is indeed composed of ZnO and Fe_3O_4 composites. In particular, the bright spot may come from the interference of other elements of the testing instrument.

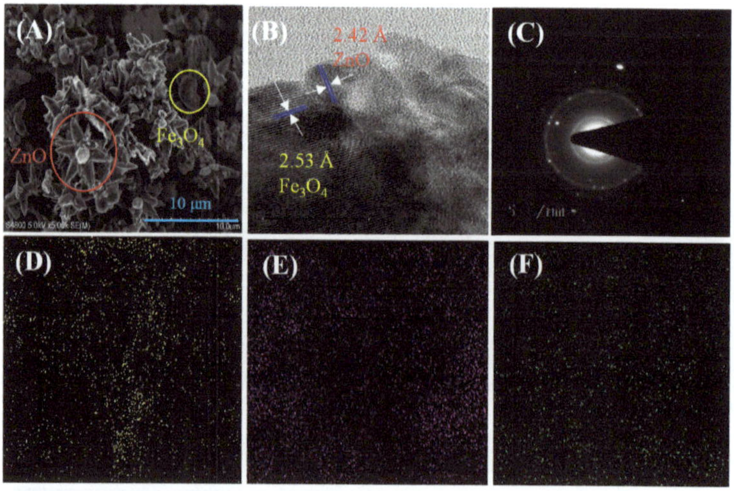

Figure 2. (**A**) SEM, (**B**) TEM, and (**C**) SAED patterns of ZnO-Fe_3O_4; EDS mapping: (**D**) Zn, (**E**) Fe, and (**F**) O.

EDS analysis was performed on the prepared composite materials. Figure 2D–F show the distribution of Zn, Fe, and O elements, respectively. It can be seen that Zn, Fe, and O are evenly distributed on the surface of ZnO-Fe_3O_4 composites, and the content of Fe and O is higher than that of Zn, which is also consistent with the SEM image of the composite. In summary, ZnO-Fe_3O_4 composite catalyst samples can be successfully prepared.

XRD was used to detect the crystal phases of ZnO-Fe_3O_4 composite materials with different doping levels. Figure 3 shows two sets of diffraction peaks of ZnO-Fe_3O_4 composite material. One group is the diffraction peaks of ZnO, (100), (002), (101), (102), (110), (103), and (112) crystal planes [21], corresponding to 2θ = 31.7°, 34.4°, 36.2°, 47.5°, 56.5°, 62.8°, and 67.8°, respectively. The other group is the diffraction peak of Fe_3O_4, the (331) crystal plane, which corresponds to 2θ = 35.2°. From Figure 3, it can be seen that the XRD patterns of ZnO-Fe_3O_4 all show the 35.2° diffraction peak belonging to the (331) crystal plane of Fe_3O_4 [20], indicating that the material was successfully synthesized from ZnO and Fe_3O_4. Figure 4 shows the XPS spectra of Zn, Fe, and O elements. The peaks at 1021.0 eV and 1044.0 eV represent Zn 2p3/2 and Zn 2p1/2, respectively (Figure 4A) [22]. Fe^{2+}2p3/2, Fe^{3+}2p3/2, Fe^{2+}2p1/2, and Fe^{3+}2p1/2 have four peaks at 710.2 eV, 712.1 eV, 723.7 eV, and 725.3 eV, respectively (Figure 4B) [23]. The peak of O element is composed of ZnO and Fe_3O_4, and there is an overlap phenomenon between the two peaks (Figure 4C), indicating that a phase separation has occurred [24].

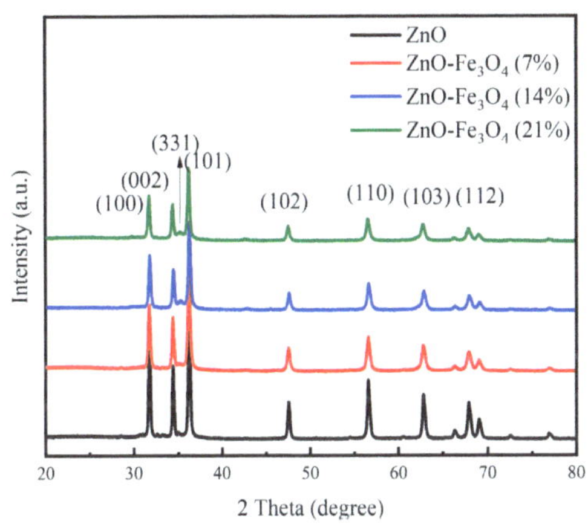

Figure 3. XRD patterns of ZnO-Fe_3O_4 composites.

3.2. Catalytic Degradation Performance of ZnO-Fe_3O_4 Composites

The catalytic effects of different catalysts in NTP system were investigated, and the results were shown in Figure 5. It can be seen that the addition of catalysts can improve the removal efficiency of CIP compared to a separate NTP system. The addition of ZnO alone can increase the removal efficiency of CIP from 80.1% to 88.2% (as Figure 5A). The addition of ZnO-Fe_3O_4 can further improve the removal efficiency of CIP. With the increase in Fe_3O_4 doping amount, the removal first increased and then decreased. When the doping amounts were 7%, 14%, and 21%, the removal rates reached 95.6%, 99.6%, and 95.9%, respectively. Obviously, when the Fe_3O_4 doping amount was 14%, the removal efficiency of CIP reached its highest. Figure 5B shows the effect of different catalysts on the first-order kinetic constant. It can be seen that the addition of ZnO can enhance the kinetic constant. The doping of Fe_3O_4 can further enhance the kinetic constant, and a 14% doping amount achieves the maximum kinetic constant. In the NTP, NTP/ZnO, and NTP/ZnO-Fe_3O_4 (14%) systems alone, the kinetic constants can reach 0.065 min^{-1}, 0.088 min^{-1}, and 0.202 min^{-1}, respectively.

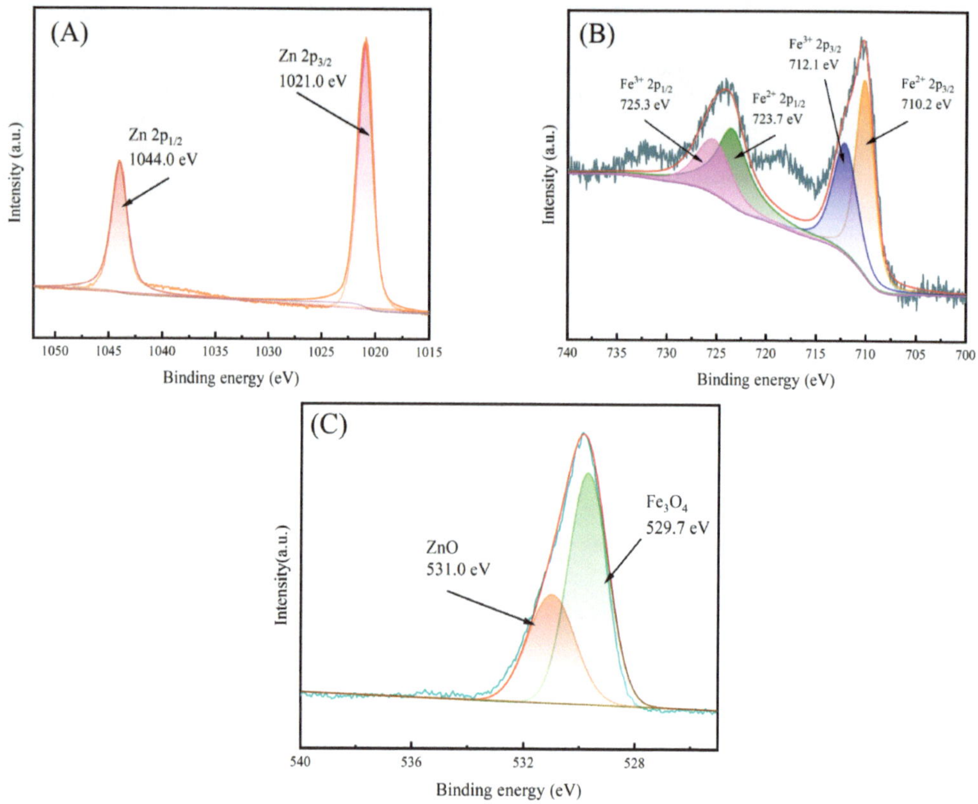

Figure 4. XPS spectra of (**A**) Zn, (**B**) Fe, (**C**) O.

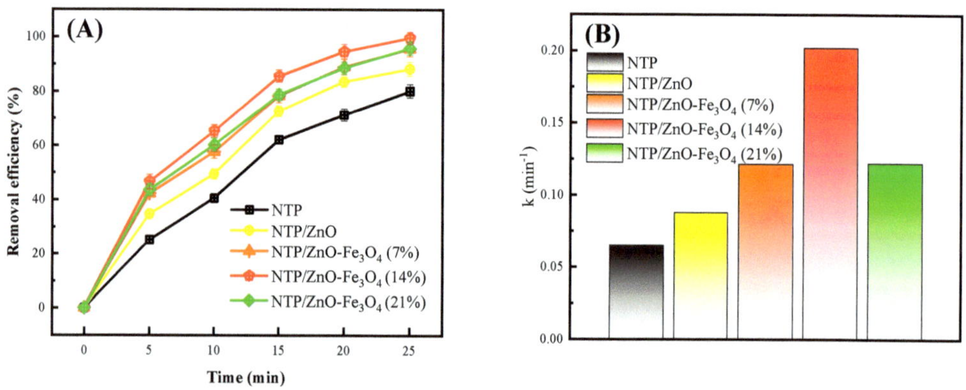

Figure 5. Effect of Fe_3O_4 doping amount: (**A**) performance and (**B**) kinetic constant. Experimental conditions: peak voltage of 19 kV; air flow rate of 4 L/min; electrode gap of 10 mm; and catalyst dosage of 0.28 g/L.

In the NTP system, the light and electronic effects generated by discharge can stimulate the photocatalytic reaction of ZnO (Equations (1)–(4)), thereby improving the removal efficiency of CIP. As for the individual ZnO system, the doping of Fe_3O_4 can shorten the band gap of ZnO and enhance its range corresponding to visible light [25]. At the same time, Fe_3O_4 can capture the conduction band electrons of ZnO, suppress the recombination of electron holes, and improve photocatalytic performance [26]. In addition, it is worth noting that Fe_3O_4 itself can also undergo heterogeneous Fenton reactions in the plasma system, catalyzing H_2O_2 to produce more ·OH (Equation (5)) [20]. However, excessive Fe_3O_4 doping actually makes it a composite center for electrons and holes of ZnO, thereby reducing the photocatalytic effect [27]. In the end, the removal efficiency of CIP decreases.

$$\text{plasma} + ZnO \rightarrow e^- + h^+ \quad (1)$$

$$O_2 + e^- \rightarrow O_2\cdot^- \quad (2)$$

$$h^+ + H_2O \rightarrow H^+ + \cdot OH \quad (3)$$

$$h^+ + OH^- \rightarrow \cdot OH \quad (4)$$

$$Fe^{2+}{}_{surf.} + H_2O_2 \rightarrow Fe^{3+}{}_{surf.} + OH^- + \cdot OH \quad (5)$$

The addition amounts of different catalysts are shown in Figure 6. It can be seen that as the addition amount increased, the removal efficiency first increased and then decreased. When the addition amount was 0.28 g/L, the removal efficiency reached its optimal value, which was 99.6%. Similarly, the kinetic constants also increased first and then decreased. When the addition amount was 0.12 g/L, 0.20 g/L, 0.28 g/L, and 0.36 g/L, the kinetic constants reached 0.073 min^{-1}, 0.100 min^{-1}, 0.202 min^{-1}, and 0.113 min^{-1}, respectively. As the addition amount increased, the catalytic effect became more significant. However, when the addition amount reached a certain value, the catalysts covered each other, and the effective contact area decreased. In addition, when too much catalyst was present, the transparency of aqueous solution decreased, resulting in low light energy utilization [28]. Therefore, an optimal amount of catalyst added, as a higher amount of catalyst was not necessarily better for the reaction.

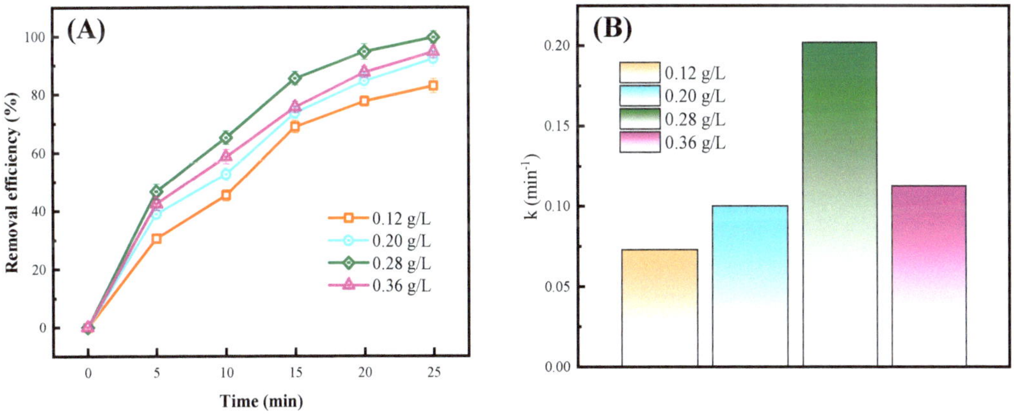

Figure 6. Effect of ZnO-Fe_3O_4 dosage: (**A**) performance and (**B**) kinetic constant. Experimental conditions: peak voltage of 19 kV; air flow rate of 4 L/min; and electrode gap of 10 mm.

3.3. Role of Active Species

To clarify the role of free radicals, capture agent experiments were conducted. IPA, BQ, and TEDA were selected to capture ·OH, $\cdot O_2^-$, and $^1O_2$, respectively [29]. As shown in Figure 7, it can be seen that the addition of free radicals leads to a decrease in removal efficiency. The addition of IPA reduced the removal rate of CIP from 99.6% to 68.5%. When BQ was present in the solution, the CIP removal rate decreased from 99% to 80.2%.

Similarly, TEDA reduced the CIP removal rate from 99.6% to 75.6%. The addition of IPA, BQ, and TEDA reduced the removal rate of CIP by 31.1%, 19.4%, and 24.0%, respectively, indicating that $\cdot OH$, $\cdot O_2^-$, and $^1O_2$ all played a certain role.

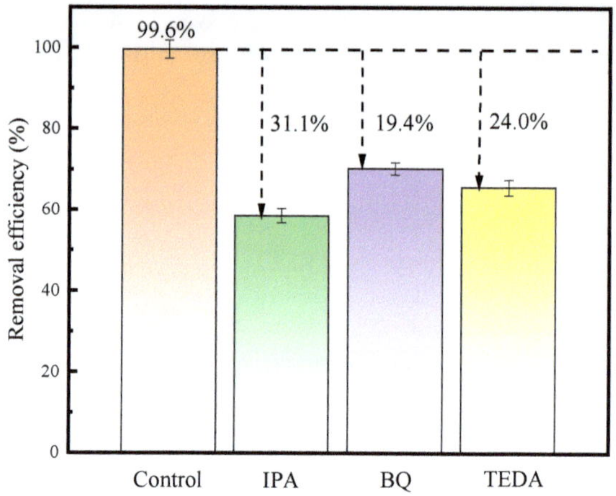

Figure 7. Role of various active species. Experimental conditions: peak voltage of 19 kV; air flow rate of 4 L/min; electrode gap of 10 mm; and catalyst dosage of 0.28 g/L.

3.4. Degradation Process and Toxicity Variation

The degradation process of CIP was thoroughly identified using three-dimensional fluorescence spectrometry, and the results are shown in Figure 8. There was one main peak in CIP solution, which was located at Ex/Em 400–500/300–400 nm in untreated CIP solution (Figure 8A), indicating that CIP molecules had the structural properties of humic acid [30]. With the progress of the reaction, the peak obviously decreased, indicating that its corresponding structure was destroyed (Figure 8B,C). After 30 min of treatment, the original main peaks of CIP basically disappeared (Figure 8D), indicating that CIP molecules can be decomposed completely.

In order to clarify the degradation pathway of CIP, we analyzed the degradation products. There were 12 intermediate products, which were detected through LC-MS analysis. Based on the identified intermediates, three specific degradation pathways were analyzed and are depicted in Figure 9, mainly involving processes such as hydroxyl addition, hydroxyl substitution, and piperazinyl ring destruction. Pathway 1: Hydroxyl groups first attacked the piperazine ring of CIP to generate C1 [31]. The hydroxyl group further attacked the C-N bond between the piperazine ring and the benzene ring, causing the piperazine ring to detach and generate C2. The cyclopropyl fragment from the quinolone moiety on C2 was further attacked by free radicals, resulting in its detachment and the generation of C3. C3 was transformed into C12 through decarboxylation. Pathway 2: CIP firstly undergoes decarboxylation reaction to generate C4 [32], which is then converted into C5 through the oxidation of piperazine by hydroxyl radicals. The hydroxyl groups on the benzene ring of C6 detached to form C7, and its F atom was replaced by hydroxyl groups to form C8. C8 was further converted into C12 through hydroxyl shedding. Pathway 3: The F atom on the CIP benzene ring was first attacked by hydroxyl groups and underwent a substitution process to form C8 [33]. The piperazine ring of C8 was further attacked by free radicals, causing its C-N bond to break and generate C9. The residual C-N bond of C9 piperazine ring was further attacked by free radicals, causing it to detach and generate C10. The carboxyl group on C10 was further replaced by hydroxyl groups to form C11, which was then removed and ultimately formed C12. These intermediate products were finally mineralized, resulting in the production of F^-, NO_3^-, CO_2, and H_2O [34,35].

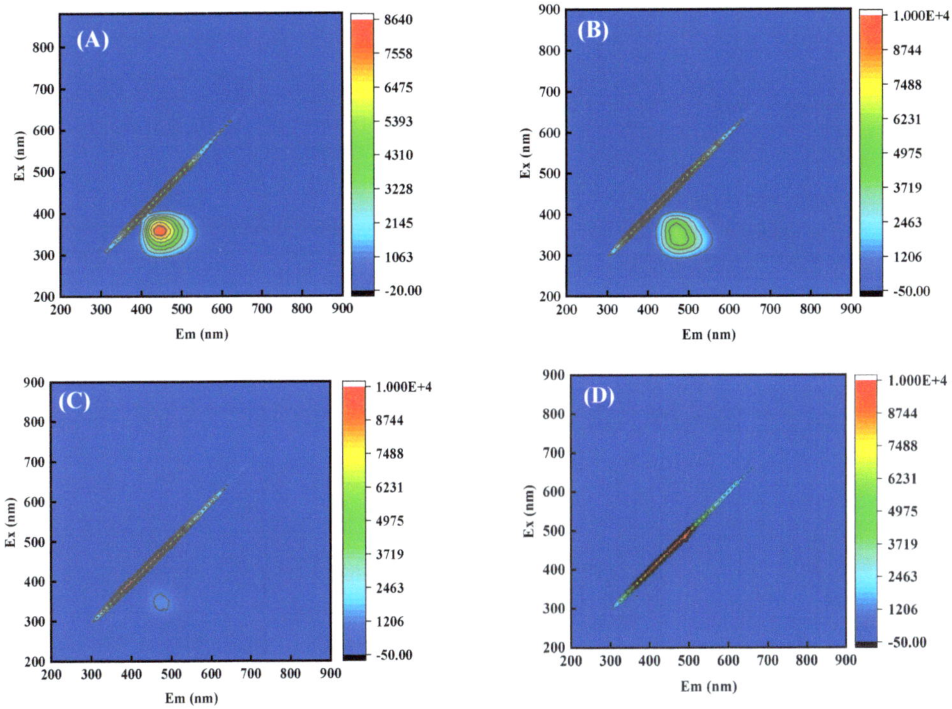

Figure 8. Three-dimensional fluorescence spectra of CIP solution after different treatment times: (**A**) 0 min; (**B**) 10 min; (**C**) 20 min; and (**D**) 30 min.

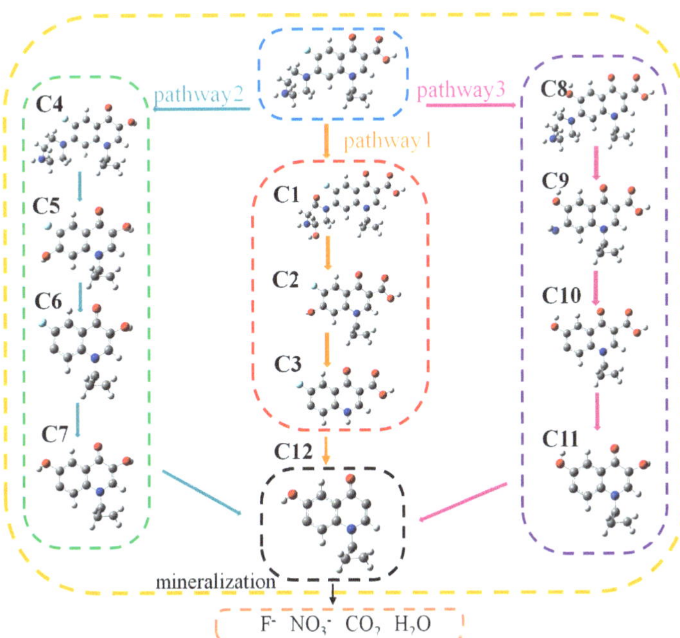

Figure 9. Proposed degradation pathway.

In addition, the toxicities of intermediates were analyzed using ECOSAR software, and the data are depicted in Figure 10. The toxicity of the corresponding substance is represented by the thickness of the string plot. From Figure 10, it can be seen that the toxicities of C1, C2, and C3 in pathway 1 are lower than that of CIP. The toxicities of C4 and C6 in pathway 2 are slightly higher, but the overall toxicity still decreases. The toxicity in pathway 3 is also significantly reduced, especially in product C8. Overall, most of the intermediate products in the NTP/ZnO-Fe_3O_4 system for CIP treatment did not exhibit a high toxicity, and their toxicities were far below international standards. Therefore, treatment with the NTP/ZnO-Fe_3O_4 system can reduce the toxicities of pollutants.

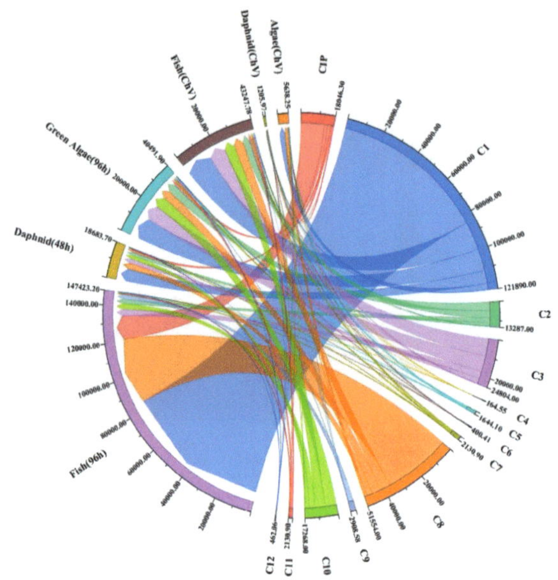

Figure 10. Toxicity of degradation intermediates.

In addition to its superior photocatalytic performance, the magnetic properties facilitate recycling, which is also one of the significant advantages of ZnO-Fe_3O_4 composite materials. In order to investigate the cycling performance of ZnO-Fe_3O_4 composite catalyst, multiple magnetic recovery cycling tests were conducted on the catalyst. The results are shown in Figure 11. From Figure 11, it can be seen that after four experiments, the removal rate of CIP decreased from the original 99.6% to 97.4%, indicating that the catalytic performance of ZnO-Fe_3O_4 composite material has hardly decreased. This indicates that the ZnO-Fe_3O_4 composite nanomaterials prepared in this study have good stability and can be reused in the NTP/ZnO-Fe_3O_4 system while maintaining a high pollutant degradation rate.

3.5. *Reuse Performance and Actual Wastewater Treatment*

The actual water quality situation is complex and variable, often containing various organic pollutants that are difficult to degrade. Most wastewater treatment methods have the problem of low mineralization rate and poor biodegradability of treated wastewater. In order to investigate the biodegradability of pollutant solutions treated with the NTP/ZnO-Fe_3O_4 system, CIP wastewater and actual pharmaceutical wastewater were selected as target pollutants, and their chemical oxygen demand (COD) after reaction was measured and plotted as shown in the bar graph in Figure 12. As shown in Figure 12, the addition of ZnO in the NTP system increases the COD removal rates of CIP wastewater and pharmaceutical wastewater by approximately 13.0% and 14.0%, respectively. The addition of ZnO-Fe_3O_4 in the NTP system increases the COD removal rates of CIP wastewater and pharmaceutical wastewater by approximately 27.0% and 19.0%, respectively. Overall, the

COD removal efficiency of the NTP/ZnO-Fe_3O_4 system for wastewater is around 70.0%, indicating that the system can effectively degrade pollutants, improve the biodegradability of wastewater solutions, and has good practical application value.

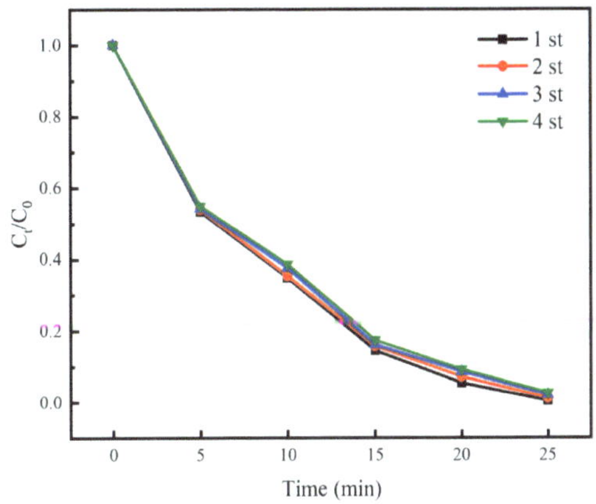

Figure 11. Recycle performance of ZnO-Fe_3O_4.

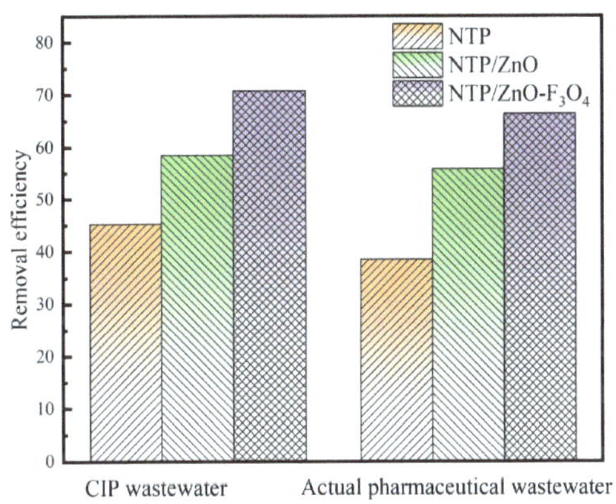

Figure 12. COD removal of CIP wastewater and actual pharmaceutical wastewater.

4. Conclusions

In this study, an emergency response scheme for the treatment of chemical wastewater by NTP/ZnO-Fe_3O_4 system was established with CIP as the target pollutant. Firstly, ZnO-Fe_3O_4 composite materials were successfully prepared using hydrothermal method and characterized with SEM, TEM, XRD, XPS, etc. Secondly, the experiment found that the removal rate of CIP with the NTP/ZnO-Fe_3O_4 system (99.6%) increased by 19.5% and 11.4% compared to that of the single NTP and NTP/ZnO systems, respectively. The optimal doping amount of ZnO-Fe_3O_4 is 14%, and the optimal dosage is 0.28 g/L. Secondly, the capture agent experiment found that active substances such as ·OH, $\cdot O_2^-$, and $^1O_2$ all play an important role in the degradation of CIP. Then, the specific degradation pathways of CIP were proposed through three-dimensional fluorescence spectroscopy and LC-MS analysis, including hydroxyl addition, hydroxyl substitution, and piperazine ring disruption. In addition, after treatment with the NTP/ZnO-Fe_3O_4 system, the overall toxicity of the product is reduced. Finally, cyclic experiments and actual wastewater treatment experiments have shown that ZnO-Fe_3O_4 can maintain a high catalytic effect even after repeated use, and the NTP/ZnO-Fe_3O_4 system can improve the biodegradability of wastewater and is suitable for various practical wastewater treatments. In summary, the NTP/ZnO-Fe_3O_4 system not only provides new ideas for the green and efficient treatment of chemical wastewater but also further enriches the knowledge of the relevant reaction mechanisms, providing reference for subsequent research by scholars.

Author Contributions: Methodology, A.L.; Software, A.L. and C.W.; Formal analysis, C.W.; Investigation, A.L.; Data curation, C.W.; Writing – original draft, A.L.; Writing – review & editing, H.G.; Visualization, C.Q.; Supervision, C.Q., J.W. and H.G.; Project administration, J.W. and H.G.; Funding acquisition, H.G. All authors have read and agreed to the published version of the manuscript.

Funding: This research was funded by Natural Science Research of Jiangsu Higher Education Institution of China (No. 20KJB610015).

Institutional Review Board Statement: Not applicable.

Informed Consent Statement: Not applicable.

Data Availability Statement: The data presented in this study are available on request from the corresponding author.

Acknowledgments: We deeply appreciate the financial support provided by Natural Science Research of Jiangsu Higher Education Institution of China (No. 20KJB610015).

Conflicts of Interest: The authors declare no conflict of interest.

References

1. Wang, Q.; Wang, P.L.; Yang, Q.X. Occurrence and diversity of antibiotic resistance in untreated hospital wastewater. *Sci. Total Environ.* **2018**, *621*, 990–999. [CrossRef] [PubMed]
2. Mishra, S.; Singh, A.K.; Cheng, L.; Hussain, A.; Maiti, A. Occurrence of antibiotics in wastewater: Potential ecological risk and removal through anaerobic–aerobic systems. *Environ. Res.* **2023**, *226*, 115678. [CrossRef] [PubMed]
3. Shi, H.M.; Ni, J.; Zheng, T.L.; Wang, X.N.; Wu, C.F.; Wang, Q.H. Remediation of wastewater contaminated by antibiotics. A review. *Environ. Chem. Lett.* **2020**, *18*, 345–360. [CrossRef]
4. Shanmugavel, S.P.; Kumar, G.; Yogalakshmi, K.N.; Gunasekaran, M.; Banu, R.J. Recent progress in mineralization of emerging contaminants by advanced oxidation process: A review. *Environ. Pollut.* **2024**, *341*, 122842.
5. Liu, W.; Lu, Y.R.; Dong, Y.B.; Jin, Q.; Lin, H. A critical review on reliability of quenching experiment in advanced oxidation processes. *Chem. Eng. J.* **2023**, *466*, 143161. [CrossRef]
6. Priyadarshini, M.; Das, I.; Ghangrekar, M.M.; Blaney, L. Advanced oxidation processes: Performance, advantages, and scale-up of emerging technologies. *J. Environ. Manag.* **2022**, *316*, 115295. [CrossRef] [PubMed]
7. Guo, H.; Pan, S.J.; Hu, Z.X.; Wang, Y.W.; Jiang, W.X.; Yang, Y.X.; Wang, Y.C.; Han, J.G.; Wu, Y.F.; Wang, T.C. Persulfate activated by non-thermal plasma for organic pollutants degradation: A review. *Chem. Eng. J.* **2023**, *470*, 144094. [CrossRef]
8. Sanito, R.C.; You, S.J.; Wang, Y.F. Degradation of contaminants in plasma technology: An overview. *J. Hazard. Mater.* **2022**, *424*, 127390. [CrossRef]
9. Guo, H.; Su, Y.Y.; Yang, X.Y.; Wang, Y.W.; Li, Z.; Wu, Y.F.; Ren, J.Y. Dielectric Barrier Discharge Plasma Coupled with Catalysis for Organic Wastewater Treatment: A Review. *Catalysts* **2023**, *13*, 10. [CrossRef]
10. Yan, X.; Yi, C.W.; Wang, Y.H.; Cao, W.D.; Mao, D.N.; Ou, Q.Q.; Shen, P.Y.; Wang, H.J. Multi-catalysis of nano-zinc oxide for bisphenol A degradation in a dielectric barrier discharge plasma system: Effect and mechanism. *Sep. Purif. Technol.* **2020**, *231*, 115897. [CrossRef]
11. Hamid, S.B.A.; The, S.J.; Lai, C.W. Photocatalytic Water Oxidation on ZnO: A Review. *Catalysts* **2017**, *7*, 93. [CrossRef]
12. Abdel-Fattah, E.; Alotibi, S. Synergistic Effect of Nonthermal Plasma and ZnO Nanoparticles on Organic Dye Degradation. *Appl. Sci.* **2023**, *13*, 10045. [CrossRef]
13. Farzinfar, B.; Qaderi, F. Synergistic degradation of aqueous p-nitrophenol using DBD plasma combined with ZnO photocatalyst. *Process Saf. Environ. Prot.* **2022**, *168*, 907–917. [CrossRef]
14. Li, Z.; Jiang, W.X.; Huang, J.W.; Wang, Y.W.; Guo, H. ZnO Promoted Persulfate Activation in Discharge Plasma System for Ofloxacin Degradation. *Catalysts* **2023**, *13*, 847. [CrossRef]
15. Wang, X.P.; Chen, Y.X.; Fu, M.; Chen, Z.H.; Huang, Q.H. Effect of high-voltage discharge non-thermal plasma on g-C3N4 in a plasma-photocatalyst system. *Chin. J. Catal.* **2018**, *39*, 1672–1682. [CrossRef]
16. Elshypany, R.; Selim, H.; Zakaria, K.; Moustafa, A.H.; Sadeek, S.A.; Sharaa, S.I.; Raynaud, P.; Nada, A.A. Elaboration of Fe_3O_4/ZnO nanocomposite with highly performance photocatalytic activity for degradation methylene blue under visible light irradiation. *Environ. Technol. Innov.* **2021**, *23*, 101710. [CrossRef]
17. Długosz, O.; Szostak, K.; Krupiński, M.; Banach, M. Synthesis of Fe_3O_4/ZnO nanoparticles and their application for the photodegradation of anionic and cationic dyes. *Int. J. Environ. Sci. Technol.* **2021**, *18*, 561–574. [CrossRef]

18. Faramawy, A.; Elsayed, H.; Scian, C.; Mattei, G. Structural, Optical, Magnetic and Electrical Properties of Sputtered ZnO and ZnO:Fe Thin Films: The Role of Deposition Power. *Ceramics* **2022**, *5*, 1128–1153. [CrossRef]
19. Li, F.F.; Sun, G.H.; Fan, T.J.; Zhang, N.; Zhao, L.J.; Zhong, R.G.; Peng, Y.Z. Ecotoxicological QSAR modelling of the acute toxicity of fused and non-fused polycyclic aromatic hydrocarbons (FNFPAHs) against two aquatic organisms: Consensus modelling and comparison with ECOSAR. *Aquat. Toxicol.* **2023**, *255*, 106393. [CrossRef]
20. Guo, H.; Li, Z.; Zhang, Y.; Jiang, N.; Wang, H.J.; Li, J. Degradation of chloramphenicol by pulsed discharge plasma with heterogeneous Fenton process using Fe_3O_4 nanocomposites. *Sep. Purif. Technol.* **2020**, *253*, 117540. [CrossRef]
21. Guo, H.; Yang, H.; Huang, J.W.; Tong, J.; Liu, X.Y.; Wang, Y.W.; Qiao, W.C.; Han, J.G. Theoretical and experimental insight into plasma-catalytic degradation of aqueous p-nitrophenol with graphene-ZnO nanoparticles. *Sep. Purif. Technol.* **2022**, *295*, 121362. [CrossRef]
22. Nekouei, S.; Nekouei, F.; Kargarzadeh, H. Synthesis of ZnO photocatalyst modified with activated carbon for a perfect degradation of ciprofloxacin and its secondary pollutants. *Appl. Organomet. Chem.* **2018**, *32*, e4198. [CrossRef]
23. Wan, Z.; Wang, J. Degradation of sulfamethazine using Fe_3O_4-Mn3O4/reduced graphene oxide hybrid as fenton-like catalyst. *J. Hazard. Mater.* **2017**, *324*, 653–664. [CrossRef] [PubMed]
24. Erim, B.; Ciğeroğlu, Z.; Şahin, S.; Vasseghian, Y. Photocatalytic degradation of cefixime in aqueous solutions using functionalized SWCNT/ZnO/Fe_3O_4 under UV-A irradiation. *Chemosphere* **2022**, *291*, 132929. [CrossRef] [PubMed]
25. Pham, L.H.; Nguyen, V.Q.; Nguyen, H.; Ha, T.H.; Duong, A.T.; Manh, T.T.; Quang, V.T.; Ta, N.B.; Nguyen, T.; Van, D.D. Efficiency enhancement of photocatalytic activity under UV and visible light irradiation using ZnO/Fe_3O_4 heteronanostructures. *Sol. Energy* **2023**, *249*, 712–724.
26. Zhang, P.; Su, Q.; Han, L.J.; Lin, J.H.; Wei, X.H.; Meng, S.J.; Wang, Y.B. Facile fabrication of magnetic Ag/ZnO/Fe_3O_4 composite and the photocatalytic performance under simulated sunlight irradiation. *Mol. Catal.* **2021**, *508*, 111606. [CrossRef]
27. Guo, H.; Wang, Y.W.; Yao, X.; Zhang, Y.T.; Li, Z.; Pan, S.J.; Han, J.G.; Xu, L.J.; Qiao, W.C.; Li, J.; et al. A comprehensive insight into plasma-catalytic removal of antibiotic oxytetracycline based on graphene-TiO_2-Fe_3O_4 nanocomposites. *Chem. Eng. J.* **2021**, *425*, 130614. [CrossRef]
28. Zhang, G.Y.; Sun, Y.; Zhang, C.; Yu, Z. Decomposition of acetaminophen in water by a gas phase dielectric barrier discharge plasma combined with TiO_2-rGO nanocomposite: Mechanism and degradation pathway. *J. Hazard. Mater.* **2017**, *323*, 719–729. [CrossRef]
29. Wang, Y.W.; Xiang, L.R.; Li, Z.; Han, J.G.; Guo, H. Sulfite activation by water film dielectric barrier discharge plasma for ibuprofen degradation: Efficiency, comparison of persulfate, mechanism, active substances dominant to pathway, and toxicity evaluation. *Sep. Purif. Technol.* **2024**, *330*, 125531. [CrossRef]
30. Chen, W.; Westerhoff, P.; Leenheer, J.A.; Booksh, K. Fluorescence Excitation—Emission Matrix Regional Integration to Quantify Spectra for Dissolved Organic Matter. *Environ. Sci. Technol.* **2003**, *37*, 5701–5710. [CrossRef]
31. Fan, G.D.; Cai, C.J.; Yang, S.W.; Du, B.H.; Luo, J.; Chen, Y.F.; Lin, X.; Li, X.; Wang, Y.M. Sonophotocatalytic degradation of ciprofloxacin by Bi_2MoO_6/$FeVO_4$ heterojunction: Insights into performance, mechanism and pathway. *Sep. Purif. Technol.* **2022**, *303*, 122251. [CrossRef]
32. Yu, X.J.; Zhang, J.; Zhang, J.; Niu, J.F.; Zhao, J.; Wei, Y.C.; Yao, B.H. Photocatalytic degradation of ciprofloxacin using Zn-doped Cu2O particles: Analysis of degradation pathways and intermediates. *Chem. Eng. J.* **2019**, *374*, 316–327. [CrossRef]
33. Hu, X.; Hu, X.J.; Peng, Q.Q.; Zhou, L.; Tan, X.F.; Jiang, L.H.; Tang, C.F.; Wang, H.; Liu, S.H.; Wang, Y.Q.; et al. Mechanisms underlying the photocatalytic degradation pathway of ciprofloxacin with heterogeneous TiO_2. *Chem. Eng. J.* **2020**, *380*, 122366. [CrossRef]
34. Anjali, R.; Shanthakumar, S. Simultaneous degradation of amoxicillin, ciprofloxacin and acetaminophen in a mixture by ozonation: Kinetics and mechanisms pathway. *J. Clean. Prod.* **2022**, *378*, 134509. [CrossRef]
35. Guo, H.; Jiang, N.; Wang, H.J.; Shang, K.F.; Lu, N.; Li, J.; Wu, Y. Pulsed discharge plasma induced WO_3 catalysis for synergetic degradation of ciprofloxacin in water: Synergetic mechanism and degradation pathway. *Chemosphere* **2019**, *230*, 190–200. [CrossRef]

toxics

MDPI

Article

Degradation of Ciprofloxacin in Water by Magnetic-Graphene-Oxide-Activated Peroxymonosulfate

Xiaoping Wang [1,*], Yulan Li [1], Jiayuan Qin [1], Ping Pan [2], Tianqing Shao [1], Xue Long [1] and Debin Jiang [1]

[1] Chongqing Key Laboratory of Catalysis and Environmental New Material, College of Environment and Resources, Chongqing Technology and Business University, Chongqing 400067, China; liyulan201713@163.com (Y.L.); 18883248697@163.com (J.Q.); shao_tian_qing@163.com (T.S.); jansev1331@163.com (X.L.); jiangdebin@ctbu.edu.cn (D.J.)

[2] Chongqing Ecological Environment Monitoring Center, No. 252, Qishan Road, Ranjiaba, Yubei District, Chongqing 401147, China; cqjczgb@126.com

* Correspondence: xpwang@ctbu.edu.cn; Tel.: +86-152-1320-1592

Abstract: Antibiotics are extensively applied in the pharmaceutical industry, while posing a tremendous hazard to the ecosystem and human health. In this study, the degradation performance of ciprofloxacin (CIP), one of the typical contaminants of antibiotics, in an oxidation system of peroxymonosulfate (PMS) activated by magnetic graphene oxide (MGO) was investigated. The effects of the MGO dosage, PMS concentration and pH on the degradation of CIP were evaluated, and under the optimal treatment conditions, the CIP degradation rate was up to 96.5% with a TOC removal rate of 63.4%. A kinetic model of pseudo-secondary adsorption indicated that it involves an adsorption process with progressively intensified chemical reactions. Furthermore, the MGO exhibited excellent recyclability and stability, maintaining strong catalytic activity after three regenerative cycles, with a CIP removal rate of 87.0%. EPR and LC-MS experiments suggested that •OH and SO_4^{-}• generated in the MGO/PMS system served as the main reactants contributing to the decomposition of the CIP, whereby the CIP molecule was effectively destroyed to produce other organic intermediates. Results of this study indicate that organic pollutants in the aqueous environment can be effectively removed in the MGO/PMS system, in which MGO has excellent catalytic activity and stabilization for being recycled to avoid secondary pollution, with definite research value and application prospects in the field of water treatment.

Keywords: magnetic graphene oxide; heterogeneous activation; peroxymonosulfate activation; sulfate radical; ciprofloxacin

Citation: Wang, X.; Li, Y.; Qin, J.; Pan, P.; Shao, T.; Long, X.; Jiang, D. Degradation of Ciprofloxacin in Water by Magnetic-Graphene-Oxide-Activated Peroxymonosulfate. *Toxics* **2023**, *11*, 1016. https://doi.org/10.3390/toxics11121016

Academic Editors: Zunyao Wang and Tiecheng Wang

Received: 12 November 2023
Revised: 10 December 2023
Accepted: 11 December 2023
Published: 13 December 2023

1. Introduction

Quinolone antibiotics are a class of synthetic antibacterial drugs, which are widely used in health care and aquaculture due to the advantages of the antibacterial effect, structural simplicity and low price [1,2]. Ciprofloxacin (CIP) is a widely used quinolone antibiotic, primarily for the treatment of various bacterial infections. In recent years, CIP has been detected in surface water, groundwater, sewage treatment plants, medical wastewater, aquaculture, river sediments, etc. [3,4]. The complex structure and high antibacterial activity of CIP, as well as the difficulty of biodegradation and certain biological toxicity, make it difficult to be removed by conventional wastewater treatment processes [5]. Consequently, it is particularly crucial to develop an economical and efficient method for removing CIP in water.

A variety of methods are currently available for treating the CIP organic pollutants in an aqueous environment, including biological, physical and chemical methods [6–8]. Advanced oxidation technology based on a sulfate radical (SO_4^{-}•, oxidation potential = 2.5~3.1 V) generated by activation of persulfate has been developed in recent years for degradation of refractory organic pollutants in water. Compared with the hydroxyl radical (•OH, oxidation potential = 1.9~2.7 V) generated by catalytic hydrogen peroxide, the sulfate radical (SO_4^-•) exhibits the features of a long lifespan, broad spectrum of action, wide application range of pH and strong selectivity and can be generated by activation of peroxodisulfate (PDS) or peroxymonosulfate (PMS) in various methods [9,10]. PMS is a structurally asymmetric oxidant, which is easier to excite or activate than PDS, generally producing •OH and SO_4^-•with the activation of transition metals (homogeneous and heterogeneous) [11], ultraviolet light [12], ultrasound [13], heating [14], a carbon catalyst [15], etc. It has been demonstrated in abundant studies that the utilization of peroxymonosulfate advanced oxidation technology provides positive results for the treatment of CIP in wastewater [16–19].

Transition metal ion activation serves as the primary option for the homogeneous activation of PMS on account of its simplicity of operation, low cost and high removal efficiency. Nevertheless, the high dosage of transition metal ions leads to an increase in the amount of residual sludge, which is prone to cause secondary contamination [20]. In comparison, heterogeneous activated PMS provides the advantages of a wide pH range and high recovery rate of the catalyst. As Co has been identified to be the most effective activator for PMS [21], currently, it is preferred by many scholars regarding choosing multi-phase cobalt oxides for the activation of PMS. However, the leaching of cobalt poses a new health hazard, which can easily lead into cobalt toxicity when exposed to the environment containing cobalt for a long period of time [22]. Certainly, it has been found that the stability and catalytic properties of cobalt composites can be enhanced as a result of modifying the preparation method [23]. Graphene oxide (GO) [24], an oxidized derivative of graphene, is a novel carbon material. An abundant amount of oxygen-containing energy groups were found at both the basal surface and the edges of GO, allowing GO with a sufficiently many sorption sites, as well as being easily dispersed in water [25]. However, owing to the fact that GO separates poorly in water and is prone to cause new environmental problems, it is commonly used as a supporting material, attracting extensive attention on supporting metals and metal oxides [26]. The distinct characteristics of GO in terms of specific surface area size, oxygen-containing functional groups, adsorption capability and stability performance make GO an essential candidate for solid carrier materials in heterogeneous catalysts [25]. In addition, during compounding with metals, metal oxides, polymers and other materials, GO can effectively disperse the adhering materials and prevent agglomeration owing to its large specific surface area [27]. Several investigations revealed that GO, reduced graphene oxide and GO metal composites allowed the absorption and degradation of CIP through the activation of peroxomonosulfate [28,29].

In this paper, GO was prepared by the modified Hummers method [30] where there is an introduction of a significant number of oxygen-containing groups providing active sites for surface modification as well as a large specific surface area in the synthesis of GO-based materials [31,32]. Moreover, MGO composites were prepared with Fe_3O_4 particles and GO compounded by the ultrasonic chemical co-precipitation method. In this study, the feasibility of MGO-catalyzed PMS for the degradation of CIP was investigated. The effects of various influential factors (MGO content, PMS concentration, pH, etc.) on CIP removal were explored, the kinetic study of CIP removal was carried out and the possible degradation mechanisms and degradation pathways of CIP were proposed. A significant advantage lies in the superior catalytic activity and high stability of MGO composites, which can be recycled upon use, avoiding secondary pollution. It is expected that the outcomes of this study will offer a certain theoretical basis and fundamental support for the practical implementation of MGO/PMS-based advanced oxidation technology in the treatment of antibiotic wastewater.

2. Materials and Methods

2.1. Reagents

The reagents mainly consisted of natural graphite (99%), sulfuric acid (H_2SO_4, 98%), potassium permanganate ($KMnO_4$), hydrogen peroxide (H_2O_2, 30%), ferric chloride hexahydrate ($FeCl_3{\cdot}6H_2O$), ferrous sulfate heptahydrate ($FeSO_4{\cdot}7H_2O$), anhydrous ethanol (CH_3CH_2OH, 99.5%), potassium hydrogen persulfate ($KHSO_5$). Chromatographic-grade acetonitrile and methanol were used as the mobile phase, and ciprofloxacin (CIP) was purchased from MACKLIN (Shanghai, China). All reagents with which the study was carried out were of an analytically pure grade without specific instructions. In addition, all solutions used for the experiments as prepared with deionised water.

2.2. Synthesis of MGO Catalysts

Figure 1 shows the preparation process for the synthesis of MGO using ultrasonic chemical precipitation. Firstly, 1.000 g of GO was transferred into 100 mL of deionized water. After that, it was sonicated for 30 min, and 1.000 g of Fe_3O_4 was added and mixed well and continued to be sonicated for 1 h. Finally, the filtered reactants were collected and then dried in an oven (60 °C), resulting in MGO.

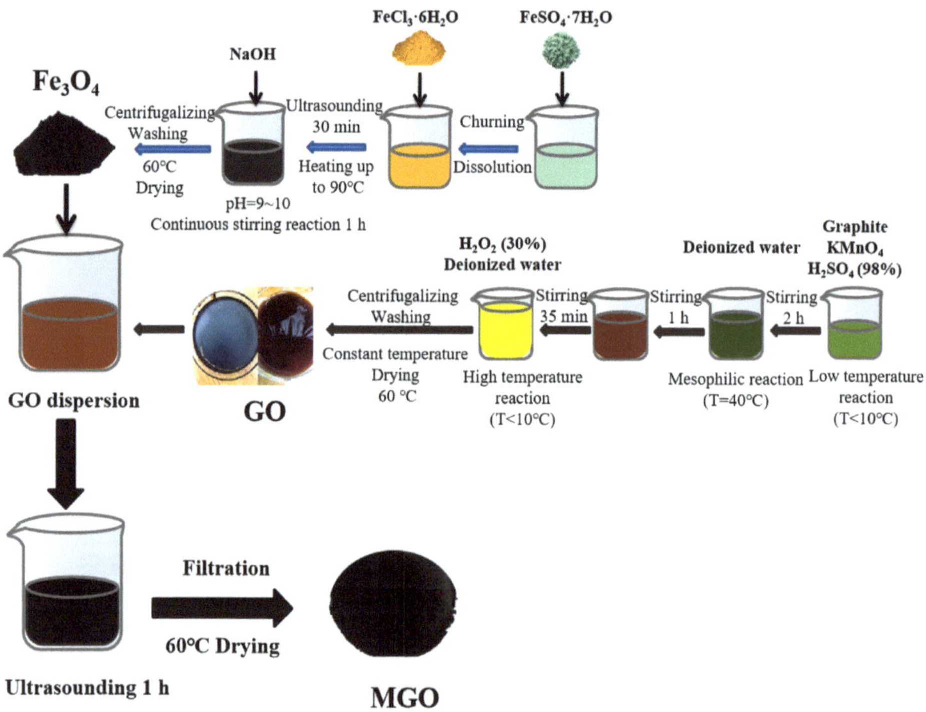

Figure 1. The fabrication process of MGO.

2.3. Catalysts' Characterization

The physical properties of the materials prepared in the study were mainly determined with a scanning electron microscope (SEM, thermo scientific Apreo 2C, ThermoFisher Scientific, Waltham, MA, USA), transmission electron microscopy (TEM, FEI Talos F200X, ThermoFisherscientifi USA) and a specific surface and porosity analyzer (BET, ASAP 2020, Mike instruments, Norcross, GA, USA). The materials in anhydrous ethanol were dispersed using ultrasonication, and scanned with an SEM and TEM, respectively, from which the morphological information of the activated materials was clarified, structural defects were analyzed and the composite of the materials was determined, whereas the specific surface area and the pore structure were identified with BET.

In addition, X-ray diffraction (XRD, XRD-D8, Bruker, Karlsruhe, Baden Wurttemberg, Germany), Fourier transform infrared (FT-IR, IR Prestige-21, Shimazu, Japan) and X-ray photoelectron spectroscopy (XPS, ESCALAB Xi+, ThermoFisher Scientific, USA) were employed to analyze the crystal structure and chemical characteristics of the activated materials. The powdered materials were pressed into flakes, and then XRD was employed to identify the crystal structure and phase transition, with the scanning range of 0~80°, the scanning speed of 5°/min and the temperature maintained at 25 °C. Similarly, the materials in powder were homogeneously mixed with potassium bromide in the ratio of 1:100, which was pressed and scanned for the infrared absorption spectrum to observe the types of functional groups contained in the materials. And finally, XPS was performed to test the whole and fractional spectra of the materials, so as to analyze the elemental composition of the samples, with the AI target as the test target.

2.4. Degradation Experiment of CIP

The aqueous solution of CIP (20 mg/L) configured in the experiment was stored protected from light and used up within 1 week at a time, for a study on the degradation of CIP by MGO-composite-activated PMS, in which the CIP removal was used as a criterion for the catalytic performance of the material. The specific experimental steps were as follows: the activators (GO or MGO) and PMS with a certain concentration were added to a conical flask containing 100 mL of 20 $mg \cdot L^{-1}$ CIP-simulated wastewater, and then placed in a thermostatic culture oscillator (ZWYR-240, Shanghai Zhicheng Analytical Instrument Manufacturing Co., Ltd., Shanghai, China) to perform the oscillation reaction for 180 min at 25 °C and a rotational speed of 150 rpm. In addition, the effects of the initial pH of the CIP solution (3, 5, 7, 9, 11), PMS concentration (0.5, 1.0, 1.5, 2.0, 3.0 g/L) and catalyst content (0.5, 1.0, 1.5, 2.0, 3.0 g/L) on the removal efficiency of CIP were investigated.

In all cases, it was possible to repeat all experiments three times under the same conditions to facilitate the subsequent error analysis and calculations, and the removal efficiency of CIP was calculated by Formula (1):

$$W = \frac{C_0 - C_t}{C_0} \times 100\% \tag{1}$$

where W is the removal efficiency (%), and C_0 and C_t denote the initial concentration of CIP and the concentration ($mg \cdot L^{-1}$) of CIP at time t, respectively.

2.5. Samples' Analysis

The full wave of the CIP standard solution was scanned using a UV spectrophotometer (UV1102II, Shanghai, China), and 277 nm as a characteristic peak of CIP was determined. The concentration of CIP was detected by high-performance liquid chromatography (LC, LC-100 plus, Shanghai, China) with a Pronaos EP-C18 column (4.6 × 260 mm, 5 μm) and a UV detection wavelength of 277 nm. The mobile phase consisted of 0.1% formic acid and acetonitrile in a volume ratio of 80:20, which flowed for 0.30 $mL \cdot min^{-1}$, and the injection volume was 20 μL.

The main active species in the CIP degradation by MGO-composite-activated PMS were detected with electron paramagnetic resonance (EPR, EMX nano, Bruker, Karlsruhe, Germany), in which DMPO acted as a free radical trapping agent. The mineralization rate of the reaction solution was measured with a total organic carbon analyzer (TOC, TOC-L CPH/CPN, Shimadzu, Kyoto, Japan), and the TOC removal rate as the basis for the oxidative degradation of CIP.

The intermediates of the CIP degradation were analyzed by liquid chromatography–mass spectrometry (LC-MS, Agilent LC1290-QQQ-6470, City of Santa Clara, CA, USA) with an ACE, UltraCore 2.5 μm C18 (2.1 × 75 mm) column of 35 °C. In addition, the mobile phase was acetonitrile and formic acid (0.1%) in a flow rate of 0.40 $mL \cdot min^{-1}$. The ESI was performed in the positive ionization mode and the scanning range of 200–400 m/z, as well as the mass spectrometry analysis being conducted in a gradient elution mode.

3. Results and Discussion

3.1. Characterization of Catalysts

3.1.1. Analysis of Surface Physical Structure

The SEM and TEM characterizations of GO and MGO materials were shown in Figure 2. As from Figure 2a,b, GO exhibited a translucent thin film sheet structure with a flat and smooth surface, visible wrinkles toward the edges of the sheet layer, as well as a low degree of polymerization among the sheets. Such a special structure effectively increased the specific surface area of GO, providing sufficient adsorption sites. At the same time, in the process of the powerful oxidation of GO, an abundant amount of oxygen-containing functional groups were inserted between the lamellae, with additional defects across the GO layers and expanded structural irregularities, resulting in the dispersion of GO in the solution, which facilitated the preparation of the composite materials. The SEM images of MGO are shown in Figure 2c, from which a large variety of Fe_3O_4 particles were well distributed on the surface of the MGO material. Additionally, the TEM image (Figure 2d) further demonstrated that the MGO material was successfully prepared by Fe_3O_4 particles and GO. Nano-Fe_3O_4 particles had a prismatic cone shape. By chemical precipitation, MGO uniformly loaded Fe_3O_4 particles between the surface and the sheet while retaining the original structure of GO.

The specific surface area is considered as one of the significant parameters to characterize the adsorption properties of the materials, whereas the pore size distribution may affect the catalytic function to which the materials are subjected. The N_2 adsorption–desorption experiments of GO, Fe_3O_4 and MGO materials were performed so that the information of adsorption isotherms and pore size distribution of the materials were obtained. Figure 2e shows that the adsorption isotherms of GO, Fe_3O_4 and MGO were all type IV isotherms and presented an adsorption hysteresis phenomenon with an H3-type hysteresis loop, indicating that GO, Fe_3O_4 and MGO were all pore structures of layered materials. As shown in Figure 2f, the pore size of Fe_3O_4 was mainly in the range of 3–200 nm, and the void sizes of GO and MGO to be distributed were mainly between 3 and 100 nm. Table 1 illustrates that the specific surface areas of Fe_3O_4, GO and MGO were 70, 140 and 163 m^2/g, respectively. Compared with Fe_3O_4 and GO, the specific surface area and pore volume of MGO composites increased, while the pore size decreased. Possibly, on account of the fact that the crystallinity of the sample reduced by the continuous ultrasonic reaction in the composite process of the material, where small molecules might be released from the GO carbon skeleton, generating a new pore structure, thus the specific surface area and void volume increased. However, the massive accumulation of Fe_3O_4 particles would lead to the reduction in pore size.

Table 1. The Surface Area, Pore Volume and Pore Size of Fe_3O_4, GO and MGO.

	Surface Area (m^2/g)	Pore Volume (cm^3/g)	Pore Size (nm)
Fe_3O_4	70	0.2796	158.00
GO	140	0.3337	94.49
MGO	163	0.3921	95.48

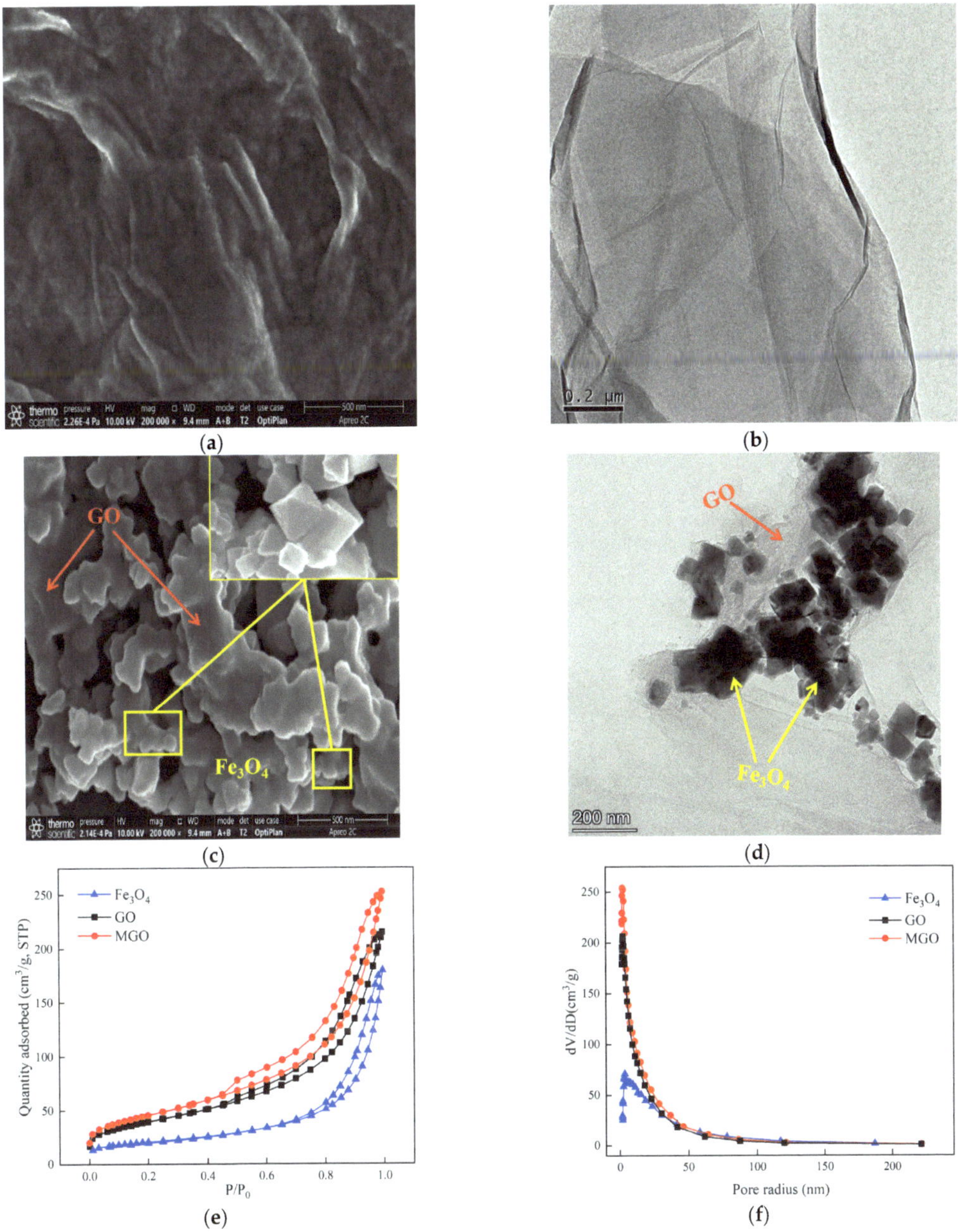

Figure 2. SEM of GO ((**a**): 500 nm) and MGO ((**c**): 500 nm); TEM of GO ((**b**): 200 nm) and MGO ((**d**): 200 nm); N_2 absorption–desorption isotherms (**e**) of GO, Fe_3O_4 and MGO; and the pore-size distribution curves (**f**) of GO, Fe_3O_4 and MGO.

3.1.2. Analysis of Crystal Structure and Chemical Characteristics

The crystal structure characteristics of GO, Fe_3O_4 and MGO were analyzed by XRD patterns at the range of 0–80° (2θ). In Figure 3a, the diffraction pattern of GO showed an obvious diffraction peak at 2θ = 10.8°, corresponding to the XRD peak of the standard

GO nanosheet (001), and the absence of the characteristic diffraction peak of graphite, indicating that graphite has been successfully oxidized to make GO. In addition, it was observed that significant characteristic peaks appeared in the XRD spectra of Fe_3O_4 at 2θ of 30.10°, 35.42°, 43.05°, 53.39°, 56.94° and 62.52°, which correspond to the diffraction peaks (220), (311), (400), (422), (511) and (440) for the pure cubic-spinel structure of Fe_3O_4 (Fe_3O_4 standard card PDF#196-0629), respectively. The diffraction peaks in the XRD spectra of MGO were identical to that of Fe_3O_4 and without the appearance of new diffraction peaks, demonstrating that an MGO composite material was successfully prepared. However, the intensity of the diffraction peaks with MGO were weaker than that of Fe_3O_4, which might be due to other lattice variations occurring within the compounding process, causing its lower average grain size and poorer crystallinity. In addition, no significant GO diffraction peak appeared at 2θ = 10.8°, which illustrated that the GO stacking in the composite presented a disordered state.

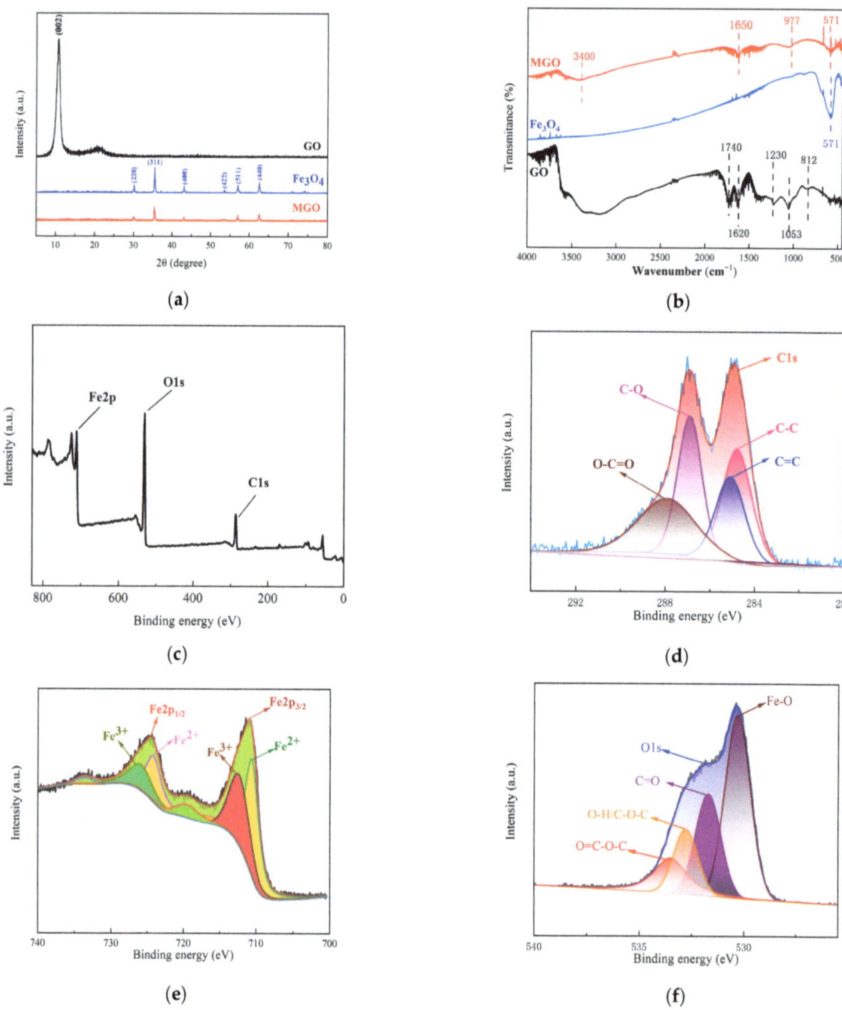

Figure 3. XRD patterns (**a**) and FT–IR spectra (**b**) of GO, Fe_3O_4 and MGO, and XPS spectra of MGO (**c**), C1s (**d**), Fe2p (**e**) and O1s (**f**).

Figure 3b showed the infrared spectrograms of GO, Fe_3O_4 and MGO for analyzing the variations observed among the functional groups. Comparing the infrared spectrum of GO and MGO, it could be seen that several characteristic peaks similar to GO were observed in MGO, yet the stretching vibration peaks containing C=O and C-O were significantly shifted (1740 cm^{-1}→1650 cm^{-1}, 1053 cm^{-1}→977 cm^{-1}), with significantly weakened absorption peaks, which probably occurred as a result of the group change in the process of the material composite. Beyond that, the characteristic peak of Fe_3O_4 could be clearly observed at 571 cm^{-1}, as a stretching vibration of the Fe-O bond, indicating that Fe_3O_4 particles were effectively introduced into the surface and lamellae of GO. The above FT-IR analysis results demonstrated the successful fabrication of Fe_3O_4-GO(MGO) composites, which contained an abundance of oxygen-containing functional groups.

The elemental composition and valence distribution of the MGO composites were characterized using X-ray photoelectron spectroscopy (XPS). The XPS whole spectrum (Figure 3c) revealed that MGO primarily consisted of just three elements, including C, O and Fe, with elemental percentages corresponding to 35.60%, 46.82% and 17.58%, respectively, as shown in Table 2. With the peak points observed at 284.0–288.0 eV, 710.0–725.5 eV and 531.0 eV for C1s, Fe2p and O1s, respectively [33–35], this suggested that Fe_3O_4 and GO compounded successfully.

Table 2. Elemental analysis of MGO.

Material	C (%)	O (%)	Fe (%)	Total (%)	C/O
MGO	35.60	46.82	17.58	100.00	0.76

The characteristic peaks of C1s (Figure 3d) were mainly distributed in the range of 284.0 to 288.0 eV, in which the peaks at 284.2, 285.0, 287.0 and 288.0 eV were identified as C-C, C=C, C-O and O-C=O, respectively. This indicated that the MGO composite material included chemical components such as cyclic hydrocarbons and olefins, and carboxyl groups and hydroxyl groups were contained on the surface, suggesting that MGO had a relatively complete structure. The fitting of the main peaks of Fe2p (Figure 3e) revealed that the characteristic peaks of $Fe2p_{1/2}$ and $Fe2p_{2/3}$ were 724.1 eV and 710.1 eV, respectively, with a difference in binding energies of approximately 14 ev. In this case, the Fe^{2+} peaks were located at 710.5 eV and 722.6 eV, whereas that of 711.0 eV and 725.3 eV belonged to the Fe^{3+} [36]. This indicated that the Fe element in the MGO composite mainly existed in the form of Fe_3O_4. Three relatively distinct peaks were observed in the energy spectrum of O1s (Figure 3f), in which the peaks around 530.0 eV and 532.8 eV corresponded to Fe-O in Fe_3O_4 and hydroxyl oxygen (O-H) on the surface of the material, respectively, as well as the characteristic peak at 530.8 eV belonging to carboxyl oxygen or carbonyl oxygen (C=O) [33,34,37]. Simultaneously, the large energy spectrum area could demonstrate that the MGO composites contained abundant carboxyl and hydroxyl functional groups, manifesting the structural integrity of the material, which was conducive to the adsorption and catalytic degradation of pollutants.

3.2. Degradation of CIP by MGO-Activated PMS

3.2.1. Analysis of Crystal Structure and Chemical Characteristics

The removal effects of Fe_3O_4, GO and MGO-activated PMS on CIP are compared in Figure 4. As shown in Figure 4a, with PMS alone, only a 31.74% removal of CIP was achieved after 120 min of treatment, which was much lower than that in the presence of Fe_3O_4, GO or MGO. This indicated that Fe_3O_4, GO and MGO all have a certain activation effect on PMS, improving the CIP removal rate to 60.61%, 76.47% and 81.39% within 120 min, respectively.

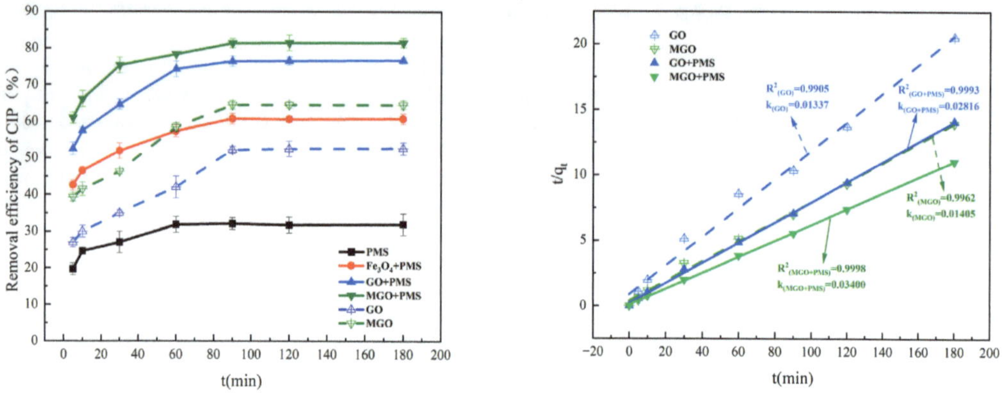

(**a**) (**b**)

Figure 4. Removal of CIP in PMS alone, Fe_3O_4/PMS, GO/PMS and MGO/PMS (**a**). Pseudo-second-order adsorption kinetic model (**b**). Experimental conditions: [CIP] = 20 mg/L, [PMS] = 1.0 g/L, [Fe_3O_4, GO, MGO] = 0.5 g/L, T = 25 °C.

In the Fe_3O_4/PMS system, for one thing, Fe_3O_4 nanoparticles adsorbed a certain concentration of CIP owing to the large specific surface area. For another, $SO_4^{-}\bullet$ generated from PMS activated by Fe^{2+} on the surface of Fe_3O_4 nanoparticles had a powerful oxidation effect and could remove CIP in water as shown in Reaction (2) [38,39]. However, it is possible that the specific surface area, the stability of dispersion and the catalytic activity of the catalysts may be decreased by the ease of aggregation of Fe_3O_4 nanoparticles in the solution [40].

$$Fe^{2+} + HSO_5^{-} \rightarrow SO_4^{-}\bullet + Fe^{3+} + OH^{-} \quad (2)$$

Several studies demonstrated that carbon-based materials could also activate PMS [41–43]. Gao et al. [44] reviewed the study on the degradation of antibiotic contaminants by activation of PMS with different carbon materials (activated carbon, biochar, carbon nanotube, other carbon materials or doped-carbon materials, etc.) and found that the majority of carbon materials with a low cost and chemical stability were commonly applied to activate PMS to produce active species, but with a low activation efficiency. In the GO/PMS system, •OH and $SO_4^{-}\bullet$ were generated from the activation of PMS with the abundance of oxygen-containing functional groups contained in GO. In addition, GO performed the role of a bridge of electron transfer to transfer electrons from organic compounds to PMS [13,45], with the following reactions, Reactions (3) and (4):

$$HSO_5^{-} + e \rightarrow SO_4^{-}\bullet + OH^{-} \quad (3)$$

$$HSO_5^{-} + e \rightarrow SO_4^{2-} + \bullet OH \quad (4)$$

As compared with the other two catalysts (Fe_3O_4 and GO), MGO displayed a better catalytic activation effect, which lead to a 80% removal of CIP within 120 min. By contrast, with the absence of PMS, the removal of CIP was just about 64% for the same reaction time by adsorption of MGO alone, which accounts for the synergistic catalytic effect of Fe_3O_4 and GO on PMS.

Kinetic investigations of the above reaction systems revealed that none of them conformed to the degradation kinetic model (Figure S1a,b), nor to the pseudo-first-order adsorption kinetic model (Figure S1c). The adsorption kinetic studies of GO, MGO, GO/PMS and MGO/PMS are presented in Figure 4b, and the linear correlation coefficients R^2 = 0.9905, 0.9962, 0.9993 and 0.9998, respectively, which reached a significant level, with the adsorption rate constants k being 0.01337, 0.01405, 0.02816 and 0.03400 mg· (g·min)$^{-1}$, approximately, demonstrating that the pseudo-second-order adsorption kinetic model better proportioned the dynamic process of CIP removal from the aqueous solution by GO, MGO, GO/PMS and MGO/PMS. As a heterogeneous reaction, the removal of CIP by MGO/PMS occurred as an adsorption process with a progressively enhanced chemical reaction, in no way suggesting that the removal was realized exclusively by adsorption of MGO alone. Actually, the degradation of CIP by MGO-activated PMS is a reaction process involving adsorption and degradation simultaneously, not a singular adsorption or degradation reaction. The degradation of CIP occurred while PMS and CIP were adsorbed onto the surface of MGO, in which the consumed PMS and the degraded CIP contributed new adsorption sites for MGO, leading to the circular reaction with adsorption and degradation being performed again. Enhanced adsorption capacity as indicated by an increase in the value of the adsorption rate constant k was not caused by improving the adsorption capacity of the MGO material itself; on the contrary, it is attributed to the fact that the oxidation reaction occurring on the MGO surface intensified the adsorption reaction processes of MGO.

3.2.2. Effects of Key Factors on CIP Degradation

The initial pH effect of the solution between 3 and 11 on CIP adsorption and degradation is explored in Figure 5a. As can be seen, the adsorption of CIP showed to be weaker than the degradation reaction of CIP. The neutral conditions were favorable for the adsorption and degradation of CIP, in which case the CIP adsorbed and degraded by MGO was 64.53% and 82.46%, respectively, significantly higher than that of the acidic and alkaline conditions.

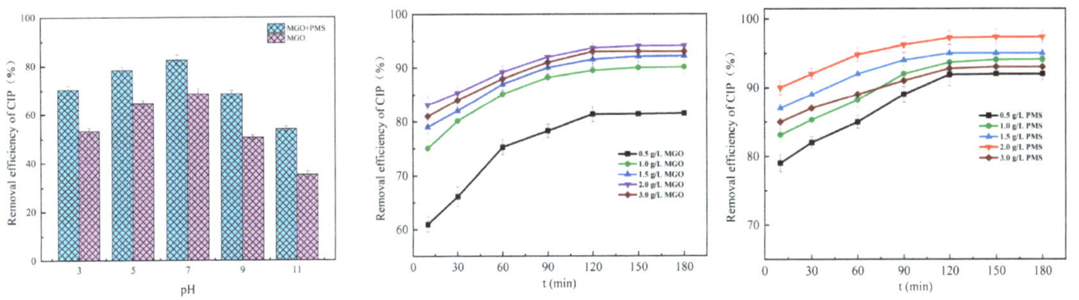

(a) (b) (c)

Figure 5. Effect of initial pH of solution (**a**), MGO dosage (**b**) and PMS concentration (**c**). Experimental conditions: [CIP] = 20 mg/L, T = 25 °C.

Typically, the initial pH of the solution would affect the existence form of PMS, the species and presence of free radicals as well as the process of the charge transferred [46], in which PMS was mainly in the form of HSO_5^- in acidic and neutral conditions, while existing primarily as SO_5^{2-} in the alkaline condition [47]. The more acidic the solution, the more prone HSO_5^- was to form •OH with weaker oxidation capacity than SO_4^-•, which inhibited the degradation reaction of CIP. Meanwhile, the CIP, which was mainly in the form of cations in acidic conditions, repelled with the positive charges on the MGO surface, diminishing the electrostatic binding force between CIP and MGO, resulting in weakened adsorption of CIP. It was also found that the removal of CIP gradually weakened as the solution pH was increased up to 7 from 11. On the one hand, the more alkaline the solution, the more negative charges accumulated on the surface of the MGO composite, which leads to the greater repulsion of the PMS negative ions and results in less PMS accumulating around the catalyst surface [48], and especially the pH value exceeding 9. On the other hand, the dominant form of CIP is the anion in strong alkali solutions, which also causes a repulsive force with the surface of the catalyst, and thereby consequently contributing to the tendency toward a reduction for both adsorption and degradation of CIP.

The effect of the concentration of MGO on CIP removal is revealed in Figure 5b, in which the concentration of PMS is controlled at 1.0 g/L. As can be seen, the final removal of CIP after 180 min treatment rose with the increase in the MGO dosage, and thereafter, it decreased as the dosage continued to increase. Specifically, as the concentration of MGO increased up to 2.0 g/L from 0 g/L, the CIP removal rate was significantly enhanced from 31.74% up to 94.11%. After that, it decreased to 93.04% for the MGO dosage of 3.0 g/L. This result indicated that moderately increasing the dosage of MGO was beneficial for the removal of CIP. On one hand, the adsorption of CIP on MGO could be improved by the increased MGO dosage. On the other, the increase in MGO could also provide additional active sites to promote the catalytic activation of PMS to generate oxidative radicals such as $SO_4^{-}\bullet$ and $\bullet OH$, which would increase the degradation of CIP. However, the excessive MGO dosage decreased the removal of CIP, and this could be ascribed to the explosive generation of oxidative free radicals, which increased the invalid consumption of free radicals by other substances in a solution like Fe^{2+}, as represented in the following reactions, Reactions (5) and (6) [13]:

$$Fe^{2+} + SO_4{}^{-}\bullet \rightarrow Fe^{3+} + SO_4{}^{2-} \tag{5}$$

$$Fe^{2+} + \bullet OH \rightarrow Fe^{3+} + OH^{-} \tag{6}$$

The contribution of PMS concentration for CIP removal was investigated (Figure 5c), where the MGO dosage was controlled as 2.0 g/L with the CIP initial concentration of 20 mg/L. As can be observed, the final removal rate of CIP after 180 min treatment increased with the addition of the PMS dosage, and thereafter, it declined with the addition of the dosage. Concretely, the CIP removal rate increased from 79.00 to 97.33% as the PMS dosage increased from 0.5 g/L to 2.0 g/L. Afterwards, it declined to 93.01% for the PMS dosage of 3.0 g/L. This result revealed that moderately increasing the amount of PMS contributed to the removal of CIP. $SO_4^{-}\bullet$ and other reactive species in the water increased as a result of the increasing concentration of PMS, which accelerated the degradation of CIP. Nevertheless, the excessive PMS dosage decreased the removal of CIP, which might be caused by the fact that the excessive generated free radicals would react with other substances in the solution, in particular with HSO_5^{-}, reacting with $SO_4^{-}\bullet$ and $\bullet OH$ to form $SO_5^{-}\bullet$, being of less activity as well as a weak oxidative capacity [49]. This not only led to the unnecessary consumption of a large amount of oxygen-containing radicals but also reduced the effective utilization of PMS, resulting in a lower removal of CIP. The main reactions were as follows [45], Reactions (7)–(9):

$$HSO_5{}^{-} + \bullet OH \rightarrow SO_5{}^{-}\bullet + H_2O \tag{7}$$

$$HSO_5{}^{-} + SO_4{}^{-}\bullet \rightarrow HSO_4{}^{-} + SO_5{}^{-}\bullet \tag{8}$$

$$SO_4{}^{-}\bullet + SO_4{}^{-}\bullet \rightarrow S_2O_8{}^{2-} \tag{9}$$

3.2.3. Recyclability of MGO

The stability and recyclability of the materials were considered as a vital condition for heterogeneous reactions. A magnetic test experiment (Figure S2) indicated that MGO displayed magnetic properties and allowed for magnetic separation. To evaluate the stability and recyclability of MGO, the used MGO was separated and collected magnetically, and washed repeatedly using a HCl solution (0.1 mol/L) and deionized water; after that, the regenerated MGO was ready for reuse. As shown in Figure 6, after 360 min treatments, the final removal rates of CIP were 96.5%, 92.2% and 87.0% for the first, second and third uses, respectively, which showed a gradually decreasing trend. This may be mainly due to the decrease in Fe^{2+} released into the solution, the incomplete elution of the adsorbed contaminants and the loss of the active functional groups on the surface of MGO. However, the catalytic performance of MGO was not significantly decreased and can meet the requirements of wastewater treatment, which may be accounted for regarding the protective effect of the multi-layer GO layers on Fe_3O_4, as shown in SEM and TEM images.

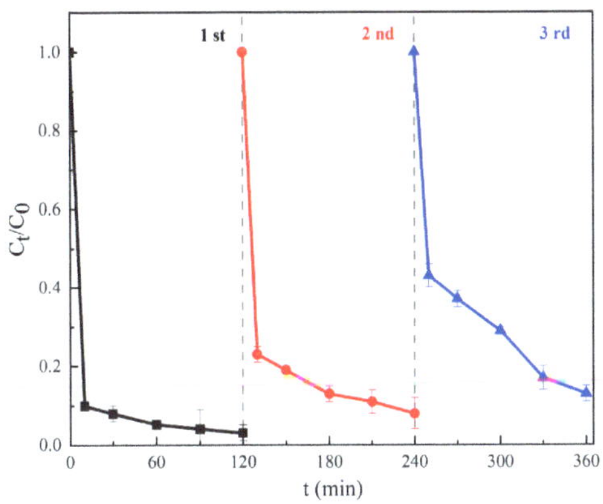

Figure 6. Stability and recyclability of MGO. Experimental conditions: [CIP] = 20 mg/L, [PMS] = 1.0 g/L, [MGO] = 0.5 g/L, pH = 7, T = 25 °C.

3.3. Mechanism of CIP Degradation by MGO-Activated PMS

A consensus is that a number of oxidative radicals can be generated during the activation of PMS [50,51]. To verify the activation effect of MGO on PMS, an EPR detection was carried out utilizing DMPO as the spin trapping agent, which is presented in Figure 7a. As can be found, an obvious 1:2:2:1 four-peak DMPO-OH signal and a relatively weaker 1:1:1:1:1 six-peak DMPO-SO_4 signal can be detected [52], which confirms the simultaneous presence of •OH and SO_4^-• in the MGO/PMS reaction system.

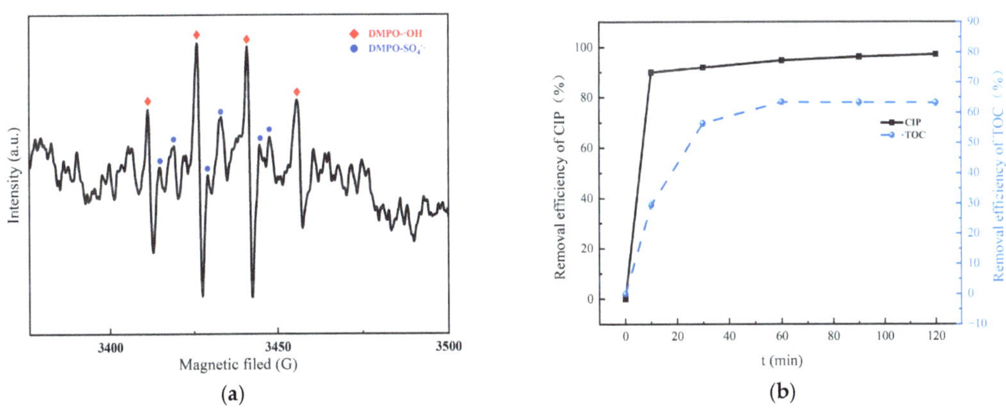

Figure 7. EPR spectra of DMPO–OH adduct and DMPO–SO_4 in MGO/PMS system for 10 min (**a**); removal rate of CIP and TOC (**b**). Experimental conditions: 10 mL of deionized water, 2 mL of DMPO, pH = 7, T = 25 °C.

To investigate the mineralization ability of the MGO/PMS reaction system on organic pollutants, the CIP and TOC removal are compared in Figure 7b. As can be seen, after a 60 min reaction, the removal for CIP and TOC was 94.8% and 63.4%, correspondingly. What this result demonstrated is that both CIP and TOC could be effectively removed by the MGO/PMS reaction. Another fact was that the removal of TOC was significantly lower than that of CIP, which suggested the generation of organic intermediates during the degradation process.

On the basis of the results of free radical trapping and CIP mineralization, as well as the combination of the FT-IR and XPS analysis, the mechanism of CIP degradation for MGO-activated PMS was speculated as follows. As shown in Figure 8, in the MGO/PMS system, in which both Fe_3O_4 and GO were involved in the reaction, the CIP degradation primarily proceeded with free radical processes. In one way, Fe^{2+} reacted with PMS, generating $SO_4^-\bullet$ and $\bullet OH$, thereby oxidizing to Fe^{3+} with loss of electrons (Reactions (2) and (10)). Simultaneously, the Fe^{3+} obtained electron would initiate the regeneration of Fe^{2+} (Reaction (11)), thus forming a cyclic reaction chain that promoted free radical generation. For the other one, the modified GO might contain an abundance of free electrons, and the sp_2 active carbon sites in GO, especially π electrons, exhibit a significantly higher electron transfer capacity, which induce the production of $SO_4^-\bullet$ with PMS (Reactions (12)–(15)) [44]. In addition, with the above reaction, in which some of the $SO_4^-\bullet$ will react with water molecules and OH^- in the solution to form a large amount of $\bullet OH$ (Reactions (16) and (17)) [52], the species and contents of the free radicals are affected dependent upon the pH of the solution, which is also in accordance with the results of the investigations in Section 3.2.2.

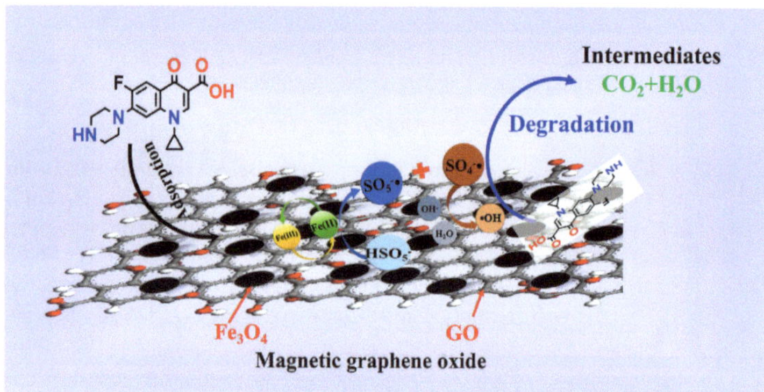

Figure 8. The mechanism of CIP degradation.

The reactive species such as $SO_4^-\bullet$ and $\bullet OH$ generated in the MGO/PMS system feature relatively high redox potentials, by which the CIP molecules may be attacked and transformed into smaller molecule intermediates that are eventually mineralized to CO_2 and H_2O with the redox reactions involving dehydroxylation, decarboxylation, etc. [44] (Reaction (18)).

$$Fe^{2+} + HSO_5{}^- \rightarrow \bullet OH + Fe^{3+} + SO_4{}^{2-} \quad (10)$$

$$Fe^{3+} + HSO_5{}^- \rightarrow Fe^{2+} + SO_5{}^-\bullet + H^+ \quad (11)$$

$$C - \pi + HSO_5{}^- \rightarrow SO_4{}^-\bullet + C - \pi^+ + OH^- \quad (12)$$

$$C - \pi^+ + HSO_5{}^- \rightarrow SO_4{}^-\bullet + C - \pi + H^+ \quad (13)$$

$$C = C = O + HSO_5{}^- \rightarrow SO_4{}^-\bullet + C = C = O^+ + OH^- \quad (14)$$

$$C = C = O^+ + HSO_5{}^- \rightarrow SO_4{}^-\bullet + C = C = O + H^+ \quad (15)$$

$$SO_4{}^-\bullet + H_2O \rightarrow \bullet OH + SO_4{}^{2-} \quad (16)$$

$$SO_4^{-}\bullet + OH^- \rightarrow \bullet OH + H^+ \quad (17)$$

$$SO_4^{-}\bullet/\bullet OH + CIP \rightarrow \text{Intermediates} + CO_2 + H_2O \quad (18)$$

The intermediates were generated from the reaction between CIP and oxidative radicals. To identify the intermediates, the LC-MS spectrum CIP solutions before and after 120 min treatment were compared in Figure S3a,b, and the corresponding intermediates speculated on based on the mass-to-charge ratios are listed in Table S1. On the basis of the detection of intermediates and oxidized radicals, the degradation mechanism of CIP in the MGO/PMS reaction system could be proposed as shown in Figure 9, which mainly has three degradation pathways.

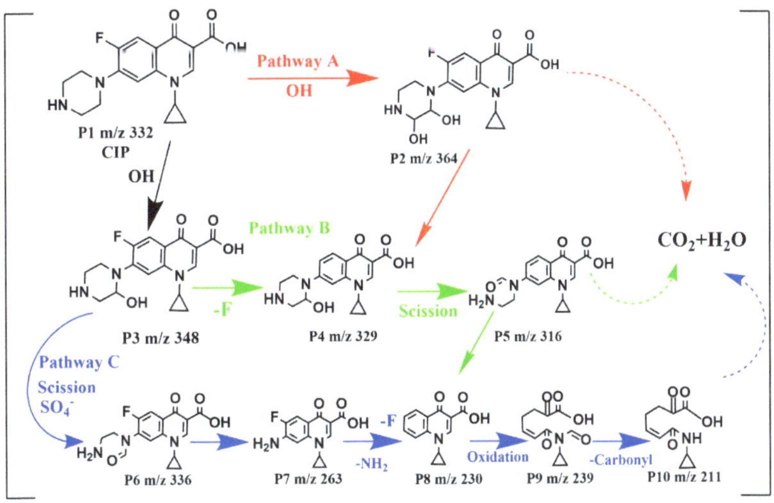

Figure 9. Proposed degradation pathways for CIP by MGO-activated PMS.

For pathway A, the piperazine ring of CIP attacked with •OH formed either P2 (m/z = 364) or P3 (m/z = 348). In pathway B, P2 (m/z = 348) was directly defluorinated to generate P4 (m/z = 329), which formed P5 (m/z = 316) with the direct piperazine ring opening and '-CH3' loss. Pathway C initiated with an oxidation reaction process in which an electron in the piperazine group was substituted by $SO_4^{-}\bullet$, followed by what seemed to be the side chain of the piperazine ring of P3 (m/z = 348) fracturing and opening for the formation of P6 (m/z = 336) [53,54]. Following further oxidation, the piperazine ring was completely removed to generate P7 (m/z = 263), which was transformed to P8 (m/z = 230) as a direct defluorination ion and amino group. As the quinoline ring of P8 continued to be opened by oxidation and a decarbonylation reaction occurred, P9 (m/z = 239) and P10 (m/z = 211) were successively generated. After that, P2, P5 and P10 produced from pathway A, B and C, respectively, would be further degraded into final products (small organic compounds, CO_2 and H_2O, etc.). In degrading CIP, the quinolone ring in the CIP molecule appears to be the weakest in oxidability, followed by the benzene ring, while the piperazine ring shows the highest oxidability [55]. Consequently, pathways B and C were the dominant degradation pathways for CIP, while pathway A was the secondary one.

4. Conclusions

In this study, an MGO composite was synthesized by the ultrasonic chemical precipitation method and characterized with SEM, TEM, BET, XRD, FT-IR and XPS in terms of morphological features, structure, functional groups and elemental groups. As compared with the degradation of CIP by PMS alone, GO/PMS and Fe_3O_4/PMS, the MGO/PMS system exhibited the better performance of the CIP removal, which could be attributed to the catalytic activation of PMS with MGO. For the solution with an initial CIP concentration of 20 mg/L, a maximum removal could reach up to 96.5% at the optimum reaction conditions, which corresponds to a TOC removal rate of 63.4%. A kinetic model of pseudo-secondary adsorption indicated a dynamic process influenced by chemisorption. After three cycles of reuse of MGO, the CIP removal rate can be maintained at 87.0%, which demonstrates the stability and recyclability of MGO. The EPR results confirmed that oxidative radicals generated in the MGO/PMS system, particularly •OH and SO_4^{-}• play a dominant role for CIP degradation. The TOC removal rate and LC-MS spectrum indicated that CIP was partially degraded into organic intermediates, and the degradation mechanism of CIP was proposed as a process with three degradation pathways. It is thus assumed based on our investigation results that MGO serves as a potential catalyst for PMS activation, which presents promising research and application prospects in the field of wastewater treatment.

Supplementary Materials: The following supporting information can be downloaded at: https://www.mdpi.com/article/10.3390/toxics11121016/s1, Figure S1: Pseudo-first-order degradation kinetic model (a), Pseudo-sencond-order degradation kinetic model (b) and Pseudo-first-order adsorption kinetic model (c); Figure S2: Magnetic properties test of MGO. (magnetic separation of solid states (a), magnetic separation of MGO in aqueous solution ((b): 1 min; (c): 30 min)); Figure S3: LC-MS detection results for CIP before treatment (a) and after 120 min of treatment (b); Table S1: Analysis of intermediates from degradation of CIP.

Author Contributions: Conceptualization (research ideas and goals), X.W.; Methodology, X.W.; Writing—review and editing, X.W.; Funding acquisition and Resources, X.W.; Experimental investigation, Y.L. and J.Q.; Data treatment, Y.L.; Formal analysis, Y.L., J.Q., T.S. and X.L.; Software, Y.L.; Writing—original draft, Y.L.; Data curation, J.Q.; Supervision, P.P. and D.J.; Validation, P.P., D.J., T.S. and X.L. All authors have read and agreed to the published version of the manuscript.

Funding: This work was supported by Chongqing Science and Technology Bureau (No. CSTB2022NSCQ-MSX0549), Scientific Research Project of Chongqing Technology and Business University (No. 2152004) and Science and Technology Research Program of Chongqing Education Commission of China (No. KJQN202300843).

Institutional Review Board Statement: Not applicable.

Informed Consent Statement: Not applicable.

Data Availability Statement: Data are available from the corresponding author by request.

Conflicts of Interest: The authors declare no conflict of interest.

References

1. Bisacchi, G.S. Origins of the quinolone class of antibacterials:an expanded "discovery story". *J. Med. Chem.* **2015**, *58*, 4874–4882. [CrossRef]
2. Yao, S.J.; Ye, J.F.; Yang, Q.; Hu, Y.R.; Cui, C.Z. Occurrence and removal of antibiotics, antibiotic resistance genes, and bacterial communities in hospital wastewater. *Environ. Sci. Pollut. Res.* **2021**, *28*, 57321–57333. [CrossRef] [PubMed]
3. Igwegbe, C.A.; Oba, S.N.; Aniagor, C.O.; Adeniyi, A.G.; Joshhua, O. Adsorption of ciprofloxacin from water: A comprehensive review. *J. Ind. Eng. Chem.* **2021**, *93*, 57–77. [CrossRef]
4. Yang, W.J.; Feierkate, F.; Ma, J.W.; Cui, Y.B. Research Progress of Antibiotic Pollution and Treatment Technologies in China. *E3S Web Conf.* **2020**, *194*, 04004. [CrossRef]
5. Le-minh, N.; Khan, S.J.; Drewes, J.E.; Stuetz, R. Fate of antibiotics during municipal water recycling treatment processes. *Water Res.* **2010**, *44*, 4295–4323. [CrossRef] [PubMed]
6. Pan, L.J.; Li, J.; Li, C.X.; Tang, X.D.; Yu, G.W.; Wang, Y. Study of ciprofloxacin biodegradation by a *Thermus* sp. isolated from pharmaceutical sludge. *J. Hazard. Mater.* **2018**, *343*, 59–67. [CrossRef] [PubMed]
7. Toli, K.; Mutavdi Pavlovi, D.; Stankir, N.; Runje, M. Biosorbents from Tomato, Tangerine, and Maple Leaves for the Removal of Ciprofloxacin from Aqueous Media. *Water Air Soil Pollut.* **2021**, *232*, 218. [CrossRef]
8. Asadi, Z.; Ahmadi, S. Investigation of the Efficiency of Coagulation Process for Ciprofloxacin Antibiotic Removal from Aqueous Solution. *J. Health Res. Community* **2019**, *5*, 38–48.

9. Kiejza, D.; Kotowska, U.; Polińska, W.; Karpińska, J. Peracids-New oxidants in advanced oxidation processes: The use of peracetic acid, peroxymonosulfate, and persulfate salts in the removal of organic micropollutants of emerging concern—A review. *Sci. Total Environ.* **2021**, *790*, 148195. [CrossRef]
10. Guerra-Rodríguez, S.; Rodríguez, E.; Singh, D.N.; Rodríguez-Chueca, J. Assessment of Sulfate Radical-Based Advanced Oxidation Processes for Water and Wastewater Treatment: A Review. *Wastewater Treat. Reuse* **2018**, *10*, 1828. [CrossRef]
11. Tang, X.J.; Hashmi, M.Z.; Zeng, B.; Yang, J.W.; Shen, C.F. Application of iron-activated persulfate oxidation for the degradation of PCBs in soil. *Chem. Eng. J.* **2015**, *279*, 673–680. [CrossRef]
12. Verma, S.; Nakamura, S.; Sillanpää, M. Application of UV-C LED activated PMS for the degradation of anatoxin-a. *Chem. Eng. J.* **2016**, *284*, 122–129. [CrossRef]
13. Li, J.Q.L.; Sun, M.T.; Ao, H.Q.; Wang, Z.M.; Liu, S.B.; Shao, M. Understanding of the Oxidation Behavior of Benzyl Alcohol by Peroxymonosulfate via Carbon Nanotubes Activation. *ACS Catal.* **2020**, *10*, 3516–3525. [CrossRef]
14. Antoniou, M.G.; De La Cruz, A.A.; Dionysiou, D.D. Degradation of microcystin-LR using sulfate radicals generated through photolysis, thermolysis and e-transfer mechanisms. *Appl. Catal. B Environ.* **2010**, *96*, 290–298. [CrossRef]
15. Zhang, B.T.; Zhang, Y.; Teng, Y.G. Electrospun magnetic cobalt–carbon nanofiber composites with axis-sheath structure for efficient peroxymonosulfate activation. *Appl. Surf. Sci.* **2018**, *452*, 443–450. [CrossRef]
16. Zhu, Z.; You, Z.D. Co_3O_4/C-PC composite derived from pomelo peel-loaded ZIF-67 for activating peroxymonosulfate (PMS) to degrade ciprofloxacin. *J. Water Process Eng.* **2022**, *49*, 103043.
17. Pu, M.J.; Ye, D.Q.; Wan, J.Q. Zinc-based metal-organic framework nanofibers membrane ZIF-65/PAN as efficient peroxymonosulfate activator to degrade aqueous ciprofloxacin. *Sep. Purif. Technol.* **2022**, *299*, 121716. [CrossRef]
18. Li, Z.Z.Q.; Ning, S.Y.; Hu, F.T.; Zhu, H.; Zeng, L.D.; Chen, L.F.; Wang, X.P.; Fujita, T.; Wei, Y.Z. Preparation of VCo-MOF@MXene composite catalyst and study on its removal of ciprofloxacin by catalytically activating peroxymonosulfate: Construction of ternary system and superoxide radical pathway. *J. Colloid Interface Sci.* **2022**, *629*, 97–110. [CrossRef]
19. He, B.; Song, L.X.; Zhao, Z.X.; Liu, W.B.; Zhou, Y.T.; Shang, J.W.; Cheng, X.W. $CuFe_2O_4$/CuO magnetic nano-composite activates PMS to remove ciprofloxacin: Ecotoxicity and DFT calculation. *Chem. Eng. J.* **2022**, *446*, 137183. [CrossRef]
20. Miklos, D.B.; Remy, C.; Jekel, M.; Linden, K.G.; Drewes, J.E.; Hübner, U. Evaluation of advanced oxidation processes for water and wastewater treatment—A critical review. *Water Res.* **2018**, *139*, 118–131. [CrossRef]
21. Guo, W.L.; Su, S.N.; Yi, C.L.; Ma, Z.M. Degradation of Antibiotics Amoxicillin by Co_3O_4-Catalyzed Peroxymonosulfate System. *Environ. Prog. Sustain. Energy* **2013**, *32*, 193–197. [CrossRef]
22. Oh, W.D.; Dong, Z.L.; Lim, T.T. Generation of sulfate radical through heterogeneous catalysis for organic contaminants removal:Current development, challenges and prospects. *Appl. Catal. B Environ.* **2016**, *194*, 169–201. [CrossRef]
23. Vakros, J. The Influence of Preparation Method on the Physicochemical Characteristics and Catalytic Activity of Co/TiO_2 Catalysts. *Catalysts* **2020**, *10*, 88. [CrossRef]
24. Fathy, M.; Gomaa, A.; Taher, F.A.; El-Fass, M.M.; Kashyout, A.E.H.B. Optimizing the preparation parameters of GO and rGO for large-scale production. *J. Mater. Sci.* **2016**, *51*, 5664–5675. [CrossRef]
25. Ersan, G.; Apul, O.G.; Perreault, F. Adsorptionof organic contaminants by graphene nanosheets: A review. *Water Res.* **2017**, *126*, 385–398. [CrossRef]
26. Wang, F.; Yang, B.S.; Wang, H.; Song, Q.X.; Tan, F.J.; Cao, Y.N. Removal of ciprofloxacin from aqueous solution by a magnetic chitosan grafted graphene oxide composite. *J. Mol. Liq.* **2016**, *222*, 188–194. [CrossRef]
27. Kumar, M.; Gholamvand, Z.; Morrissey, A.; Nolan, K.; Ulbricht, M.; Lawler, J. Preparation and characterization of low fouling novel hybrid ultrafiltration membranes based on the blends of $GO-TiO_2$ nanocomposite and polysulfone for humic acid removal. *J. Membr. Sci.* **2016**, *506*, 38–49. [CrossRef]
28. Wang, X.L.; Wang, Q.D. Preparation of graphite oxide (GO) and the thermal stability of silicone rubber/GO nanocomposites. *Thermochim. Acta* **2012**, *529*, 25–28. [CrossRef]
29. Tang, Y.; Guo, H.; Xiao, L.; Yu, S.L.; Gao, N.Y.; Wang, Y.L. Synthesis of reduced graphene oxide/magnetite composites and investigation of their adsorption performance of fluoroquinolone antibiotics. *Colloid Surf. A* **2013**, *424*, 74–80. [CrossRef]
30. Shraban, K.S.; Jitentra, K.S.; Gagan, K.P.; Dillip, K.P.; Aninda, S.R.; Animesh, L. Preparation of Graphene Oxide from Bio-soot wastes: As an efficient adsorbent forhighly noxious Congo red dye. *FlatChem* **2020**, *24*, 100198.
31. Cao, N.; Zhang, Y. Study of reduced graphene oxide preparation by hummers' method and related characterization. *J. Nanomater.* **2015**, *2015*, 168125. [CrossRef]
32. Song, J.G.; Wang, X.Z.; Chang, C.T. Preparation and Characterization of Graphene Oxide. *J. Nanomater.* **2014**, *2014*, 276143. [CrossRef]
33. Long, J.P.; Li, S.X.; Liang, J.M.; Wang, A.G.; Liang, B. Preparation and characterization of graphene oxide and it application as a reinforcement in polypropylene composites. *Polym. Compos.* **2019**, *40*, 723–729. [CrossRef]
34. Mullet, M.; Khare, V.; Ruby, C. XPS study of Fe(II)–Fe(III) (oxy)hydroxycarbonate green rust compounds. *Anal. Sci. J.* **2008**, *40*, 323–328. [CrossRef]
35. Stoch, J.; Gablankowska-Kukucz, J. The Effect of Carbonate Contaminations on the XPS O1s Band Structure in Metal Oxides. *Anal. Sci. J.* **1991**, *17*, 165–167.
36. Grosvenor, A.P.; Kobe, B.A.; Biesinger, M.C.; Mcintyre, N.S. Investigation of multiplet splitting of Fe2p XPS spectra and bonding in iron compounds. *Surf. Interface Anal.* **2004**, *36*, 1564–1574. [CrossRef]

37. Lopez, G.P.; Castner, D.G.; Ratner, B.D. XPS O1s Binding Energies for Polymers Containing Hydroxyl, Ether, Ketone and Ester Groups. *Surf. Interface Anal.* **1991**, *17*, 267–272. [CrossRef]
38. Zou, J.; Ma, J.; Chen, L.W.; Li, X.C.; Guan, Y.H.; Xie, P.C.; Pan, C. Rapid Acceleration of Ferrous Iron/Peroxymonosulfate Oxidation of Organic Pollutants by Promoting Fe(III)/Fe(II) Cycle with Hydroxylamine. *Environ. Sci. Technol.* **2013**, *47*, 11685–11691. [CrossRef]
39. Luo, Y.; Liu, C.; Zhao, M.Q. CoFe-LDO nanoparticles as a novel catalyst of peroxymonosulfate (PMS) for histidine removal. *Environ. Sci. Pollut. Res.* **2022**, *29*, 16517–16528. [CrossRef]
40. Yin, F.; Wang, C.; Lin, K.Y.A.; Tong, S. Persulfate activation for efficient degradation of norfloxacin by a rGO-Fe_3O_4 composite. *J. Taiwan Inst. Chem. Eng.* **2019**, *102*, 163–169. [CrossRef]
41. Wang, A.W.; Ni, J.X.; Wang, W.; Liu, D.M.; Xue, B.X.; Chang, C.C.; Ma, J.; Zhao, Y. MOF Derived Co-Fe nitrogen doped graphite carbon@crosslinkedmagnetic chitosan Micro-nanoreactor for environmental applications: Synergy enhancement effect of adsorptio−PMS activation. *Appl. Catal. B Environ.* **2022**, *319*, 121926. [CrossRef]
42. Hou, S.Y.; Zhang, H.Z.; Wang, P.; Zhang, M.; Yang, P. Multiwall carbon nanotube decorated hemin/Mn-MOF towards BPA degradation through PMS activation. *J. Environ. Chem. Eng.* **2022**, *10*, 108426. [CrossRef]
43. Fu, J.Y.; Feng, L.; Liu, Y.Z.; Zhang, L.Q.; Li, S. Electrochemical activation of peroxymonosulfate(PMS) by carbon cloth anode for sulfamethoxazole degradation. *Chemosphere* **2022**, *287*, 132094. [CrossRef]
44. Gao, Y.; Wang, Q.; Ji, G.Z.; Li, A.M. Degradation of antibiotic pollutants by persulfate activated with various carbon materials. *Chem. Eng. J.* **2022**, *429*, 132387. [CrossRef]
45. Wang, C.; Du, J.Y.; Liang, Z.J.; Zhan, Z.W.; Cui, F.Y.; Shi, W.X. High-efficiency oxidation of fluoroquinolones by the synergistic activation of peroxymonosulfate via vacuum ultraviolet and ferrous iron. *J. Hazard. Mater.* **2022**, *422*, 126884. [CrossRef]
46. Ntzoufra, P.; Vakros, J.; Frontistis, Z.; Tsatsos, S.; Mantzavinos, D. Effect of sodium persulfate treatment on the physicochemical properties and catalytic activity of biochar prepared from spent malt rootlets. *J. Environ. Chem. Eng.* **2021**, *9*, 105071. [CrossRef]
47. Tan, C.Q.; Gao, N.Y.; Deng, Y.; Deng, J.; Zhou, S.; Li, J.; Xin, X.Y. Radical induced degradation of acetaminophen with Fe_3O_4 magnetic nanoparticles as heterogeneous activator of Peroxymonosulfate. *J. Hazard. Mater.* **2014**, *5*, 452–460. [CrossRef]
48. Zhang, T.; Zhu, H.; Croue, J.P. Production of sulfate radical from peroxymonosulfate induced by a magnetically separable $CuFe_2O_4$ spinel in water: Efficiency, stability, and mechanism. *Environ. Sci. Technol.* **2013**, *47*, 2784–2791. [CrossRef]
49. Guan, Y.H.; Ma, J.; Ren, Y.M.; Liu, Y.L.; Xiao, J.Y.; Lin, L.Q.; Zhang, C. Efficient degradation of atrazine by magnetic porous copper ferrite catalyzed peroxymonosulfate oxidation via the formation of hydroxyl and sulfate radicals. *Water Res.* **2013**, *47*, 5431–5438. [CrossRef]
50. He, Y.X.; Qian, J.; Wang, P.F.; Wu, J.; Lu, B.H.; Tang, S.J.; Gao, P. Acceleration of levofloxacin degradation by combination of multiple free radicals via MoS_2 anchored in manganese ferrite doped perovskite activated PMS under visible light. *Chem. Eng. J.* **2022**, *431*, 133933. [CrossRef]
51. Hua, L.C.; Huang, C.; Huang, C.P. Insights on free radical oxidation and in-situ coagulation in PMS/Fe(II) process for the removal of algogenic organic matter precursors. *Chem. Eng. J.* **2022**, *446*, 136986. [CrossRef]
52. Lai, L.D.; Ji, H.D.; Zhang, H.; Liu, R.; Zhou, C.Y.; Liu, W.; Ao, Z.M.; Li, N.W. Activation of peroxydisulfate by V-Fe concentrate ore for enhanced degradation of carbamazepine: Surface≡V(III) and≡V(IV) as electron donors promoted the regeneration of≡Fe(II). *Appl. Catal. B Environ.* **2021**, *282*, 119559. [CrossRef]
53. Zhou, Y.F.; Yu, M.C.; Liang, H.J.; Chen, J.; Xu, L.; Niu, J.F. Novel dual-effective Z-scheme heterojunction with g-C_3N_4, Ti_3C_2 mxene and black phosphorus for improving visible light-induced degradation of ciprofloxacin. *Appl. Catal. B* **2021**, *291*, 120105. [CrossRef]
54. Mahdi-Ahmed, M.; Chiron, S. Ciprofloxacin oxidation by UV-C activated peroxymonosulfate in wastewater. *J. Hazard. Mater.* **2014**, *265*, 41–46. [CrossRef]
55. Deng, J.; Ge, Y.J.; Tan, C.Q.; Wang, H.Y.; Li, Q.S.; Zhou, S.Q.; Zhang, K.J. Degradation of ciprofloxacin using α-MnO_2 activated peroxymonosulfate process: Effect of water constituents, degradation intermediates and toxicity evaluation. *Chem. Eng. J.* **2017**, *330*, 1390–1400. [CrossRef]

toxics

MDPI

Article

A Theoretical Study on the Degradation Mechanism, Kinetics, and Ecotoxicity of Metronidazole (MNZ) in •OH- and $SO_4^{\bullet-}$-Assisted Advanced Oxidation Processes

Jingyu Sun [1,*], Ruijun Chu [1] and Zia Ul Haq Khan [2]

1 Hubei Key Laboratory of Pollutant Analysis & Reuse Technology, College of Chemistry and Chemical Engineering, Hubei Normal University, Cihu Road 11, Huangshi 435002, China; 15845304247@163.com

2 Department of Chemistry, COMSATS University Islamabad, Park Road, Islamabad 45550, Pakistan; ziaulhaqkhan11@gmail.com

* Correspondence: sunjy@hbnu.edu.cn; Tel./Fax: +86-0714-6515602

Abstract: Metronidazole (MNZ), a typical example of nitroimidazole antibiotics, is widely used in the treatment of infectious diseases caused by anaerobic bacteria. The degradation mechanism and kinetics of MNZ in the presence of HO• and $SO_4^{\bullet-}$ were studied using density functional theory (DFT). It was confirmed that both HO• and $SO_4^{\bullet-}$ easily added to the carbon atom bonded to the NO_2 group in the MNZ molecule as the most feasible reaction channel. This study shows that subsequent reactions of the most important product (M-P) include the O_2 addition, hydrogen abstraction and bond breakage mechanisms. The rate constants of HO• and $SO_4^{\bullet-}$-initiated MNZ in the aqueous phase were calculated in the temperature range of 278–318 K. The total rate constants of MNZ with HO• and $SO_4^{\bullet-}$ were determined to be 8.52×10^9 and 1.69×10^9 $M^{-1}s^{-1}$ at 298 K, which were consistent with experimental values of $(3.54 \pm 0.42) \times 10^9$ and $(2.74 \pm 0.13) \times 10^9$ $M^{-1}s^{-1}$, respectively. The toxicity of MNZ and its degradation products to aquatic organisms has been predicted. The results proposed that the toxicity of the initial degradation product (M-P) was higher than that of MNZ. However, further degradation products of MNZ induced by HO• were not harmful to three aquatic organisms (fish, daphnia, and green algae). This study provides a comprehensive theoretical basis for understanding the degradation behavior of MNZ.

Keywords: metronidazole (MNZ); quantum chemical calculations; degradation mechanism; rate constants; ecotoxicity assessment

check for updates

Citation: Sun, J.; Chu, R.; Khan, Z.U.H. A Theoretical Study on the Degradation Mechanism, Kinetics, and Ecotoxicity of Metronidazole (MNZ) in •OH- and $SO_4^{\bullet-}$-Assisted Advanced Oxidation Processes. *Toxics* **2023**, *11*, 796. https://doi.org/10.3390/toxics11090796

Academic Editors: Zunyao Wang and Tiecheng Wang

Received: 9 August 2023
Revised: 18 September 2023
Accepted: 18 September 2023
Published: 20 September 2023

1. Introduction

The general use of antibiotics in human and veterinary medicine indirectly introduces their traces into the water and soil environments [1,2]. Metronidazole (MNZ) is a common type of anti-anaerobic and anti-protozoal antibiotic [3]. It is extensively used as an additive in poultry and fish feeds [4]. In addition, MNZ is an effective radiosensitizer which can be used in radiotherapy to overcome the radio-resistance of hypoxic tumor cells [5–7]. Due to its widespread use, MNZ has been detected in drinking water, ground water, surface water, waste water, and soil sediment at a range from one to several ng L^{-1} [8]. In order to investigate the environmental effect of MNZ, removal of MNZ from the water environment is necessary.

Different degradation techniques like adsorption [9], biodegradation [10], and advanced oxidation processes (AOPs) [4,11,12], including UV, UV/H_2O_2, Fenton reaction, granular activated carbon (GAC)/persulfate (PS), and gamma radiolysis, have been applied to remove metronidazole from wastewater. The literature has shown that advanced oxidation processes show good performance in the removal of antibiotics [13–15]. The AOPs based on gamma-ray, X-ray, and electron beams from the electron accelerators are also very important techniques which are broadly applied in the decomposition of various pollutants in water and waste water [16]. It has been reported that the UV/peroxide (UV/H_2O_2) and UV/persulfate (UV/PS) have better degradation efficiencies for some pharmaceuticals [17,18]. During the advance oxidation process (UV/H_2O_2) and (UV/PS), highly reactive species, i.e., HO• and $SO_4^{\bullet-}$ radicals, are generated, initiating the degradation of antibiotics [19,20]. The HO• radicals are very reactive and react rapidly with many pharmaceuticals, with a rate constant of 10^8–10^{10} M^{-1} s^{-1} [21]. The redox potential of $SO_4^{\bullet-}$ is 2.5–3.1 V, which is higher than that of HO• (1.9–2.7 V) [22]. Thus, AOPs based on $SO_4^{\bullet-}$ are gaining more interest [23–26]. Wojnárovits tabulated and discussed the rate constants of $SO_4^{\bullet-}$ radicals with about 230 organic molecules. They found that the highest rate constants for $SO_4^{\bullet-}$ radicals with aromatic molecules were in the ~10^9 $M^{-1}s^{-1}$ range [27].

Shemer et al., in 2006, investigated the degradation rates and removal efficiencies of MNZ through advance oxidation process (UV, UV/H_2O_2, H_2O_2/Fe^{2+}, and UV/H_2O_2/Fe^{2+}) in de-ionized water [28]. Ammar et al. studied sono-Fenton and solar photo-Fenton processes and found that they were effective for the degradation of MNZ [29,30]. Cheng et al. also investigated the degradation of MNZ through advanced oxidation processes [31]. Kinetic models for fifteen pharmaceuticals and personal care products (PPCPs), including metronidazole, were built to predict the degradation of PPCPs in the advanced oxidation processes (AOPs) of HO• and $SO_4^{\bullet-}$. The secondary reaction rate constants of metronidazole with HO• and $SO_4^{\bullet-}$ radicals were $(3.54 \pm 0.42) \times 10^9$ and $(2.74 \pm 0.13) \times 10^9$ M^{-1} s^{-1}, respectively [32]. Whillans et al. measured the rate constant for HO• with MNZ by means of the pulse radiolysis technique, and its value was $(5.5 \pm 0.2) \times 10^9$ M^{-1} s^{-1} [33]. These studies were based on advanced oxidation processes which mainly focused on accelerating the degradation of MNZ by optimizing the reaction conditions and analyzing the kinetics, but without providing the specific MNZ degradation mechanism or products.

Recently, only one study has obtained the main cleavage products of MNZ based on DFT using a statistical molecular fragmentation (SMF) model [34]. However, the detailed reaction potential energy surfaces for •OH-initiated MNZ degradation were not provided. Yao et al. used DFT calculations to give the detailed degradation pathways and possible degradation products for the reactions of dimetridazole (DMZ) and ornidazole (ONZ) with HO• [35], which is helpful in exploring the degradation of MNZ initiated by HO•. Recently, quantum chemical calculations have become an effective method by which to investigate the mechanism and kinetics of HO• and $SO_4^{\bullet-}$ with some organic contaminants [36–40]. For example, the degradation mechanism and products of acetaminophen and *N,N*-diethyl-m-toluamide (DEET), initiated by HO• and $SO_4^{\bullet-}$ in the aqueous phase, have been studied using the quantum chemical method by Xu et al. [38,39]. Theoretical studies give us more information regarding the structures, thermodynamic data, and microscopic reaction mechanism.

To our knowledge, there have been no theoretical studies on MNZ with HO• and $SO_4^{\bullet-}$. Thus, in the current paper, the quantum chemical method was used to investigate the degradation mechanism of HO• and $SO_4^{\bullet-}$ with MNZ in aqueous solution reactions. On the basis of the reaction mechanism, the rate constants of all initial reaction pathways were calculated in the temperature ranges of 278–318 K. In addition, the ecotoxicity of MNZ and its degradation products to aquatic organisms was predicted in order to assess their exposure risk.

2. Computational Methods

2.1. Electronic Structure and Kinetic Calculations

The geometric optimization of all structures was performed at the M06-2X/6-31+G(d, p) level using Gaussian 16 software [41]. Previous studies have confirmed the reliability of the density functional M06-2X method for optimizing the geometries of transition states, intermediates, and products involved in the reactions of nitroimidazole drugs with HO• [34,35]. To obtain more accurate energies, single-point energies were calculated using the more expensive 6-311++G(3df, 2p) basis set. In addition, intrinsic reaction coordinate (IRC) calculations [42] were performed to verify whether each transition state was connected to the corresponding reactants and products. The solvent effect was considered using a universal solvation model of SMD [43] that includes the bulk electrostatic contribution and the cavity-dispersion-solvent-structure term. The SMD has been successfully applied in some aqueous reactions involving HO• and $SO_4^{\bullet-}$ [38,39,44]. Based on the thermodynamic properties of these reactions, the bimolecular reaction rate constants in the aqueous phase were performed using the conventional transition state theory (TST) [45], with the Wigner tunneling correction [46], using the KiSThelP program [47]. The calculated formula is given below.

$$k = \kappa(T)\sigma\frac{k_B T}{h}\left(\frac{RT}{P^0}\right)^{\Delta n} e^{\frac{-\Delta G^{\neq}}{k_B T}} \quad (1)$$

$$\Delta G^{\neq} = G_{TS} - G_R \quad (2)$$

In the formula, k is the calculated rate constant. k_B, h and R are the Boltzmann constant, Planck constant, and gas constant, respectively. The Gibbs free energy of activation ($\Delta G^{\neq}$) was equal to the Gibbs free energy of the transition state minus the total Gibbs free energy of the reactants. σ represents the reaction path degeneracy, which accounts for the number of equivalent reaction paths. P^0 and T are pressure and temperature, respectively. Δn was calculated to be $(n-1)$ for the n-molecular reaction. For instance, Δn was 1 for the bimolecular reaction studied herein. The tunneling correction factor $\kappa(T)$ was calculated using the Wigner model:

$$\kappa(T) = 1 + \frac{1}{24}\left(\frac{h\mathrm{lm}(V^{\neq})}{k_b T}\right)^2 \quad (3)$$

where $\mathrm{lm}(V^{\neq})$ is the transition state's imaginary frequency. Some studies have used the KiSThelP program to calculate rate constants, with good results [48–50].

2.2. Ecotoxicity Assessment

In this research work, the ecotoxicity levels of MNZ and its degradation products were measured by adopting the Ecological Structure–Activity Relationship Model (ECOSAR V2.0) [51]. The procedure measured the acute and chronic toxicity to three aquatic organisms, including green algae, daphnia, and fish, as a basis for assessment. The median lethal concentration (LC50) was used to calculate the acute toxicity for fish and daphnia. Median effect concentration (EC50) was used to calculate the acute toxicity for green algae. The chronic toxicity value (ChV) was used to reflect the chronic toxicity for three aquatic organisms. The lowest toxicity value was chosen as the most conservative estimate.

3. Results and Discussion

Figure 1 shows the optimized structure of MNZ, HO•, and $SO_4^{\bullet-}$. The geometries of the transition states involved in the initial reactions of HO• and $SO_4^{\bullet-}$ are presented in Figure 2. The reactions of MNZ with HO• and $SO_4^{\bullet-}$ can be classified into the following types: (a) radical adduct formation (RAF) at the C1, C2, C3 and N1 of imidazole ring (RAF-1 to RAF-4); (b) hydrogen atom transfer (HAT) on the imidazole ring (HAT-1 and HAT-6) and in the methyl group (HAT-2 and HAT-7) and the $-CH_2CH_2OH$ group (HAT-3 to HAT5 and HAT-8 to HAT-10). The intramolecular hydrogen bond would appear in some transition states, as indicated by Qu et al. [52].

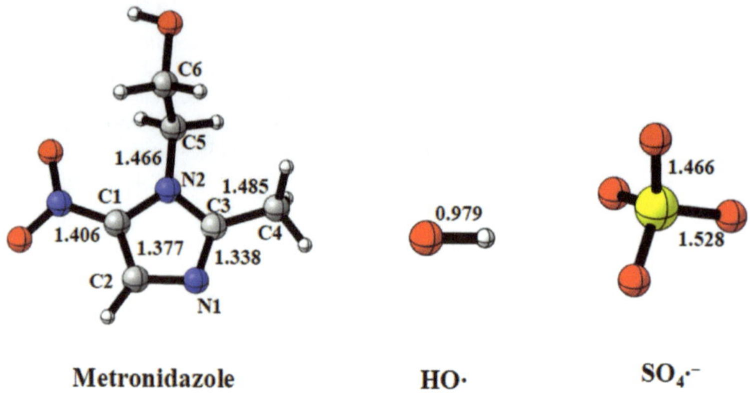

Figure 1. Optimized structures of MNZ, HO•, and $SO_4^{\bullet -}$ at the M06-2X/6-31+G(d, p) level. Bond lengths are given in angstroms S C N O H.

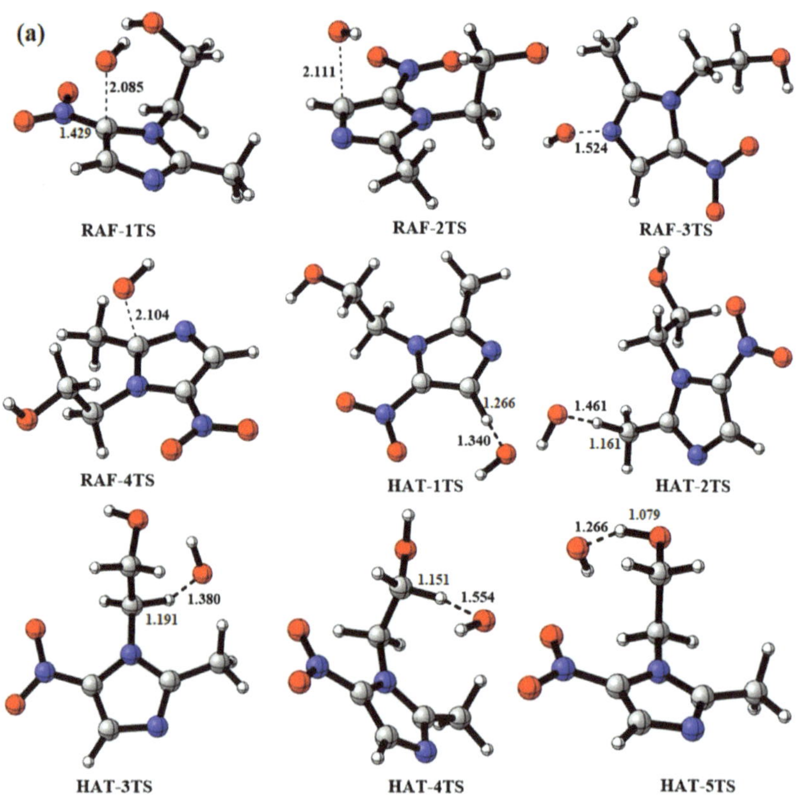

Figure 2. *Cont.*

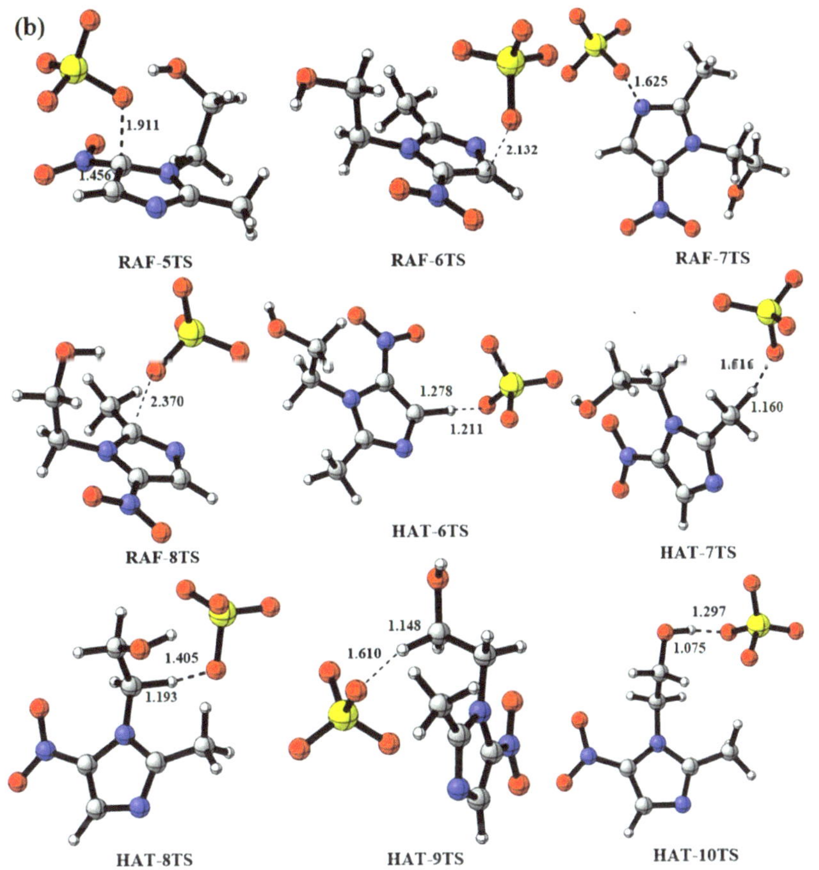

Figure 2. Optimized geometries of the transition states (TSs) involved in the initial reactions of MNZ with HO• and $SO_4^{\bullet-}$ (**a**) MNZ with HO• reactions; (**b**) MNZ with $SO_4^{\bullet-}$ reactions. Bond lengths are given in angstroms S C N O H .

3.1. Initial Reaction of MNZ with HO• and $SO_4^{\bullet-}$

3.1.1. Radical Adduct Formation Reactions

Figure 3 shows that HO• and $SO_4^{\bullet-}$ could add to the –C=C– and –C=N bonds of the imidazole ring to form intermediates (IM1 to IM3 and IM1-S to IM3-S) and products (M-P and M-P-S). Furthermore, the thermodynamic data of all channels were calculated and were listed in Table S2. The addition of both HO• and $SO_4^{\bullet-}$ to the imidazole ring was exothermic, except for when added to the N1 atom of the imidazole ring. Meanwhile, HO• and $SO_4^{\bullet-}$ addition to the N1 atom of the imidazole ring resulted in the highest Gibbs energy barriers, 26.92 and 37.12 kcal mol^{-1}, respectively, which possibly have been associated with the greater electronegativity of the N atom. The C2 and C3 addition reactions for HO• and $SO_4^{\bullet-}$ had relatively lower barriers and higher exothermicity. It was notable that HO• and $SO_4^{\bullet-}$ were added to the C1 position in the MNZ molecule with the lowest Gibbs free energy barriers of 5.89 and 6.78 kcal mol^{-1}, respectively, releasing 29.01 and 24.06 kcal mol^{-1}. The addition of HO• and $SO_4^{\bullet-}$ caused the removal of the $-NO_2$ group (RAF-1 and RAF-5) and generated by-products like the hydroxyimidazole compound (M-P) and the sulfate anion substance (M-P-S). As shown in Figure 2, the bond lengths of the oxygen atoms in the HO• and $SO_4^{\bullet-}$ radicals approaching C1 atom were

2.085 and 1.911 Å in RAF-1TS and RAF-5TS, respectively. Meanwhile, the C–N bond lengths in RAF-1TS and RAF-5TS were 1.429 and 1.456 Å, which exceeded MNZ by 0.023 and 0.050 Å. The HO• radicals with dimetridazole (DMZ) and ornidazole (ONZ) caused the addition to the C1 position, followed by the products of D-P/O-P and NO_2, with barriers of 4.94 and 6.32 kcal mol^{-1}, respectively [35], which were the most important reaction channels. The same mechanisms were suggested for the reaction of HO• with MNZ, DMZ, and ONZ. However, the results obtained by Xu et al. showed that nitro-elimination reactions included two steps where a single HO• or double HO• was added, then $-NO_2$ was leaved [34]. Therefore, M-P as a representative product was selected in order to discuss its subsequent reactions.

3.1.2. Hydrogen Abstraction Reactions

In the structure of MNZ, there are five different hydrogen-involved imidazole ring and branching chains. Intermediates (IM4–IM8) and H_2O were generated by five hydrogen atom abstraction reaction pathways of MNZ with HO•. They were all exothermic reactions (ΔG = −3.08 to −30.76 kcal mol^{-1}), with $\Delta G^{\neq}$ values of 8.33 to 43.77 kcal mol^{-1}. For the reactions of $SO_4^{\bullet-}$ and MNZ, hydrogen abstraction from imidazole ring (HAT-6) via a higher Gibbs free energy barrier of 42.26 kcal mol^{-1} was endothermic by 5.37 kcal mol^{-1}. However, the others were both exothermic (ΔG = −6.45 to −22.31 kcal mol^{-1}), with Gibbs free energy barriers of 12.00–31.88 kcal mol^{-1}. From the calculated thermodynamic data, for either HO• or $SO_4^{\bullet-}$, it was confirmed that the ordering of the difficulty of the hydrogen abstraction in the MNZ molecule could be derived as: HAT-1(HAT-6) > HAT-5(HAT-10) > HAT-3(HAT-8) > HAT-2(HAT-7) ≈ HAT-4(HAT-9). It was noted that the H attached to the C2 atom of the imidazole ring was the most inaccessible for HO• and $SO_4^{\bullet-}$ abstraction. However, hydrogen abstraction from the C6 atom in the $-CH_2OH$ branched ring was the most important, because the relative free energies were 8.33 and 12.00 kcal mol^{-1}, and they were exothermic, 21.12 and 12.67 kcal mol^{-1}, for the HO• and $SO_4^{\bullet-}$ reactions, respectively. The abstraction of hydrogen on C atom in the $-CH_2OH$ group was investigated by Xu et al. [34]. They calculated that this step needed to overcome a barrier of 5.3 kcal mol^{-1}. In HAT-4TS and HAT-9TS, the breaking bond (C–H) lengths were 1.151 and 1.148 Å, and the forming O–H bond lengths were 1.554 and 1.610 Å for HO• and $SO_4^{\bullet-}$ hydrogen abstraction, respectively.

All the hydrogen abstraction pathways were required to overcome many more energy barriers than the RAF-1 pathway, which is consistent with the reaction mechanism of dimetridazole (DMZ) and ornidazole (ONZ) with HO• radicals. Therefore, for the initial reaction pathways of MNZ with HO• and $SO_4^{\bullet-}$ radicals, addition at the position of C1 in the MNZ molecule was the most feasible reaction channel. The reaction mechanism of MNZ degraded by HO• and $SO_4^{\bullet-}$ was able to guide other new nitroimidazoles to be removed, which represents the novelty and significance of this study.

In order to investigate the products of MNZ degradation in wastewater in detail, the secondary reaction pathways of M-P obtained from reacting MNZ with HO• were investigated at the double level of M06-2X/6-31+G(3df, 2p)//M06-2X/6-31+G(d, p).

3.1.3. Further Conversion Reactions of M-P

Based on the above results, we concluded that M-P is the most important degradation product for the initial reaction of MNZ with HO•. And M-P can undergo a series of subsequent transformation reactions to produce other degradation products. Therefore, it is essential to study the subsequent reactions of this product in order to assess the risk of these products in an aquatic environment.

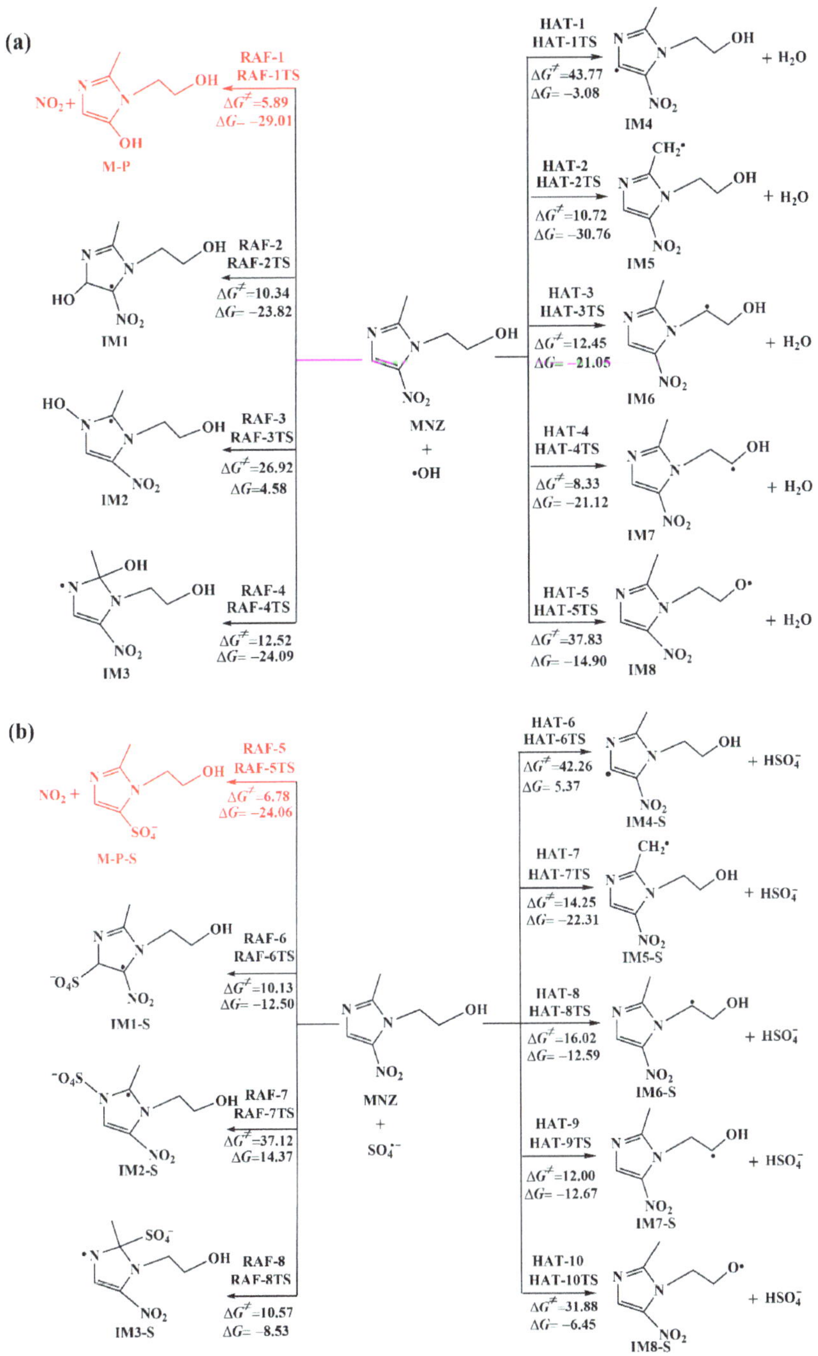

Figure 3. The detailed initial reaction pathways of MNZ with HO• and $SO_4^{\bullet-}$ including Gibbs free energy of activation ($\Delta G^{\neq}$) and Gibbs free energy of reaction (ΔG). All values are in kcal mol^{-1}. (**a**) MNZ with HO•; (**b**) MNZ with $SO_4^{\bullet-}$. The lowest energy barriers pathways are marked in red.

The Fukui function based on DFT was successfully adopted in order to predict the attacking site of the reactant molecules [53–55]. For the Fukui function, the following three situations were as follows:

$$\text{Nucleophilic attack}: \ f_k^+ = q_k(N+1) - q_k(N)$$

$$\text{Electrophilic attack}: \ f_k^- = q_k(N) - q_k(N-1)$$

$$\text{Radical attack}: \ f_k^0 = \frac{1}{2}[q_k(N+1) - q_k(N-1)]$$

where f_k^+, f_k^- and f_k^0 represent nucleophilic, electrophilic, and radical attacks, respectively. $q_k(N+1)$, $q_k(N)$ and $q_k(N-1)$ are the electronic populations of the $(N+1)$, N, and $(N-1)$ electron states, respectively. A high concentrated Fukui index (f^0) indicates the most likely site to be attacked by free radicals. As shown in Table S1, the C1 site had the largest Fukui index (f^0) value of 0.1038. Thus, the C1 site is the most favorable attack site for the subsequent reaction of M-P.

Based on the Fukui function prediction, HO• was preferentially added to the C1 site in the M-P molecule. The Gibbs free energy barrier of 2.09 kcal mol^{-1} needed to be overcome in order to obtain IM9. As shown in Figure 4, the IM9 was able to perform three types of reactions: (I) branched chain breakage; (II) intramolecular H-transfer and C–C bond breakage; and (III) reaction with O_2. The corresponding transition states were shown in Figure S1. Through comparing the thermodynamic data of the three reactions, it was revealed that P1 and IM10, as unstable degradation products, were obtained from the branched chain breakage and intramolecular H transfer, accompanied with C–C bond breakage. The above two types of reactions were endothermic, and their Gibbs free energy barriers ($\Delta G^{\neq}$) were 31.36 kcal mol^{-1} and 36.06 kcal mol^{-1}, respectively. For the reaction of IM9 with O_2, IM9 firstly reacted with O_2 via H abstraction from the C–H/O–H bond to produce P2, P3, and HO_2. The P2 and P3 formation were exothermic by 29.40 and 37.95 kcal mol^{-1}. P1 and P2 are double •OH imidazole products that have been found in subsequent reactions of •OH with DMZ and ONZ. It is important that the O_2 can associate with the C atom of an unpaired electron of IM9 from the anti- or syn-position with respect to the •OH. The peroxyl radicals IM11 and IM12 were generated from spontaneous exothermic reactions, with the corresponding $\Delta G = -25.19$ kcal mol^{-1} and $\Delta G = -26.15$ kcal mol^{-1} between IM9 and O_2, respectively. Subsequently, the toxicity of the final products was investigated and analyzed.

3.2. Kinetic Study

The bimolecular rate constants of the elementary aqueous reactions induced by HO• and $SO_4^{\bullet-}$ for MNZ were calculated using transition state theory (TST) in a temperature range from 278 to 318 K, as shown in Tables S3 and S4. Aqueous phase temperatures of 278–318 K were considered in order to avoid ice and water vapor. The rate constants in the solid phase were difficult to be calculated, as reactions involving ice are extremely complex and different from reactions in the gas and aqueous phases. The transition state theory is no longer applicable for calculating the solid phase rate constant. However, the reaction rate constants involving water vapor could be calculated by TST. Previous studies have shown that the kinetic calculations for gas-phase reactions could provide better results [40,49,54]. And gas-phase rate constants could be changed to aqueous-phase ones. However, in this paper, gas-phase rate constants were not calculated, because the reactions of MNZ with HO• and $SO_4^{\bullet-}$ mainly occurred in the aqueous phase. The calculated reaction rate constants and branching ratio of MNZ with HO• and $SO_4^{\bullet-}$ were summarized in Table 1. The total rate constant of the aqueous phase oxidation reaction of HO• with MNZ at 298 K was 8.52×10^9 M^{-1} s^{-1}, in agreement with the experimental rate constants of $(3.54 \pm 0.42) \times 10^9$ M^{-1} s^{-1} and $(5.5 \pm 0.2) \times 10^9$ M^{-1} s^{-1} [32,33]. In our previous study, the calculated rate constants for DMZ + HO• and ONZ + HO• reactions were 4.32×10^9 M^{-1} s^{-1} and 4.42×10^9 M^{-1} s^{-1}, respectively, which were also close to the experimental values $(3.65 \pm 0.01) \times 10^9$ and $(2.66 \pm 0.26) \times 10^9$ M^{-1} s^{-1}, respectively. It is indicated that kinetics of the three reactions would be influenced by the substituent groups ($-CH_2CH_2OH$, $-CH_3$, $-CH_2CHOHCH_2Cl$). The calculated total rate constant initiated by $SO_4^{\bullet-}$ was 1.69×10^9 M^{-1} s^{-1} at 298 K, which was consistent with a previous reported value $(2.74 \pm 0.13) \times 10^9$ M^{-1} s^{-1} [32]. It was indicated that the degradation efficiencies of HO• and $SO_4^{\bullet-}$ for MNZ were essentially equal throughout the AOPs.

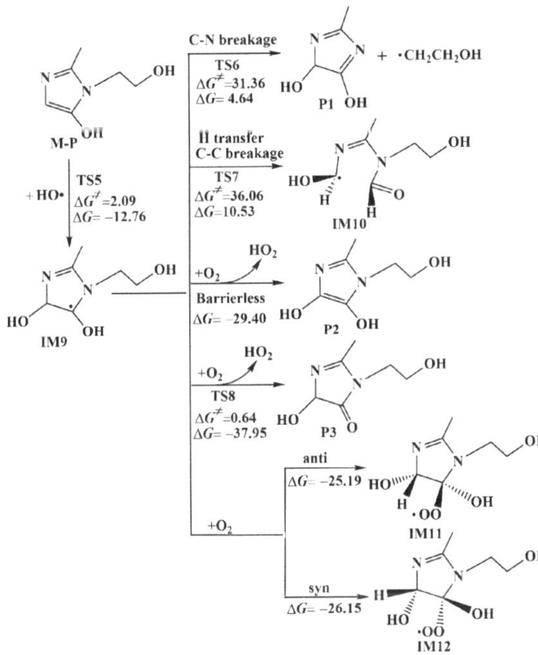

Figure 4. Subsequent reactions of M-P with Gibbs free energy of activation ($\Delta G^{\neq}$, kcal mol^{-1}) and Gibbs free energy of reaction (ΔG, kcal mol^{-1}).

Table 1. The rate constants and branching ratios for the initial reaction channels of MNZ with HO• and $SO_4^{\bullet-}$ at 298 K.

Reactions	HO• k_{aq}	(%)	Reactions	$SO_4^{\bullet-}$ k_{aq}	(%)
RAF-1	8.10×10^9	95	RAF-5	1.68×10^9	99.5
RAF-2	4.49×10^6		RAF-6	5.78×10^6	0.3
RAF-3	4.32×10^6		RAF-7	3.35×10^{-13}	
RAF-4	1.36×10^5		RAF-8	2.72×10^6	0.2
$k_{total\text{-}RAF}$	8.10×10^9	95	$k'_{total\text{-}RAF}$	1.69×10^9	100
HAT-1	2.44×10^{-18}		HAT-6	6.08×10^{-17}	
HAT-2	1.38×10^7		HAT-7	2.80×10^4	
HAT-3	6.46×10^5		HAT-8	1.20×10^3	
HAT-4	3.98×10^8	5	HAT-9	6.85×10^5	
HAT-5	1.22×10^{-13}		HAT-10	2.70×10^{-9}	
$k_{total\text{-}HAT}$	4.13×10^8	5	$k'_{total\text{-}HAT}$	7.15×10^5	
k_{total}	8.52×10^9		k'_{total}	1.69×10^9	
$k_{experiment}$	$(3.54 \pm 0.42) \times 10^9$; $(5.5 \pm 0.2) \times 10^9$		$k'_{experiment}$	$(2.71 \pm 0.13) \times 10^9$	

Figure 5 shows the rate constants of MNZ reaction, with HO• and $SO_4^{\bullet-}$ as a function of temperature from 278 to 318 K. In the temperature range of 278–318 K, the total rate constants (k_{total} and k'_{total}) of the reactions of MNZ with HO• and $SO_4^{\bullet-}$ decreased as the temperature increased. The total addition rate constants ($k_{total-RAF}$ and $k'_{total-RAF}$) had negative temperature dependence. However, the total hydrogen abstraction rate constants ($k_{total-HAT}$ and $k'_{total-HAT}$) had weakly positive temperature dependence. It is worth noting that the total rate constant was mainly determined by the radical adduct formation channel. As shown in Table 1, the total rate constants of radical adduct formation channels of the reactions of MNZ with HO• and $SO_4^{\bullet-}$ were 8.10×10^9 and 1.69×10^9 M^{-1} s^{-1}, accounting for 95% and 100% for the corresponding total reactions. It was indicated that hydrogen abstraction reactions can be neglected and radical adduct formation reactions are the most important during the aqueous phase. The above kinetic calculations are consistent with the conclusions obtained from thermodynamics.

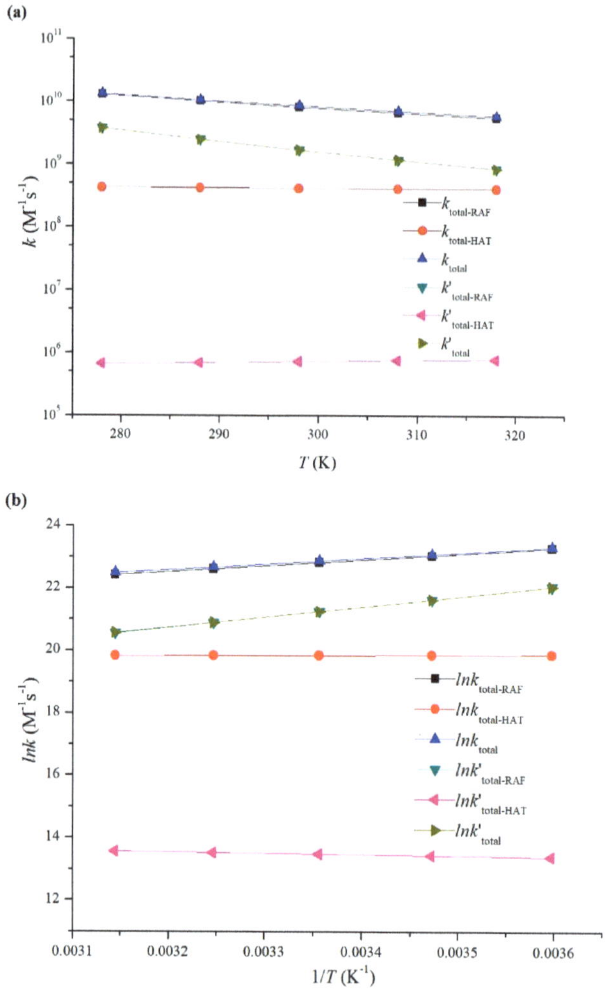

Figure 5. The calculated total rate constants (k_{total} and k'_{total}), the total addition rate constants ($k_{total-RAF}$ and $k'_{total-RAF}$), and the total hydrogen abstraction rate constants ($k_{total-HAT}$ and $k'_{total-HAT}$) for the corresponding reactions of MNZ with HO• and $SO_4^{\bullet-}$ in the temperature range of 278–318 K. **(a)** k vs. T; **(b)** lnk vs. $1/T$.

The half-life ($t_{1/2}$) of the OH-initiated transformation of MNZ can be calculated by the total rate constant and the concentration of HO•: $t_{1/2}$ = ln2/(k_{aq} × [HO•]). Here, k_{aq} is the total rate constant for the reaction of MNZ with HO•. [HO•] is the concentration of HO•. Previous studies have proven that [HO•] reached 10^{-9} to 10^{-10} mol/L in wastewater subjected to AOP treatment and 10^{-14} to 10^{-18} mol/L in natural water [56–58]. Thus, the half-life of MNZ initiated by HO• in the temperature range of 278–318 K, with the [HO•] range of 10^{-9} to 10^{-18} mol/L, was calculated and given in Table S5. The half-life ($t_{1/2}$) in AOP wastewater (10^{-9}–10^{-10}) was 0.081–0.81 s at 298 K (see Table S5). When the concentration of HO• was lower than 5.88×10^{-17} mol/L, MNZ remained in the water for more than 16 days, which indicated that the half-life of MNZ was longer in natural water. Obviously, at a certain temperature, $t_{1/2}$ increases with the decrease in the concentration of HO•. At the highest [HO•] of 10^{-9} mol/L, $t_{1/2}$ increased from 0.052s to 0.119 s when the temperature increased from 278 K to 318 K. Obviously, at a certain [HO•], $t_{1/2}$ increases with the increase in temperature. Therefore, both the temperature and the concentration of HO• influence $t_{1/2}$. Briefly, the stay time of MNZ in AOPs is very short, but it persists a long time in natural water.

3.3. Risk Predictions

Ecotoxicity assessment plays an important role in controlling organic contaminants. Bai et al. assessed the ecotoxicity of ciprofloxacin (CIP) and its products through ecological structure–activity relationships (ECOSAR). It was found that most of stable degradation products were harmless to the three tested aquatic organisms (fish, daphnia, and green algae) [59]. Recently, the ecotoxicity of mevinphos and monocrotophos and their degradation products were assessed by Tang et al. using the ECOSAR program. The ecotoxicity predictions for the three aqueous organisms implied that their toxicity levels were reduced during degradation [60]. To explore the potential risks of MNZ and its degradation products for aquatic organisms, the acute and chronic toxicities were evaluated using three different nutritional levels of aquatic organisms (daphnia, fish, and green algae) based on the ECOSAR program. The specific classification criteria for acute and chronic toxicity are shown in Table S6. The predicted toxicity values of MNZ and its degradation products are shown in Table S7. The acute toxicity values of MNZ to fish, daphnia, and green algae were 878.33, 179.75, and 6.92 mg L^{-1}, respectively. It is indicated that MNZ is not harmful to either fish or daphnia, but is toxic to green algae in terms of acute toxicity. The chronic toxicity values of MNZ were 0.95, 3.08, and 2.70 mg L^{-1} for fish, daphnia, and green algae, respectively, showing that MNZ was chronically toxic to fish and harmful to daphnia and green algae. It is indicated that MNZ is a toxic pollutant. Thus, it is necessary to pay further attention to it and its degradation products for aquatic organisms.

As shown in Table S7, the acute toxicity values of the initial degradation product of M-P were 260.97, 79.78, and 3.56 mg L^{-1} for fish, daphnia, and green algae, respectively. And the chronic toxicity values of M-P were 0.50, 1.37, and 1.45 mg L^{-1} for fish, daphnia, and green algae, respectively. The toxicity evolution of MNZ and its major degradation products is shown in Figure 6. The toxicity level of the initial degradation product of M-P was increased for fish, daphnia, and green algae. The toxic predication for DMZ and ONZ degradation also showed that the initial degradation products of D-P and O-P enhanced aquatic toxicity. The toxicity levels of P1, P2, and P3 were decreased and found not to be harmful to three aquatic organisms. Similarly, in the study of DMZ and ONZ with HO•, it was indicated that most of the secondary degradation products were not harmful to aquatic organisms. Thus, the risk of degradation of MNZ should be a concern in the formation process of M-P.

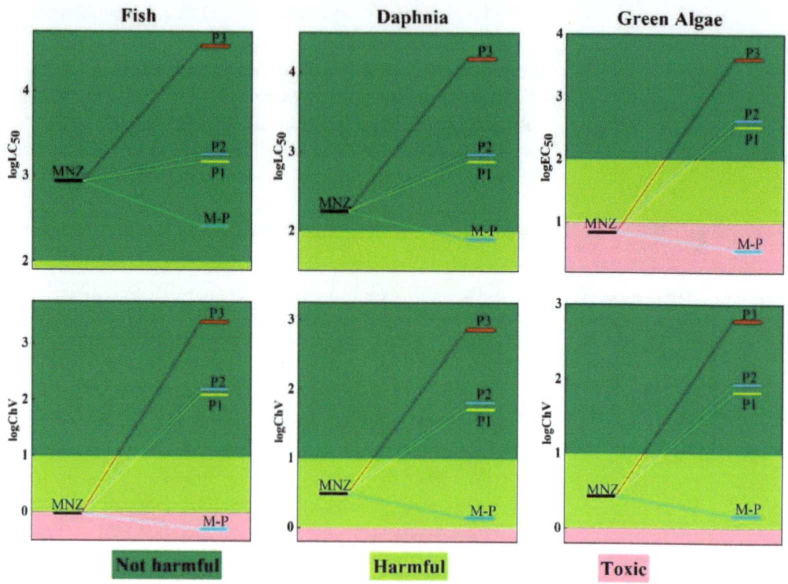

Figure 6. Acute and chronic toxicity values of MNZ and its main degradation products.

4. Conclusions

In the present research work, the degradation process of MNZ was systematically described from three aspects: the degradation mechanism, the kinetics, and a toxicity evaluation from theoretical point of view. Several findings were concluded:

(1) The main reaction channel for both HO• and $SO_4^{\bullet-}$-initiated MNZ was the RAF reaction. The main reaction site was the C1 atom on the imidazole ring of MNZ. The main by-products obtained in the subsequent hydroxylation reaction were P2 and P3.

(2) The total rate constants of MNZ with HO• and $SO_4^{\bullet-}$ were determined to be 8.52×10^9 and 1.69×10^9 M^{-1} s^{-1} at 298 K in an aqueous environment, which is in agreement with the experimental values.

(3) The toxicity analysis showed that the end products were not harmful to aquatic organisms. The toxicity level of the initial degradation product of M-P was higher than the parent of MNZ.

We are finally able to answer the questions surrounding MNZ degradation. These results suggest that HO• and $SO_4^{\bullet-}$ oxidation is effective in removing MNZ from aquatic environments. This study can guide the aquatic transformation and removal of nitroimidazole antibiotics.

Supplementary Materials: The following supporting information can be downloaded at: https://www.mdpi.com/article/10.3390/toxics11090796/s1, Figure S1: Optimized geometries of the transition states (TSs) involved subsequent degradation of M-P; Table S1: The condensed Fukui function values of M-P product; Table S2: Gibbs free energies (ΔG) and relative energies (ΔE) for reaction intermediates (IM), products, and transition states (TS) of MNZ with HO•and $SO_4^{\bullet-}$initial reaction at the double level of M06-2X/6-31+G(3df, 2p)//M06-2X/6-31+G(d, p); Table S3: The calculated rate constants ($M^{-1}s^{-1}$) between 278 and 318 K in the reaction of MNZ and HO•; Table S4: The calculated rate constants ($M^{-1}s^{-1}$) between 278 and 318 K in the reaction of MNZ and $SO_4^{\bullet-}$; Table S5: The calculated half-life ($t_{1/2}$) for MNZ depending on the concentration of •OH of 10^{-9}–10^{-18} mol/L within the temperature range of 278–318 K; Table S6: The acute and chronic toxicity class (mg L^{-1}); Table S7: Eoctoxicity values of MNZ and its transformation products to aquatic organisms (mg L^{-1}).

Author Contributions: Conceptualization, J.S. and R.C.; methodology, J.S.; software, J.S.; validation, J.S., R.C. and Z.U.H.K.; formal analysis, J.S., R.C. and Z.U.H.K.; investigation, J.S. and R.C.; resources, J.S.; data curation, J.S. and R.C.; writing—original draft preparation, J.S.; writing—review and editing, Z.U.H.K.; visualization, J.S.; supervision, J.S.; project administration, J.S.; funding acquisition, J.S. All authors have read and agreed to the published version of the manuscript.

Funding: This research was funded by the National Natural Science Foundation of China (No. 21507027) and the open fund of Hubei Key Laboratory of Pollutant Analysis & Reuse Technology (No. PA200205).

Institutional Review Board Statement: Not applicable.

Informed Consent Statement: Not applicable.

Conflicts of Interest: The authors declare no conflict of interest.

References

1. Philip, J.M.; Aravind, U.K.; Aravindakumar, C.T. Emerging contaminants in Indian environmental matrices–A review. *Chemosphere* **2018**, *190*, 307–326. [CrossRef]
2. Wilkinson, J.; Hooda, P.S.; Barker, J.; Barton, S.; Swinden, J. Occurrence, fate and transformation of emerging contaminants in water: An overarching review of the field. *Environ. Pollut.* **2017**, *231*, 954–970. [CrossRef] [PubMed]
3. Agarwal, A.; Kanekar, S.; Sabat, S.; Thamburaj, K. Metronidazole-induced cerebellar toxicity. *Neurol. Int.* **2016**, *8*, 6365. [CrossRef] [PubMed]
4. Fang, Z.Q.; Chen, J.H.; Qiu, X.H.; Qiu, X.Q.; Cheng, W.; Zhu, L.C. Effective removal of antibiotic metronidazole from water by nanoscale zero-valent iron particles. *Desalination* **2011**, *268*, 60–67. [CrossRef]
5. Asquith, J.C.; Foster, J.L.; Willson, R.L.; Ings, R.; McFadzean, J.A. Metronidazole ("Flagyl"). A radiosensitizer of hypoxic cells. *Brit. J. Radiol.* **1974**, *47*, 474–481. [CrossRef] [PubMed]
6. Willson, R.L.; Cramp, W.A.; Ings, R.M.J. Metronidazole ("Flagyl"): Mechanisms of radiosensitization. *Int. J. Radiat. Biol.* **1974**, *26*, 557–569. [CrossRef]
7. Wardman, P. Some reactions and properties of nitro radical-anions important in biology and medicine. *Environ. Health Persp.* **1985**, *64*, 309–320. [CrossRef]
8. Wagil, M.; Maszkowska, J.; Białk-Bielińska, A.; Caban, M.; Stepnowski, P.; Kumirska, J. Determination of metronidazole residues in water, sediment and fish tissue samples. *Chemosphere* **2015**, *119*, S28–S34. [CrossRef]
9. Ramavandi, B.; Akbarzadeh, S. Removal of metronidazole antibiotic from contaminated water using a coagulant extracted from *Plantago ovata*. *Desalin. Water. Treat.* **2015**, *55*, 2221–2228. [CrossRef]
10. Rivera-Utrilla, J.; Prados-Joya, G.; Sánchez-Polo, M.; Ferro-García, M.A.; Bautista-Toledo, I. Removal of nitroimidazole antibiotics from aqueous solution by adsorption/bioadsorption on activated carbon. *J. Hazard. Mater.* **2009**, *170*, 298–305. [CrossRef]
11. Forouzesh, M.; Ebadi, A.; Aghaeinejad-Meybodi, A. Degradation of metronidazole antibiotic in aqueous medium using activated carbon as a persulfate activator. *Sep. Purif. Technol.* **2019**, *210*, 145–151. [CrossRef]
12. Sánchez-Polo, M.; López-Peñalver, J.; Prados-Joya, G.; Ferro-García, M.A.; Rivera-Utrilla, J. Gamma irradiation of pharmaceutical compounds, nitroimidazoles, as a new alternative for water treatment. *Water. Res.* **2009**, *43*, 4028–4036. [CrossRef] [PubMed]
13. Guo, K.H.; Wu, Z.H.; Yan, S.W.; Yao, B.; Song, W.H.; Hua, Z.C.; Zhang, X.W.; Kong, X.J.; Li, X.C.; Fang, J.Y. Comparison of the UV/chlorine and UV/H_2O_2 processes in the degradation of PPCPs in simulated drinking water and wastewater: Kinetics, radical mechanism and energy requirements. *Water. Res.* **2018**, *147*, 184–194. [CrossRef]
14. Yang, Y.; Ok, Y.S.; Kim, K.-H.; Kwon, E.E.; Tsang, Y.F. Occurrences and removal of pharmaceuticals and personal care products (PPCPs) in drinking water and water/ sewage treatment plants: A review. *Sci. Total Environ.* **2017**, *596–597*, 303–320. [CrossRef] [PubMed]
15. Zhang, R.C.; Meng, T.; Huang, C.H.; Ben, W.W.; Yao, H.; Liu, R.N.; Sun, P.Z. PPCP degradation by chlorine−UV processes in ammoniacal water: New reaction insights, kinetic modeling, and DBP formation. *Environ. Sci. Technol.* **2018**, *52*, 7833–7841. [CrossRef]
16. Trojanowicz, M.; Bobrowski, K.; Szreder, T.; Bojanowska-Czajka, A. Gamma-ray, X-ray and electron beam based processes. Advanced oxidation processes for waste water treatment. In *Advanced Oxidation Processes for Waste Water Treatment*; Academic Press: Cambridge, MA, USA, 2018; pp. 257–331. [CrossRef]
17. Kanakaraju, D.; Glass, B.D.; Oelgemöller, M. Advanced oxidation processmediated removal of pharmaceuticals from water: A review. *J. Environ. Manag.* **2018**, *219*, 189–207. [CrossRef]
18. Miklos, D.B.; Remy, C.; Jekel, M.; Linden, K.G.; Drewes, J.E.; Hübner, U. Evaluation of advanced oxidation processes for water and wastewater treatment–a critical review. *Water. Res.* **2018**, *139*, 118–131. [CrossRef] [PubMed]
19. Li, A.L.; Wu, Z.H.; Wang, T.T.; Hou, S.D.; Huang, B.J.; Kong, X.J.; Li, X.C.; Guan, Y.H.; Qiu, R.L.; Fang, J.Y. Kinetics and mechanisms of the degradation of PPCPs by zero-valent iron (Fe°) activated peroxydisulfate (PDS) system in groundwater. *J. Hazard Mater.* **2018**, *357*, 207–216. [CrossRef]

20. Zhang, R.C.; Sun, P.Z.; Boyer, T.H.; Zhao, L.; Huang, C.-H. Degradation of pharmaceuticals and metabolite in synthetic human urine by UV, UV/H_2O_2, and UV/PDS. *Environ. Sci. Technol.* **2015**, *49*, 3056–3066. [CrossRef]
21. Deng, Y.; Zhao, R.Z. Advanced Oxidation Processes (AOPs) in Wastewater Treatment. *Curr. Pollut. Rep.* **2015**, *1*, 33–40. [CrossRef]
22. Xiao, R.Y.; He, L.; Luo, Z.H.; Spinney, R.; Wei, Z.S.; Dionysiou, D.D.; Zhao, F.P. An experimental and theoretical study on the degradation of clonidine by hydroxyl and sulfate radicals. *Sci. Total. Environ.* **2020**, *710*, 136333. [CrossRef] [PubMed]
23. Zou, M.T.; Qi, Y.M.; Qu, R.J.; Al-Basher, G.; Pan, X.X.; Wang, Z.Y.; Huo, Z.L.; Zhu, F. Effective degradation of 2,4-dihydroxybenzophenone by zero−valent iron powder (Fe°)-activated persulfate in aqueous solution: Kinetic study, product identification and theoretical calculations. *Sci. Total. Environ.* **2021**, *771*, 144743. [CrossRef]
24. Pan, X.X.; Chen, J.; Wu, N.N.; Qi, Y.M.; Xu, X.X.; Ge, J.L.; Wang, X.H.; Li, C.G.; Qu, R.J.; Sharma, V.K.; et al. Degradation of aqueous 2,4,4′-trihydroxybenzophenone by persulfate activated with nitrogen doped carbonaceous materials and the formation of dimer products. *Water. Res.* **2018**, *240*, 176–187. [CrossRef] [PubMed]
25. Qi, Y.M.; Wei, J.Y.; Qu, R.J.; Al-Basher, G.; Pan, X.X.; Ahmed Dar, A.; Shad, A.; Zhou, D.M.; Wang, Z.Y. Mixed oxidation of aqueous nonylphenol and triclosan by thermally activated persulfate: Reaction kinetics and formation of co-oligomerization products. *Chem. Eng. J.* **2021**, *403*, 126396. [CrossRef]
26. Wacławek, S.; Lutze, H.V.; Grübel, K.; Padil, V.V.T.; Černík, M.; Dionysiou, D.D. Chemistry of persulfates in water and wastewater treatment: A review. *Chem. Eng. J.* **2017**, *330*, 44–62. [CrossRef]
27. Wojnárovits, L.; Takács, E. Rate constants of sulfate radical anion reactions with organic molecules: A review. *Chemosphere* **2019**, *220*, 1014–1032. [CrossRef]
28. Shemer, H.; Kunukcu, Y.K.; Linden, K.G. Degradation of the pharmaceutical Metronidazole via UV, Fenton and photo-Fenton processes. *Chemosphere* **2006**, *63*, 269–276. [CrossRef]
29. Ammar, H.B. Sono-Fenton process for metronidazole degradation in aqueous solution: Effect of acoustic cavitation and peroxydisulfate anion. *Ultrason. Sonochem.* **2016**, *33*, 164–169. [CrossRef]
30. Ammar, H.B.; Brahim, M.B.; Abdelhédi, R.; Samet, Y. Enhanced degradation of metronidazole by sunlight via photo-Fenton process under gradual addition of hydrogen peroxide. *J. Mol. Catal A Chem.* **2016**, *420*, 222–227. [CrossRef]
31. Cheng, W.; Yang, M.; Xie, Y.Y.; Liang, B.; Fang, Z.Q.; Tsang, E.P. Enhancement of mineralization of metronidazole by the electro-Fenton process with a Ce/SnO_2-Sb coated titanium anode. *Chem. Eng. J.* **2013**, *220*, 214–220. [CrossRef]
32. Lian, L.S.; Yao, B.; Hou, S.D.; Fang, J.Y.; Yan, S.W.; Song, W.H. Kinetic study of hydroxyl and sulfate radical-mediated oxidation of pharmaceuticals in wastewater effluents. *Environ. Sci. Technol.* **2017**, *51*, 2954–2962. [CrossRef] [PubMed]
33. Whillans, D.W.; Adams, G.E.; Neta, P. Electron-affinic sensitization: VI. A pulse radiolysis and ESR comparison of some 2- and 5-nitroimidazoles. *Radiat. Res.* **1975**, *62*, 407–421. [CrossRef] [PubMed]
34. Xu, L.J.; Li, W.Y.; Désesquelles, P.; Van-Oanh, N.T.; Thomas, S.; Yang, J. A statistical model and DFT study of the fragmentation mechanisms of metronidazole by advanced oxidation processes. *J. Phys. Chem. A* **2019**, *123*, 933–942. [CrossRef] [PubMed]
35. Yao, J.F.; Tang, Y.Z.; Zhang, Y.J.; Ruan, M.; Wu, W.Z.; Sun, J.Y. New theoretical investigation of mechanism, kinetics, and toxicity in the degradation of dimetridazole and ornidazole by hydroxyl radicals in aqueous phase. *J. Hazard. Mater.* **2022**, *422*, 126930. [CrossRef]
36. Qu, R.J.; Li, C.G.; Liu, J.Q.; Xiao, R.Y.; Pan, X.X.; Zeng, X.L.; Wang, Z.Y.; Wu, J.C. Hydroxyl radical based photocatalytic degradation of halogenated organic contaminants and paraffin on silica gel. *Environ. Sci. Technol.* **2018**, *120*, 7220–7229. [CrossRef]
37. Xu, M.M.; Yan, S.D.; Liu, X.F.; Sun, S.M.; Khan, Z.U.H.; Wu, W.Z.; Sun, J.Y. Theoretical investigation on the degradation of sulfadiazine in water environments: Oxidation of •OH, $SO_4^{\bullet-}$ and $CO_3^{\bullet-}$ and reactivity of $(TiO_2)_n$ clusters (n = 1−6). *J. Environ. Chem. Eng.* **2023**, *11*, 109994. [CrossRef]
38. Xu, M.M.; Yao, J.F.; Sun, S.M.; Yan, S.D.; Sun, J.Y. Theoretical calculation on the reaction mechanisms, kinetics and toxicity of acetaminophen degradation initiated by hydroxyl and sulfate radicals in the aqueous phase. *Toxics.* **2021**, *9*, 234. [CrossRef]
39. Xu, M.M.; Yan, S.D.; Sun, S.M.; Ni, Z.R.; Wu, W.Z.; Sun, J.Y. *N*, *N*-diethyl-m-toluamide (DEET) degradation by •OH and $SO_4^{\bullet-}$-assisted AOPs in wastewater treatment: Theoretical studies into mechanisms, kinetics and toxicity. *J. Environ. Chem. Eng.* **2022**, *10*, 108435. [CrossRef]
40. Yao, J.F.; Tang, Y.Z.; Zhang, Y.J.; Ruan, M.; Chen, F.; Wu, W.Z.; Sun, J.Y. Quantum chemical study of the mechanisms, kinetics, and ecotoxicity assessment of OH radical-initiated reactions of 2, 2′ 4, 4′, 5, 5′-hexabrominated diphenyl ether (BDE-153) in atmosphere and wastewater. *Chem. Eng. J.* **2021**, *422*, 129916. [CrossRef]
41. Frisch, M.J.; Trucks, G.W.; Schlegel, H.B.; Scuseria, G.E.; Robb, M.A.; Cheeseman, J.R.; Scalmani, G.; Barone, V.; Mennucci, B.; Petersson, G.A. *Gaussian 09 (Revision D.01)*; Gaussian, Inc.: Wallingford, CT, USA, 2010.
42. Fukui, K. The path of chemical reactions-the IRC approach. *Acc. Chem. Res.* **1981**, *14*, 363–368. [CrossRef]
43. Marenich, A.V.; Cramer, C.J.; Truhlar, D.G. Universal solvation model based on solute electron density and on a continuum model of the solvent defined by the bulk dielectric constant and atomic surface tensions. *J. Phys. Chem. B.* **2009**, *113*, 6378–6396. [CrossRef] [PubMed]
44. Li, M.X.; Sun, J.F.; Han, D.D.; Wei, B.; Mei, Q.; An, Z.X.; Wang, X.Y.; Cao, H.J.; Xie, J.; He, M.X. Theoretical investigation on the contribution of •OH, $SO_4^{\bullet-}$ and $CO_3^{\bullet-}$ radicals to the degradation of phenacetin in water: Mechanisms, kinetics, and toxicity evaluation. *Ecotox. Environ. Safe.* **2020**, *204*, 110977. [CrossRef]
45. Truhlar, D.G.; Garrett, B.C.; Klippenstein, S.J. Current status of transition-state theory. *J. Phys. Chem.* **1996**, *100*, 12771–12800. [CrossRef]

46. Garrett, B.C.; Truhlar, D.G. Accuracy of tunneling corrections to transition state theory for thermal rate constants of atom transfer reactions. *J. Phys. Chem.* **1979**, *83*, 200–203. [CrossRef]
47. Canneaux, S.; Bohr, F.; Henon, E. KiSThelP: A program to predict thermodynamic properties and rate constants from quantum chemistry results. *J. Comput. Chem.* **2014**, *35*, 82–93. [CrossRef]
48. Wang, X.Y.; Sun, J.F.; Han, D.D.; Bao, L.; Mei, Q.; Wei, B.; An, Z.X.; He, M.X.; Yuan, S.L.; Xie, J.; et al. Gaseous and heterogeneous reactions of low-molecular-weight (LMW) unsaturated ketones with O_3: Mechanisms, kinetics, and effects of mineral dust in tropospheric chemical processes. *Chem. Eng. J.* **2020**, *395*, 125083. [CrossRef]
49. Yao, J.F.; Sun, Y.N.; Tang, Y.Z.; Zhang, Y.J.; Wu, W.Z.; Sun, J.Y. Atmospheric oxidation of 4-(2-methoxyethyl) phenol initiated by OH radical in the presence of O_2 and NO_x: A mechanistic and kinetic study. *Int. J. Quantum. Chem.* **2021**, *121*, e26650. [CrossRef]
50. Zhao, H.; Wang, S.J.; Sun, J.Y.; Zhang, Y.J.; Tang, Y.Z. OH-initiated degradation of 1, 2, 3-trimethylbenzene in the atmosphere and wastewater: Mechanisms, kinetics, and ecotoxicity. *Sci. Total. Environ.* **2023**, *857*, 159534. [CrossRef]
51. ECOSAR V2.0. Available online: https://www.epa.gov/oppt/newchems/tools/21ecosar.htm (accessed on 1 September 2012).
52. Qu, R.J.; Liu, H.X.; Feng, M.B.; Yang, X.; Wang, Z.Y. Investigation on intramolecular hydrogen bond and some thermodynamic properties of polyhydroxylated anthraquinones. *J. Chem. Eng. Data.* **2012**, *57*, 2442–2455. [CrossRef]
53. Xiao, W.K.; Yan, S.D.; Liu, X.F.; Sun, S.M.; Khan, Z.U.H.; Sun, J.Y. Theoretical study on the degradation mechanism, kinetics and toxicity for aqueous ozonation reaction of furan derivatives. *Chemosphere* **2023**, *332*, 138782. [CrossRef]
54. Xiao, W.K.; Sun, S.M.; Yan, S.D.; Wu, W.Z.; Sun, J.Y. Theoretical study on the formation of Criegee intermediates from ozonolysis of pentenal: An example of trans-2-pentenal. *Chemosphere* **2022**, *303*, 135142. [CrossRef] [PubMed]
55. Sun, Y.N.; Yao, J.F.; Tang, Y.Z.; Zhang, Y.J.; Wu, W.Z.; Sun, J.Y. Theoretical study on the atmospheric degradation mechanism and subsequent products of *E*, *E*-2, 4-hexadienal with hydroxyl radical. *Int. J. Quantum. Chem.* **2021**, *121*, e26563. [CrossRef]
56. Brezonik, P.L.; Fulkerson-Brekken, J. Nitrate-induced photolysis in natural Waters: Controls on concentrations of hydroxyl radical photo-intermediates by natural scavenging agents. *Environ. Sci. Technol.* **1998**, *32*, 3004–3010. [CrossRef]
57. Burns, J.M.; Cooper, W.J.; Ferry, J.L.; King, D.W.; DiMento, B.P.; McNeill, K.; Miller, C.J.; Miller, W.L.; Peake, B.M.; Rusak, S.A.; et al. Methods for reactive oxygen species(ROS) detection in aqueous environments. *Aquat. Sci.* **2012**, *74*, 683–734. [CrossRef]
58. Haag, W.R.; Hoigné, J. Photo-sensitized oxidation in natural water via OH radicals. *Chemosphere* **1985**, *14*, 1659–1671. [CrossRef]
59. Bai, F.Y.; Ni, S.; Tang, Y.Z.; Pan, X.M.; Zhao, Z. Ciprofloxacin transformation in aqueous environments: Mechanism, kinetics, and toxicity assessment during •OH-mediated oxidation. *Sci. Total. Environ.* **2020**, *699*, 134190. [CrossRef] [PubMed]
60. Zhao, H.; Wang, S.J.; Zhang, Y.J.; Lu, C.G.; Tang, Y.Z. Degradation of mevinphos and monocrotophos by OH radicals in the environment: A computational investigation on mechanism, kinetic, and ecotoxicity. *J. Hazard. Mater.* **2023**, *445*, 130478. [CrossRef]

toxics

MDPI

Article

Dielectric Barrier Discharge Plasma Coupled with Cobalt Oxyhydroxide for Methylene Blue Degradation

Xiaomei Yao [1], Yingbo Fang [1], Xiaochen Cui [1], Xian Cheng [1,2,*] and Zixia Cheng [1]

1 School of Electrical and Information Engineering, Zhengzhou University, Zhengzhou 450001, China; xm_yao@zzu.edu.cn (X.Y.); 16696658179@163.com (Y.F.); 13031652377@163.com (X.C.); chzx@zzu.edu.cn (Z.C.)
2 Henan Engineering Research Center of Power Transmission & Distribution Equipment and Electrical Insulation, Zhengzhou University, Zhengzhou 450001, China
* Correspondence: chengxian@zzu.edu.cn

Abstract: In this study, the coupled use of a double dielectric barrier discharge (DDBD) and CoOOH catalyst was investigated for the degradation of methylene blue (MB). The results indicated that the addition of CoOOH significantly promoted MB degradation performance compared to the DDBD system alone. In addition, both the removal rate and energy efficiency increased with an increase in CoOOH dosage and discharge voltage. After 30 min of discharge treatment in the coupled system (with CoOOH of 150 mg), the removal rate reached 97.10% when the discharge voltage was 12 kV, which was 1.92 times that in the single DDBD system. And when the discharge time was 10 min, the energy efficiency could reach 0.10 g (k·Wh)$^{-1}$, which was 3.19 times better than the one in the single DDBD system. Furthermore, the addition of CoOOH could also significantly enhance the TOC and COD removal rates of MB. In the DDBD-coupled-with-CoOOH system, TOC and COD were 1.97 times and 1.99 times those of the single DDBD system after 20 min of discharge treatment with a discharge voltage of 12 kV and 100 mg of CoOOH. The main active substances detected in the coupled system indicated the conversion of the active species H_2O_2 and O_3 into a more oxidizing ·OH was enhanced through the addition of a CoOOH catalyst, resulting in the more effective decomposition of MB and intermediate molecules.

Keywords: double dielectric barrier discharge (DDBD); MB degradation; CoOOH; degradation performance

check for updates

Citation: Yao, X.; Fang, Y.; Cui, X.; Cheng, X.; Cheng, Z. Dielectric Barrier Discharge Plasma Coupled with Cobalt Oxyhydroxide for Methylene Blue Degradation. *Toxics* **2023**, *11*, 763. https://doi.org/10.3390/toxics11090763

Academic Editors: Zunyao Wang and Tiecheng Wang

Received: 14 August 2023
Revised: 3 September 2023
Accepted: 6 September 2023
Published: 8 September 2023

1. Introduction

In the textile industry, various organic synthetic dyes are commonly used as coloring materials. Because of their variety and complex structure, synthetic dyes are a kind of complex environmental pollutant, which poses a huge ecological risk to the environment [1]. Dyes enter water systems and cause the death of aquatic species and microorganisms, posing a serious threat to ecosystems and human health [2]. Synthetic dyes are relatively stable compounds that are difficult to degrade in wastewater treatment based on physical, chemical, or/and biological treatments. Therefore, developing an effective wastewater treatment process to remove dyes has been the focus.

Advanced oxidation processes (AOPs) have gained attention in recent years in the treatment of dye contaminants, such as photochemical, catalytic oxidation, ozonation, and Fenton or photo-Fenton systems. However, some problems, such as the addition of additional chemical reagents and unsatisfactory energy efficiency, still exist in these AOP systems [3]. As a comprehensive AOP technology, plasma has become a feasible technology for decomposing these pollutants effectively. In the plasma method, energy is injected into an aqueous solution through a high-voltage discharge plasma channel between two electrodes, and it produces various active species such as ·OH, ·O, O_3, and H_2O_2 in the aqueous solution with high oxidation–redox potential for the decomposition of the target molecule [4].

Among these active substances, ·OH exhibits a high oxidation potential and is recognized as the most effective active substance for degrading pollutant molecules. Therefore, how to make full use of various active substances generated by plasma, especially those with a relatively low oxidation potential (such as O_3 and H_2O_2), and increase the ·OH yield has become the focus of research. The addition of catalysts, such as adding an iron agent into plasma system for a Fenton effect, can realize the conversion of H_2O_2 generated in situ to ·OH and then maximize the utilization of various generated active substances in the plasma. Therefore, compared to a separate discharge plasma system, the plasma–catalytic system has the advantages of both good degradation performance and high energy utilization in catalytic reactions, making it an effective technology. Commonly used heterogeneous catalysts, which are conducive to recovery, mainly include transition metal oxides, such as TiO_2 [5], ZnO [6], WO_3, etc. In recent years, studies have shown that hydroxyl metal oxide is a promising type of catalyst with good catalytic performance, abundant properties, and good stability [7]. Hydroxyl metal oxides exhibit high specific surface area, which can provide sufficient active sites for catalyst reactions, and are also beneficial for complex multi-electron and multi-phase reaction process [7,8]. Zhao et al. compared the catalytic effects of FeOOH and various iron oxides in Fenton-like reactions, and found that FeOOH can more effectively catalyze the degradation of organic pollutants in water, and the conversion efficiency of H_2O_2 to ·OH is also the highest [8]. Additionally, cobalt is an ideal metal catalyst owing to its excellent catalytic performance and low ion leakage, arousing a wide range of research interests [9,10]. Xu et al. has successfully employed cobalt hydroxide in the decomposition of ozone and p-chloronitrobenzene (pCNB) in water [11]. Ren et al. introduced a CoOOH catalyst into a plasma system for the degradation of trans-ferulic acid wastewater, and the results showed good degradation performance. It has been concluded that the H_2O_2 and O_3 generated in the discharge plasma are first decomposed into peroxide-like oxides at the active site of the catalyst and then converted into ·OH with higher oxidation ability to achieve the efficient degradation of the target [12].

Inspired by this, we introduced a CoOOH catalyst into a gas–liquid hybrid double dielectric barrier discharge (DDBD) for the degradation of dye in wastewater. DDBD devices, which usually employ an additional dielectric barrier to isolate the metal electrode from the solution to avoid corrosion, have been widely used in water treatment [13–16]. The specific DDBD system can be reflected in our previous research work [17]. In this paper, we chose methylene blue (MB) as the target, which is a model azo dye with a stable molecular structure, and it is difficult to degrade using traditional methods [18–20]. The physical and chemical properties of CoOOH in the plasma–catalytic system were characterized via XRD, EDX, XPS, and SEM. The MB degradation performance including MB degradation efficiency, energy efficiency, COD, and TOC was evaluated. Additionally, the effect of different operating parameters on MB degradation were investigated. Moreover, the catalytic mechanisms of CoOOH in this discharge plasma system were explored.

2. Materials and Methods

2.1. Materials

Cobaltous nitrate ($Co(NO_3)_2 \cdot 6H_2O$) and methylene blue (99.99%) were purchased from Shanghai Aladdin Biochemical Technology Co., Ltd., Shanghai, China, and Tianjin Zhiyuan Chemical Reagent Co., Ltd., Tianjin, China, respectively. Sodium indigo disulphonate and potassium permanganate were purchased from Sinopharm Chemical Reagent Co., Ltd., Shanghai, China. Salicylic acid, 2,3-dihydroxybenzoic acid, and 2,5-dihydroxybenzoic acid were purchased from Aladdin. Besides these, other reagents used in this study were analytical grade, and no further purification was required. Deionized water was used in all experiments.

2.2. Catalyst Preparation

The CoOOH catalyst was prepared according to the method in previous work [17]. An amount of 4.3 g of $Co(NO_3)_2 \cdot 6H_2O$ was completely dissolved in 250 mL deionized water, and then an H_2O_2 solution with a concentration of 30% was added. The mixed liquid was stirred at a constant temperature under 55 °C with a magnetic stirrer. During the stirring process, 0.5 mol L^{-1} NaOH solution was added to adjust the pH value of the mixed solution to 9~10, and then, it was stirred at a constant temperature for more than 1 h for a uniform solution. The solution containing the CoOOH catalyst was then repeatedly washed and separated at least three times using an accelerated centrifuge. Finally, it was dried at 70 °C, and the dried powder was crushed and sifted (<35 μm). Thus, CoOOH catalyst powder could be obtained.

The crystal phases of the samples were evaluated via X-ray powder diffraction (XRD, Brucker D8 Advance, Broker Scientific Instruments, Billerica, MA, USA) in the range of 10–90° (2θ), using Cu Kα radiation. The morphologies of the obtained samples were characterized using a scanning electron microscope (SEM, Hitachi, SU8020, Tokyo, Japan). X-ray photoelectron spectroscopy (XPS) was performed on a Thermo escalab 250Xi (Thermo escalab, Carlsbad, CA, USA).

2.3. Experimental Setup

The gas–liquid hybrid double dielectric barrier discharge (DDBD) reactor used in this study was the same as in our previous work [17]. A copper rod (8 mm in diameter and 220 mm in length) was used as the high-voltage electrode. Two quartz tubes were used as the insulating medium, with a gap of 1 mm. The inner one, with an inner diameter of 9 mm and an outer diameter of 12 mm, was covered outside the copper rod electrode, and the gap between them was 1.5 mm. The inner and outer diameter of the outer quartz tube were 18 mm and 20 mm, respectively. Four identical copper strips connected in series with a total length of 80 mm, which were attached tightly to the outside surface of the outer quartz tube, were used as the grounding electrode. The DDBD plasma reactor was supplied by a high-frequency AC power supply (CTP-2000K, Nanjing Suman Electronics Co., Ltd., Nanjing, China). The voltage frequency was 10.13 kHz. A high-voltage probe (Tektronix P6015A, Cincinnati, OH, USA) and a voltage probe (Rigol PVP2150, Suzhou, China) were employed to record the voltage and current delivered to the reactor. The schematic diagram of the experimental setup is shown in Figure 1.

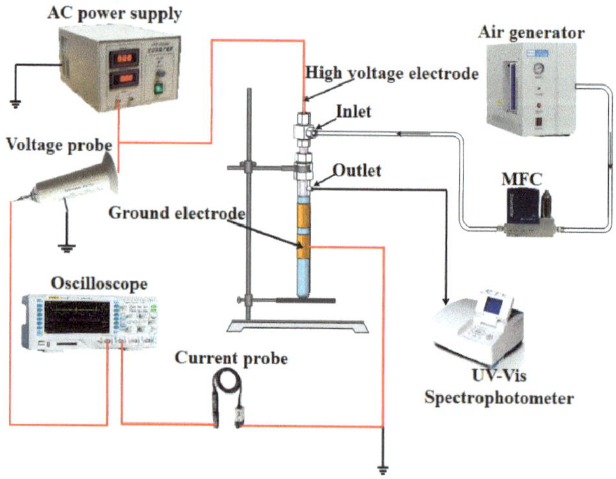

Figure 1. Schematic diagram of experimental setup.

A mass flow meter (MFC, D07, Beijing Seven star Flow Co., Ltd., Beijing, China) was employed to control the compressed air from the air generator flowing into the DDBD reactor. Compressed air was injected through the air inlet into the zone between the copper rod and the inner quartz tube, forming a gas-phase discharge under a strong electric field.

Then, the discharged gas was injected into the MB solution between two dielectric tubes with an aerator for the following chemical reaction process.

2.4. Methods and Analysis

A UV-vis spectrophotometer (TU-1810, Puxi, Beijing, China) was used to detect the concentration of the MB solution. The calculation methods of removal efficiency, energy efficiency, and the first-order reaction kinetics model were similar to our previous work [17]. The dissolved O_3 centration (mg L^{-1}) in the liquid phase was measured and calculated using indigo disulfonate spectrophotometry (IDS) [21], the H_2O_2 concentration (mg L^{-1}) in the MB solution was determined using titanium sulfate spectrophotometry, and the ·OH concentration (mol L^{-1}) in the MB solution was determined via the salicylic acid probe method. TOC and COD were determined using the Analytik Jena Mutiny N/C 2100 (Jena, Germany) and HQ40D-Multi (hach, Loveland, CO, USA), respectively. In addition, unless otherwise specified, the volume of the MB solution in the experiment was 30 mL, the air flow rate was 50 mL min^{-1}, and the concentration of the MB solution was 50 mg L^{-1}.

3. Results

3.1. Characterization of CoOOH

XRD is a crucial method for investigating the crystal structure of substances. Through analyzing the XRD diffraction pattern using Jade 9 software, the valuable information about the crystal structure of the catalyst used was obtained. In Figure 2a, it can be observed that CoOOH exhibited six diffraction peaks at specific angles: 20.2°, 36.9°, 38.8°, 50.7°, 65.3°, and 69.1°. Through a comparison of the diffraction data with the PDF4-2009 card database, it was determined that these peaks correspond to the crystal phases 003, 101, 012, 015, 110, and 113 of CoOOH, respectively.

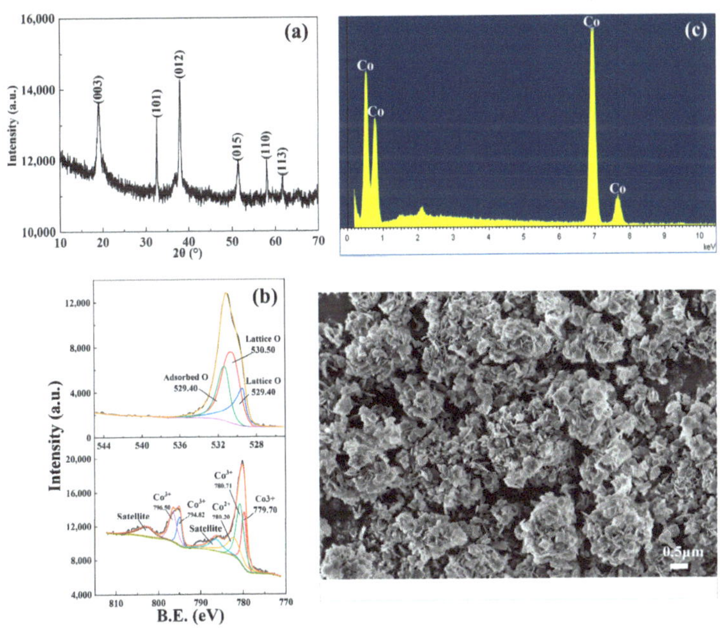

Figure 2. (**a**) XRD, (**b**) XPS, (**c**) SED, and (**d**) SEM of the CoOOH catalyst.

XPS was employed to analyze the valence distribution of each element in the catalyst. In Figure 2b, the peaks at 529.40 eV and 530.5 eV represent oxygen atoms within the lattice of the substance [22,23], and the peak at 531.28 eV indicates oxygen atoms from ·OH adsorbed on the surface of the measured substance [24–27]. It is calculated that the lattice oxygen atoms account for 33.28% of the total number of atoms, and the oxygen atoms from the adsorbed ·OH account for 13.39% of the total. This ratio is consistent with the expected proportion of lattice oxygen atoms and adsorbed oxygen atoms in CoOOH. Additionally, the XPS energy spectrum of Co2p was analyzed. The peaks of Co^{3+} appeared at 779.70 eV, 780.71 eV, 794.82 eV, and 796.50 eV, and the peak of Co^{2+} appeared at 782.20 eV [27,28]. Additionally, EDS was utilized to detect the catalyst, enabling the quantitative analysis of each element. As depicted in Figure 2c, distinct characteristic peaks of Co and O are detected, with a molecular ratio of approximately 36.52% for Co and 63.48% for O, which closely matches the theoretical ratio for CoOOH. Consequently, there is strong evidence supporting the identification of the catalyst as CoOOH.

The surface morphology of CoOOH was examined via SEM, and the particle size was calculated. Figure 2d presents an enlarged SEM image of CoOOH, revealing its relatively rough surface with a typical layered structure. The particle size distribution is fairly uniform, with the majority of particles concentrated within 4.5 μm, which could ensure that CoOOH has a considerable contact area with the active substances.

3.2. *MB Degradation Performance via DDBD with CoOOH Catalyst*

3.2.1. Effect of CoOOH Dosage on MB Degradation

In Figure 3a,b, it is evident that with the same duration of discharge treatment, both the removal rate and first-order reaction rate constant increase as the dosage of CoOOH catalyst increases. After 30 min of discharge treatment, without CoOOH, the removal rate of MB was only 50.52%. However, when the dosages of CoOOH were 100 mg and 150 mg, the removal rates increased to 89.54% and 97.10%, respectively, which were 1.77 times and 1.92 times that without CoOOH, respectively. Meanwhile, the first-order reaction rate constants rose from 0.03 min^{-1} to 0.08 min^{-1} and 0.10 min^{-1}, which were 2.33 times and 3.03 times that in the single DDBD system. Furthermore, after 40 min of discharge, the removal rate of MB reached 99.99% (12 kV, 100 mg of CoOOH). These results indicate that the inclusion of CoOOH significantly promotes the removal rate and reaction rate of MB degradation. Therefore, the DDBD-coupled-with-CoOOH system exhibits a distinct advantage over DDBD alone for the degradation of MB.

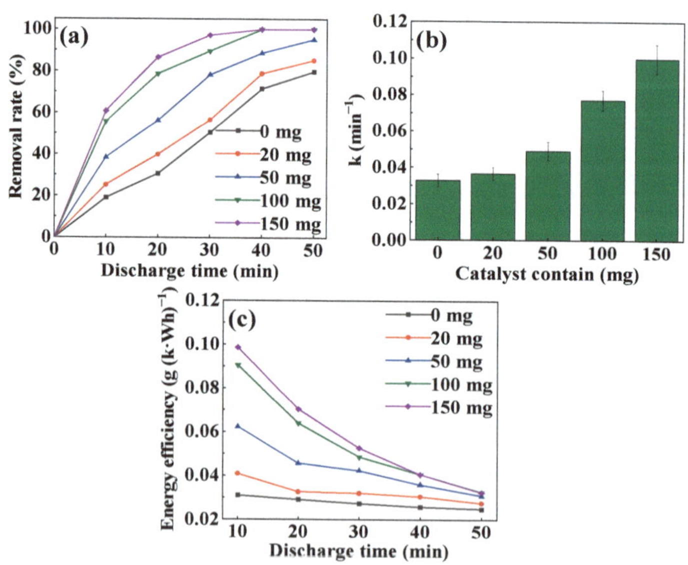

Figure 3. MB degradation under different dosages of CoOOH catalyst: (**a**) removal rate, (**b**) first-order reaction rate constant, (**c**) energy efficiency (discharge voltage of 12 kV).

From Figure 3c, it can be observed that the energy efficiency gradually increases with the increase in CoOOH dosage after the same discharge treatment time, while decreases as the discharge time prolongs. Without CoOOH, the energy efficiency was only 0.03 g $(k{\cdot}Wh)^{-1}$ after 10 min, and it decreased slowly with time. However, when 100 mg and 150 mg of CoOOH were added, respectively, the energy efficiency reached 0.09 g $(k{\cdot}Wh)^{-1}$ and 0.1 g $(k{\cdot}Wh)^{-1}$, respectively, which were 2.90 times and 3.19 times of that without CoOOH. Additionally, the energy efficiency decreased significantly as the discharge time extended.

Some researches indicated the CoOOH facilitated the formation of peroxide-like compounds on its surface by the active substances generated during the discharge process [29,30]. Additionally, the addition of CoOOH in this DDBD plasma system can significantly improve the MB degradation performance, which may attribute to the abundant ·OH on the catalyst surface and the conversion process is discussed in detail in Section 3.3. Although more active substances were generated as the discharge time extends, most MB molecules were degraded into smaller molecules in the early stage, reducing the average collision probability with active substances, which lead to a continuous decline of energy efficiency.

3.2.2. Effect of Discharge Voltage on MB Degradation

From Figure 4a,b, it is obvious that in the DDBD plasma coupled with CoOOH system, the MB removal rate and first-order reaction rate constant increase with the increase in discharge voltage. After 20 min of discharge treatment, when the discharge voltage increased from 8 kV to 14 kV, the MB removal rate rose from 35.43% to 96.61%. Meanwhile, the first-order reaction rate constant increased from 0.02 min^{-1} to 0.18 min^{-1}. This was because as the discharge voltage increased, more energy was injected into the reactor, causing MB to degrade more in a limited time. In the DDBD system alone, after 40 min of discharge treatment and the discharge voltage was 12 kV, the MB removal rate and first-order reaction rate constant were 71.62% and 0.03 min^{-1}, respectively. However, in the system comprising DDBD plasma coupled with CoOOH, the MB was almost completely degraded and the energy efficiency reached 0.09 min^{-1}, which was 1.33 times higher than that of without CoOOH. In brief, the addition of the catalyst CoOOH significantly enhances the removal rate and first-order reaction rate constant.

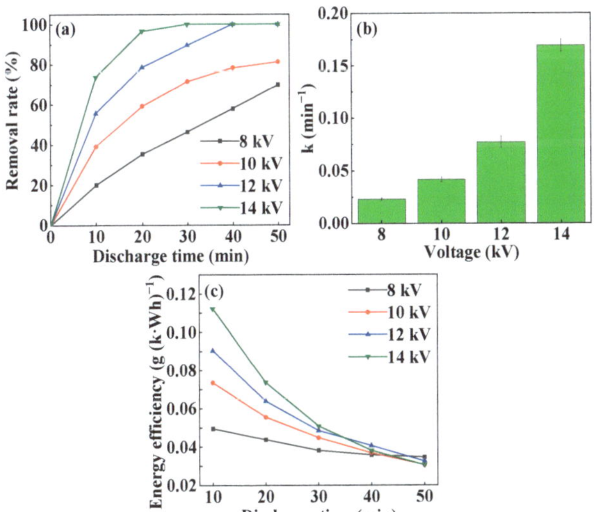

Figure 4. MB degradation under different discharge voltages: (**a**) removal rate, (**b**) first-order reaction rate constant, (**c**) energy efficiency (the dosage of CoOOH was 100 mg).

Figure 4c shows the energy efficiency of MB degradation. It can be seen that the energy efficiency decreases with the extension in time at the same voltage. At 14 kV of discharge voltage, when the discharge time increased from 10 min to 30 min, the energy efficiency decreased from 0.11 g $(k{\cdot}Wh)^{-1}$ to 0.05 g $(k{\cdot}Wh)^{-1}$. This was due to the higher concentration of MB in the initial solution, the larger average contact area between MB molecules and active particles, which lead to the higher probability of high-energy particles colliding with MB molecules. However, with the increase in discharge time, more and more MB molecules were gradually oxidized into smaller molecules, resulting in a huge decrease in the number of MB molecules that could be oxidized per unit time in the subsequent time, so the energy efficiency decreased over time. For the same discharge time, the energy efficiency increased with the increase in discharge voltage. After discharge treatment for 10 min, it was 0.11 g $(k{\cdot}Wh)^{-1}$ at a discharge voltage of 14 kV, which was 2.20 times that of 8 kV. In the discharge process, in addition to the production of more active substances, the O_3 and H_2O_2 that are generated would not only be decomposed into peroxide-like compounds on the surface of CoOOH [31], but also be converted into ·OH under the catalysis of CoOOH, resulting in an exceptionally high MB removal rate. The increase in discharge voltage greatly promoted the conversion of this process, so the energy efficiency increased with the increase in discharge voltage. After 30 min, due to the large reduction in MB molecules contained in the solution, even if a voltage of 14 kV produces more active substances, the number of MB molecules that could be degraded per unit time was greatly reduced, resulting in a decrease in energy efficiency, even less than the energy efficiency at a voltage of 8 kV.

3.2.3. COD and TOC

As shown in Figure 5, the chemical oxygen demand (COD) and total organic carbon (TOC) concentrations of MB solution under different discharge times were also measured. It can be seen that the removal rates of both COD and TOC gradually increase with the extension in discharge time, and in the DDBD-coupled-with-CoOOH system, they are significantly higher than in the single DDBD system. After 20 min of discharge treatment, the removal rates of COD and TOC in the DDBD system were 30.04% and 28.14%, respectively, while in the DDBD-coupled-with-CoOOH system, they reached 59.10% and 55.90%, respectively, which were 1.97 times and 1.99 times the former. Obviously, the addition of CoOOH significantly increases the MB mineralization rate. It may be due to the addition of CoOOH promoting the formation of ·OH, which exhibits stronger oxidation performance than H_2O_2 and O_3, so that more MB and intermediates are degraded into small molecules per unit time, which will be analyzed in more detail in the following Section 3.3.

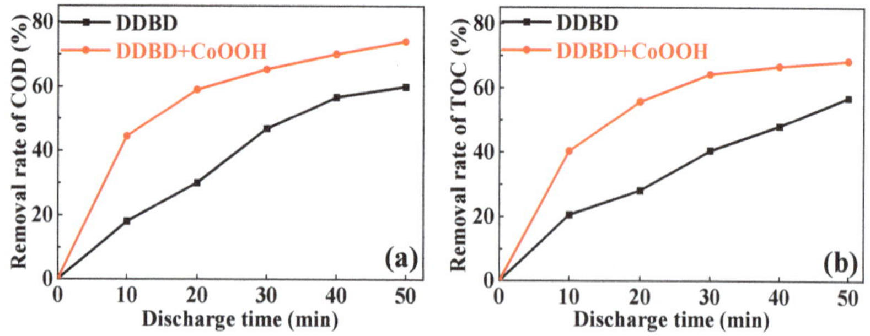

Figure 5. The removal rates of (**a**) COD and (**b**) TOC (discharge voltage of 12 kV, CoOOH dosage of 100 mg).

However, under the same conditions, the mineralization rate in the DDBD-coupled-with-CoOOH-system is still lower than the removal rate. It can be seen from Figure 5 that after 40 min of discharge treatment, the removal rates of COD and TOC in the coupled system were 70.13% and 66.90%, respectively, while the removal rate is close to 100% in Figure 3a (100 mg CoOOH). It is due to the fact that during the plasma discharge process, the MB molecules are partially converted into small organic molecules or intermediates that are difficult to be mineralized, and only a part of the MB molecules are completely oxidized and decomposed into CO_2 and H_2O. MB is an azo dye whose azo bond is more likely to be broken than those of benzene and naphthalene rings [32]. Additionally, the azo bond is the chromogenic base of the azo dye, and most of the azo bond is broken during the degradation of MB, resulting in obvious the decolorization of the MB solution [33].

3.3. Mechanism of MB Degradation via DDBD Coupled with CoOOH

In order to explore the mechanism of the synergistic effect between the DDBD and CoOOH, the content of active substances (H_2O_2, O_3, ·OH) at different discharge times was detected. As shown in Figure 6a, in both systems, the O_3 content initially increased and then decreased. It was due to the fact that in the discharge process, the air passed through the gas phase discharge region and produced active substances such as high-energy electrons and O_3, and then these active substances entered the liquid phase with air to form $e_{(aq)}$ and $O_{3(aq)}$, as shown in Formulas (1)–(3). Therefore, at the beginning of the discharge, the O_3 concentration in the solution gradually accumulated and increased with the extension in the discharge time. Until 30 min, the O_3 concentration reached the maximum, 0.41 mg L^{-1} and 0.19 mg L^{-1} in both systems, respectively. However, after 30 min, the O_3 content gradually decreased over time. In several studies, there was also an inflection point in the variation of dissolved O_3 concentration with treatment time with a high-frequency power supply. It was due to the fact that the electrical discharge could heat up the temperature of the solution in these cases, which was adverse to the dissolution of O_3 in water [5]. In contrast, it can be seen from Figure 6b,c that the H_2O_2 and ·OH contents continuously increased as the discharge time prolonged. This was mainly because the active substances dissolved in the solution underwent a series of oxidation reactions in the liquid phase to form H_2O_2 and ·OH, as shown in Formulas (4) and (5), and then these active substances (H_2O_2, ·OH) were gradually accumulated with the extension in time [34].

$$O_{2(gas)} + e_{(gas)} \rightarrow O + O \tag{1}$$

$$O + O_{2(gas)} + O_2 \rightarrow O_{3(gas)} + O_2 \tag{2}$$

$$O + O_{2(aq)} + e_{aq} \rightarrow O_{3(aq)} \tag{3}$$

$$O + H_2O + e_{aq} \rightarrow H_2O_2 \tag{4}$$

$$O + H_2O \rightarrow 2{\cdot}OH \tag{5}$$

Although the change trend of the active substance content in the two systems is similar, it can be observed from Figure 6 that the DDBD system had a higher content of O_3 and H_2O_2, and a lower content of ·OH, compared to the DDBD-coupled-with-CoOOH system. This indicates that the addition of CoOOH changes the content of active substances in the solution. In other words, the addition of CoOOH promotes the formation of ·OH, which is a highly reactive specie with superior oxidation properties compared to O_3 and H_2O_2 [29,35]. As a result, the MB removal rate, as well as COD and TOC removal rates, in the DDBD-coupled-with-CoOOH system are higher than those in the single DDBD system.

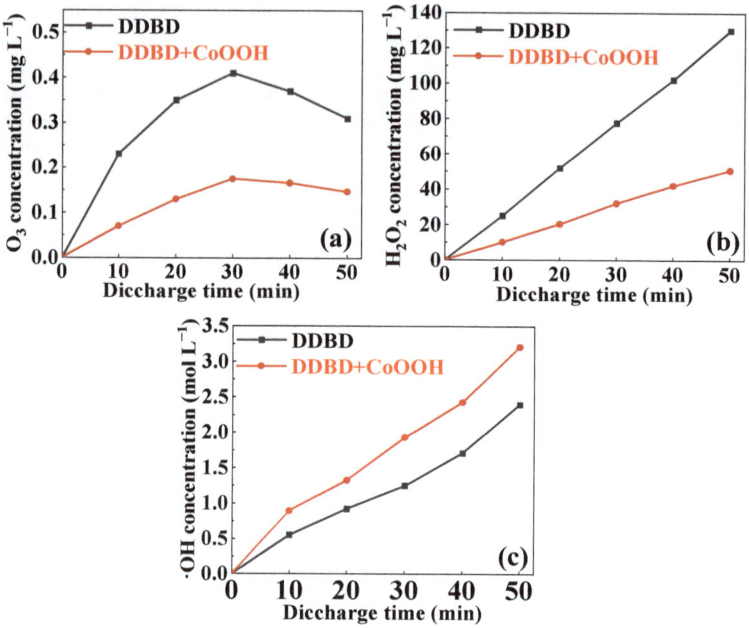

Figure 6. The dissolved concentration of (**a**) O_3, (**b**) H_2O_2, and (**c**) ·OH in MB solution (discharge voltage of 12 kV, CoOOH dosage of 100 mg).

Some studies have shown that there are abundant oxygen vacancies on the surface of CoOOH [36,37], and it is very easy for active substances such as H_2O_2 to form oxide-like compounds in the oxygen vacancies on the surface of CoOOH [38]; eventually, these oxide-like compounds react with $O_{3(aq)}$ in the solution to form ·OH, such as in (6)~(11). The catalytic mechanism is shown in Figure 7 (route (1)). In addition, CoOOH has a large number of Lewis sites on its surface, which easily adsorb H_2O molecules in contact with them and react with $O_{3(aq)}$ in solution to form a ·OH group [39], as shown in Equations (12)–(14). These formed ·OH groups eventually dissociate from the CoOOH surface, leaving the Lewis site exposed again for the next reaction. This is the second way CoOOH promotes ·OH content, as shown in Figure 7 (route (2)). Furthermore, CoOOH exhibits efficient electron transfer through the Fenton reaction with oxides. First, Co^{3+} can oxidize with the generated oxide-like intermediates to produce ·OH and Co^{2+} [40–42]. Then, the H_2O_2 produced by the discharge react with Co^{2+} to produce Co^{3+}, ·OH, and OH^-, as shown in Formulas (15)–(17). The Co atom of CoOOH participates in this reaction to form ·OH, and a cycle is formed, which is shown in Figure 7 (route (3)).

$$H_2O_2 \leftrightarrow H^+ + HO_2^- \tag{6}$$

$$HO_2^- + O_{3(aq)} \leftrightarrow HO_2^{\cdot} + O_3^{\cdot -} \tag{7}$$

$$HO_2^{\cdot} \leftrightarrow O_2^{\cdot -} + H^+ \tag{8}$$

$$O_2^{\cdot -} + O_3 \rightarrow O_2 + O_3^{\cdot -} \tag{9}$$

$$O_3^{\cdot -} + H^+ \rightarrow HO_3^{\cdot -} \tag{10}$$

$$HO_3^{\cdot} \rightarrow \cdot OH + O_2 \tag{11}$$

$$O_2^{2-}(ad) + O_3 + H_2O \rightarrow OH(ad) + HO_3^+ + O_2 \tag{12}$$

$$OH(ad) + 2O_3 \rightarrow O_2^{2-}(ad) + HO_3^- + O_2 \tag{13}$$

$$HO_3^- \rightarrow \cdot OH + O_2 \tag{14}$$

$$H_2O_2 + OH^- \rightarrow HO_2^- + H_2O \tag{15}$$

$$Co^{3+} + HO_2^- \rightarrow Co^{2+} + HO_2^+ \tag{16}$$

$$Co^{2+} + H_2O_2 \rightarrow Co^{3+} + \cdot OH + OH^- \tag{17}$$

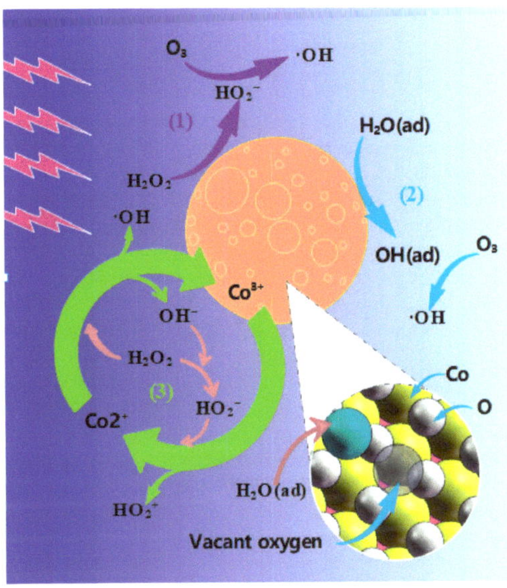

Figure 7. Mechanism of MB degradation via DDBD coupled with CoOOH.

4. Conclusions

In this study, CoOOH was added as a catalyst in a DDBD reaction system to achieve an effective removal rate and energy utilization efficiency when treating MB wastewater. The experimental results showed that, under a discharge voltage of 12 kV, an initial concentration of 50 mg L^{-1}, an air flow rate of 50 mL min^{-1}, and discharge treatment for 40 min, the removal rate of MB wastewater using the DDBD system with 100 mg CoOOH increased to 99.99%, which was 28.37% more than the DDBD system alone. Under the same conditions, the energy utilization efficiency increased to 0.04 g $(k\cdot Wh)^{-1}$, which was 1.57 times higher than that of the DDBD system alone, and the COD and TOC removal rates increased from 56.70% and 48.21% to 70.13% and 66.90%, respectively. This indicates that the addition of CoOOH could indeed achieve the purpose of improving the removal rate and energy efficiency of DDBDs and is expected to become a new efficient and environmentally friendly treatment method for MB wastewater.

Author Contributions: Conceptualization, writing—review and editing, funding acquisition, X.Y.; formal analysis, methodology, investigation, data curation, writing—original draft, Y.F.; supervision, resources, X.C. (Xian Cheng); investigation, data curation, X.C. (Xiaochen Cui); funding acquisition, resources, Z.C. All authors have read and agreed to the published version of the manuscript.

Funding: This work was supported by the National Natural Science Foundation of China (No. 52107171), the General Financial Grant from the China Postdoctoral Science Foundation (No. 2021M692900), the Key Laboratory of Engineering Dielectrics and its Application (Harbin University of Science and Technology), and the Ministry of Education (No. KFM202001).

Institutional Review Board Statement: Not applicable.

Informed Consent Statement: Not applicable.

Data Availability Statement: Data will be made available on request.

Conflicts of Interest: The authors declare no conflict of interest.

References

1. Kyere-Yeboah, K.; Bique, I.K.; Qiao, X.C. Advances of non-thermal plasma discharge technology in degrading recalcitrant wastewater pollutants. A comprehensive review. *Chemosphere* **2023**, *320*, 138061. [CrossRef]
2. Topolovec, B.; Skoro, N.; Puac, N.; Petrovic, M. Pathways of organic micropollutants degradation in atmospheric pressure plasma processing—A review. *Chemosphere* **2022**, *294*, 133606. [CrossRef] [PubMed]
3. Oturan, M.A.; Aaron, J.J. Advanced Oxidation Processes in Water/Wastewater Treatment: Principles and Applications. A Review. *Crit. Rev. Environ. Sci. Technol.* **2014**, *44*, 2577–2641. [CrossRef]
4. Zhang, Y.; Zhou, M.H.; Hao, X.L.; Lei, L.C. Degradation mechanisms of 4-chlorophenol in a novel gas-liquid hybrid discharge reactor by pulsed high voltage system with oxygen or nitrogen bubbling. *Chemosphere* **2007**, *67*, 702–711. [CrossRef] [PubMed]
5. Duana, L.J.; Jiang, N.; Lu, N.; Shang, K.F.; Li, J.; Wu, Y. Synergetic effect of TiO_2 and Fe^{3+} as co-catalysts for enhanced phenol degradation in pulsed discharge system. *Appl. Catal. B-Environ.* **2018**, *221*, 521–529. [CrossRef]
6. Li, Z.; Jiang, W.X.; Huang, J.W.; Wang, Y.W.; Guo, H. ZnO Promoted Persulfate Activation in Discharge Plasma System for Ofloxacin Degradation. *Catalysts* **2023**, *13*, 847. [CrossRef]
7. Qian, X.F.; Ren, M.; Zhu, Y.; Yue, D.T.; Han, Y.; Jia, J.P.; Zhao, Y.X. Visible Light Assisted Heterogeneous Fenton-Like Degradation of Organic Pollutant via alpha-FeOOH/Mesoporous Carbon Composites. *Environ. Sci. Technol.* **2017**, *51*, 3993–4000. [CrossRef]
8. Zhao, L.; Lin, Z.R.; Ma, X.H.; Dong, Y.H. Catalytic activity of different iron oxides: Insight from pollutant degradation and hydroxyl radical formation in heterogeneous Fenton-like systems. *Chem. Eng. J.* **2018**, *352*, 343–351. [CrossRef]
9. Abdedayem, A.; Guiza, M.; Rivas Toledo, F.J.; Ouederni, A. Nitrobenzene degradation in aqueous solution using ozone/cobalt supported activated carbon coupling process: A kinetic approach. *Sep. Purif. Technol.* **2017**, *184*, 308–318. [CrossRef]
10. Mahamallik, P.; Pal, A. Degradation of textile wastewater by modified photo-Fenton process: Application of Co(II) adsorbed surfactant-modified alumina as heterogeneous catalyst. *J. Environ. Chem. Eng.* **2017**, *5*, 2886–2893. [CrossRef]
11. Xu, Z.Z.; Chen, Z.L.; Joll, C.; Ben, Y.; Shen, J.M.; Tao, H. Catalytic efficiency and stability of cobalt hydroxide for decomposition of ozone and p-chloronitrobenzene in water. *Catal. Commun.* **2009**, *10*, 1221–1225. [CrossRef]
12. Ren, J.Y.; Jiang, N.; Shang, K.F.; Lu, N.; Li, J.; Wu, Y. Synergistic degradation of trans-ferulic acid by water falling film DBD plasma coupled with cobalt oxyhydroxide: Performance and mechanisms. *Chem. Eng. J.* **2019**, *372*, 321–331. [CrossRef]
13. Mei, S.F.; Liu, Y.A. Degradation of Reactive Black 5 in Aqueous Solution by Double-Dielectric Barrier Discharge. In Proceedings of the 2nd International Conference on Energy, Environment and Sustainable Development (EESD 2012), Jilin, China, 12–14 October 2012; pp. 1616–1619.
14. Zhang, H.B.; Li, K.; Shu, C.H.; Lou, Z.Y.; Sun, T.H.; Jia, J.P. Enhancement of styrene removal using a novel double-tube dielectric barrier discharge (DDBD) reactor. *Chem. Eng. J.* **2014**, *256*, 107–118. [CrossRef]
15. Rong, S.P.; Sun, Y.B. Degradation of TAIC by water falling film dielectric barrier discharge-Influence of radical scavengers. *J. Hazard. Mater.* **2015**, *287*, 317–324. [CrossRef]
16. Wu, J.L.; Xiong, Q.; Liang, J.L.; He, Q.; Yang, D.X.; Deng, R.Y.; Chen, Y. Degradation of benzotriazole by DBD plasma and peroxymonosulfate: Mechanism, degradation pathway and potential toxicity. *Chem. Eng. J.* **2020**, *384*, 123300. [CrossRef]
17. Yao, X.; Fang, Y.; Guo, Y.; Xu, M. Degradation of methylene blue using a novel gas-liquid hybrid DDBD reactor: Performance and pathways. *Chemosphere* **2023**, *336*, 139172. [CrossRef]
18. Jing, H.P.; Wang, C.C.; Zhang, Y.W.; Wang, P.; Li, R. Photocatalytic degradation of methylene blue in ZIF-8. *Rsc Adv.* **2014**, *4*, 54454–54462. [CrossRef]
19. Din, M.I.; Khalid, R.; Najeeb, J.; Hussain, Z. Fundamentals and photocatalysis of methylene blue dye using various nanocatalytic assemblies-a critical review. *J. Clean. Prod.* **2021**, *298*, 126567. [CrossRef]
20. Khan, I.; Saeed, K.; Zekker, I.; Zhang, B.L.; Hendi, A.H.; Ahmad, A.; Ahmad, S.; Zada, N.; Ahmad, H.; Shah, L.A.; et al. Review on Methylene Blue: Its Properties, Uses, Toxicity and Photodegradation. *Water* **2022**, *14*, 242. [CrossRef]
21. Suarasan, I.; Ghizdavu, L.; Ghizdavu, I.; Budu, S.; Dascalescu, L. Experimental characterization of multi-point corona discharge devices for direct ozonization of liquids. *J. Electrost.* **2002**, *54*, 207–214. [CrossRef]
22. Renault, O.; Gosset, L.G.; Rouchon, D.; Ermolieff, A. Angle-resolved x-ray photoelectron spectroscopy of ultrathin Al_2O_3 films grown by atomic layer deposition. *J. Vac. Sci. Technol. A* **2002**, *20*, 1867–1876. [CrossRef]

23. Rodríguez, J.L.; Poznyak, T.; Valenzuela, M.A.; Tiznado, H.; Chairez, I. Surface interactions and mechanistic studies of 2,4-dichlorophenoxyacetic acid degradation by catalytic ozonation in presence of Ni/TiO_2. *Chem. Eng. J.* **2013**, *222*, 426–434. [CrossRef]
24. Zhao, X.L.; Wang, Y.; Wu, H.Y.; Fang, L.C.; Liang, J.J.; Fan, Q.H.; Li, P. Insights into the effect of humic acid on Ni(II) sorption mechanism on illite: Batch, XPS and EXAFS investigations. *J. Mol. Liq.* **2017**, *248*, 1030–1038. [CrossRef]
25. Peng, J.L.; Lai, L.D.; Jiang, X.; Jiang, W.J.; Lai, B. Catalytic ozonation of succinic acid in aqueous solution using the catalyst of Ni/Al_2O_3 prepared by electroless plating-calcination method. *Sep. Purif. Technol.* **2018**, *195*, 138–148. [CrossRef]
26. Wang, Q.L.; Tang, M.H.; Peng, Y.Q.; Du, C.C.; Lu, S.Y. Ozone assisted oxidation of gaseous PCDD/Fs over CNTs-containing composite catalysts at low temperature. *Chemosphere* **2018**, *199*, 502–509. [CrossRef] [PubMed]
27. Zhu, M.; Miao, J.; Duan, X.; Guan, D.; Zhong, Y.; Wang, S.; Zhou, W.; Shao, Z. Postsynthesis Growth of CoOOH Nanostructure on $SrCo_{0.6}Ti_{0.4}O_{3-\delta}$ Perovskite Surface for Enhanced Degradation of Aqueous Organic Contaminants. *ACS Sustain. Chem. Eng.* **2018**, *6*, 15737–15748. [CrossRef]
28. Ren, J.Y.; Li, J.; Lv, L.; Wang, J. Degradation of caffeic acid by dielectric barrier discharge plasma combined with Ce doped CoOOH catalyst. *J. Hazard. Mater.* **2021**, *402*, 123772. [CrossRef] [PubMed]
29. Huang, J.H.; Chen, J.T.; Yao, T.; He, J.F.; Jiang, S.; Sun, Z.H.; Liu, Q.H.; Cheng, W.R.; Hu, F.C.; Jiang, Y.; et al. CoOOH Nanosheets with High Mass Activity for Water Oxidation. *Angew. Chem. Int. Ed.* **2015**, *54*, 8722–8727. [CrossRef] [PubMed]
30. Zhang, Q.H.; He, D.; Li, X.R.; Feng, W.; Lyu, C.; Zhang, Y.F. Mechanism and performance of singlet oxygen dominated peroxymonosulfate activation on CoOOH nanoparticles for 2,4-dichlorophenol degradation in water. *J. Hazard. Mater.* **2020**, *384*, 121350. [CrossRef]
31. Wang, Z.M.; Ma, H.; Zhang, C.; Feng, J.; Pu, S.Y.; Ren, Y.M.; Wang, Y. Enhanced catalytic ozonation treatment of dibutyl phthalate enabled by porous magnetic Ag-doped ferrospinel $MnFe_2O_4$ materials: Performance and mechanism. *Chem. Eng. J.* **2018**, *354*, 42–52. [CrossRef]
32. Hamada, K.; Qian, J.; Hirata, Y.; Satomura, K.; Mitsuishi, M. Effects of bolaform electrolytes on the interaction between a water-soluble polymer and sulphonated monoazo dyes. Part 3: Dyes containing two naphthalene rings. *Dye. Pigment.* **1996**, *31*, 19–30. [CrossRef]
33. Isaad, J.; El Achari, A. A novel glycoconjugated N-acetylamino aldehyde hydrazone azo dye as chromogenic probe for cyanide detection in water. *Anal. Chim. Acta* **2011**, *694*, 120–127. [CrossRef]
34. Wang, H.J.; Shen, Z.; Yan, X.; Guo, H.; Mao, D.N.; Yi, C.W. Dielectric barrier discharge plasma coupled with WO3 for bisphenol A degradation. *Chemosphere* **2021**, *274*, 129722. [CrossRef] [PubMed]
35. Yang, Y.; Jiang, J.; Lu, X.L.; Ma, J.; Liu, Y.Z. Production of Sulfate Radical and Hydroxyl Radical by Reaction of Ozone with Peroxymonosulfate: A Novel Advanced Oxidation Process. *Environ. Sci. Technol.* **2015**, *49*, 7330–7339. [CrossRef] [PubMed]
36. Ye, S.H.; Wang, J.P.; Hu, J.; Chen, Z.D.; Zheng, L.R.; Fu, Y.H.; Lei, Y.Q.; Ren, X.Z.; He, C.A.X.; Zhang, Q.L.; et al. Electrochemical Construction of Low-Crystalline CoOOH Nanosheets with Short-Range Ordered Grains to Improve Oxygen Evolution Activity. *Acs Catal.* **2021**, *11*, 6104–6112. [CrossRef]
37. Zeng, H.X.; Zhu, H.; Deng, J.; Shi, Z.; Zhang, H.J.; Li, X.Y.; Deng, L. New insight into peroxymonosulfate activation by CoAl-LDH derived CoOOH: Oxygen vacancies rather than Co species redox pairs induced process. *Chem. Eng. J.* **2022**, *442*, 136251. [CrossRef]
38. Staehelin, J.; Hoigne, J. Decomposition of ozone in water-rate of initiation by hydroxide ions and hydrogen-peroxide. *Environ. Sci. Technol.* **1982**, *16*, 676–681. [CrossRef]
39. Jain, P.; Jha, S.; Ingole, P.P. Concurrently Engineered Lewis Acid Sites and Coordination Sphere Vacancies in CoFe Prussian Blue Analogues for Boosted Bifunctional Oxygen Electrocatalysis. *ACS Appl. Energy Mater.* **2023**, *6*, 3278–3290. [CrossRef]
40. Zhang, Q.R.; Qu, G.Z.; Wang, T.C.; Li, C.G.; Qiang, H.; Sun, Q.H.; Liang, D.L.; Hu, S.B. Humic acid removal from micro-polluted source water in the presence of inorganic salts in a gas-phase surface discharge plasma system. *Sep. Purif. Technol.* **2017**, *187*, 334–342. [CrossRef]
41. Lin, Y.T.; Chen, C.H.; Lin, M.S. pH-Dependent Catalytic Behavior in Cathodic Application of Hydrogen Peroxide with Cobalt Oxide Modified Electrode and Its Application in Electrochemical Fenton Process in Alkaline Media. *J. Electrochem. Soc.* **2018**, *165*, H360–H364. [CrossRef]
42. Ren, J.Y.; Jiang, N.; Li, J.; Shang, K.F.; Lu, N.; Wu, Y. Synergistic degradation of trans-ferulic acid in aqueous solution by dielectric barrier discharge plasma combined with ozone. *Environ. Sci. Pollut. Res.* **2018**, *25*, 35479–35491. [CrossRef] [PubMed]

MDPI AG
Grosspeteranlage 5
4052 Basel
Switzerland
Tel.: +41 61 683 77 34

Toxics Editorial Office
E-mail: toxics@mdpi.com
www.mdpi.com/journal/toxics

www.ingramcontent.com/pod-product-compliance
Lightning Source LLC
Chambersburg PA
CBHW041344210926
51850CB00013B/2986